Springer Proceedings in Physics

Volume 333

Indexed by Scopus

The series Springer Proceedings in Physics, founded in 1984, is devoted to timely reports of state-of-the-art developments in physics and related sciences. Typically based on material presented at conferences, workshops and similar scientific meetings, volumes published in this series will constitute a comprehensive up to date source of reference on a field or subfield of relevance in contemporary physics. Proposals must include the following:

- Name, place and date of the scientific meeting
- A link to the committees (local organization, international advisors etc.)
- Scientific description of the meeting
- List of invited/plenary speakers
- An estimate of the planned proceedings book parameters (number of pages/articles, requested number of bulk copies, submission deadline).

Please contact:

For Americas and Europe: Dr. Zachary Evenson; zachary.evenson@springer.com
For Asia, Australia and New Zealand: Dr. Loyola DSilva; loyola.dsilva@springer.com

John Strologas · Leandros Perivolaropoulos
Editors

HEP2023

40th Conference on Recent Developments in High Energy Physics and Cosmology

Springer

Editors
John Strologas
Department of Physics
University of Ioannina
Ioannina, Greece

Leandros Perivolaropoulos
Department of Physics
University of Ioannina
Ioannina, Greece

ISSN 0930-8989 ISSN 1867-4941 (electronic)
Springer Proceedings in Physics
ISBN 978-3-031-76855-2 ISBN 978-3-031-76856-9 (eBook)
https://doi.org/10.1007/978-3-031-76856-9

This Springer imprint is published by the registered company Springer Nature Switzerland AG
The registered company address is: Gewerbestrasse 11, 6330 Cham, Switzerland

Preface

The 40th installment of the Conference on Recent Developments in High Energy Physics and Cosmology (HEP2023) took place on the campus of the University of Ioannina, on April 5–7, 2023. We had 199 participants, 31 invited speakers, and 61 presentations overall, 37 of which were plenary. Recent activities in both experimental and theoretical particle physics and cosmology were presented, with emphasis on LHC physics and instrumentation, gravity and black holes, effective field theories, and educational outreach. The conference fostered interactions and collaborations between the participants and provided a means for exposure of younger scientists.

The meeting venue was the Karolos Papoulias conference center of the University of Ioannina. In the opening ceremony, welcome speeches were delivered by the Rector of the University of Ioannina, Prof. Anna Batistatou, the Chair of the Physics Department Prof. Samuel Cohen, and the Chair of the local organizing committee Prof. Costas Fountas.

Each plenary (parallel) session speaker delivered a talk of 30 (20) minutes, which included time for questions and discussion. A dedicated convener chaired each plenary and parallel session. After-session social activities included a banquet dinner that took place at the Nautical Club of Ioannina. The local organizing committee consisted of professors A. Dedes, I. Evangelou, C. Fountas, P. Kokkas, G. Leondaris, N. Manthos, I. Papadopoulos, L. Perivolaropoulos, and J. Strologas.

This volume presents peer-reviewed proceedings articles based on oral presentations. At least two reviewers were assigned per paper. We would like to congratulate the authors for their important contributions and the reviewers for their excellent work.

Ioannina, Greece

John Strologas
Leandros Perivolaropoulos

The original version of the book has been revised. A correction to this book can be found at https://doi.org/10.1007/978-3-031-76856-9_25

Contents

Black Hole Solutions in Chern-Simons Gravity with Axion Hair

Nikos Chatzifotis

Abstract Chern-Simons gravity with axions is revisited from the point of view of studying Kerr-like black hole solutions which take into account the backreaction of the axion field onto the spacetime geometry. We extend previous results by giving analytic expressions for slowly rotating black holes, which formally include an all order expansion in the pertinent coupling constant. We investigate potentially observable effects, e.g. the black hole angular-momentum reversal in the near horizon regime by the axion cloud surrounding the black hole, which occurs for sufficiently strong interaction coupling.

Keywords Chern-Simons · Axion · Black hole solutions · Angular momentum · Strong interaction

1 Introduction

This presentation is based on work done in [1, 2]. Modified gravity theories which emanate from high order derivative corrections inspired from string theory, from the presence of high curvature terms and from scalar fields coupled to gravity, known as scalar-tensor theories have long captured the attention and the interest of the community. In all these theories, a crucial aspect to investigate is the behavior of the black hole solutions they support. It is of utmost importance to understand how black holes behave in these theories, as they have a major impact on the production of gravitational waves during non-spherical coalescence or collapse [3].

One of the fundamental aspects of these modified gravity theories is the involvement of scalar fields in the solutions of black holes. In most cases, these scalar fields interact with the geometry, a process referred to as scalarization, resulting in the black hole acquiring "hair". In the context of string theory [4–8], slowly rotating black holes of the Kerr type exhibit scalar hair due to the presence of the so-called string model

N. Chatzifotis (✉)
Physics Department, School of Applied Mathematical and Physical Sciences, National Technical University of Athens, 15780 Zografou Campus, Athens, Greece
e-mail: chatzifotisn@gmail.com

J. Strologas and L. Perivolaropoulos (eds.), *HEP2023*, Springer Proceedings in Physics 333 https://doi.org/10.1007/978-3-031-76856-9_1

independent axion field [9]. In the (3+1)-dimensional effective low-energy gravitational theory derived from string compactification, this axion field is dual to the field strength of the spin-one antisymmetric tensor field within the massless gravitational multiplet.

Chern-Simons (CS) theories are gravity theories that incorporate higher powers of curvature and belong to the family of consistent Lanczos-Lovelock gravity theories in dimensions higher than four. Within these theories, CS black holes represent special solutions with second-order field equations for the metric and well-defined asymptotic solutions in Anti-de Sitter (AdS) space [10]. Previous studies have investigated CS black holes in d-dimensions [11], exploring their stability by analytically calculating the quasinormal modes of scalar perturbations (for a comprehensive review on CS black holes, refer to [12]). By allowing the axion field to be dynamical, a solution describing a slowly rotating black hole was discovered in the context of small coupling and slow rotation limit [13, 14]. Other interesting works on the problem of extracting rotating black hole solutions in modified gravitational theories with a CS term include [15–17], where more generic gravitational theories were considered in the weak field limit and weak coupling limit respectively, while in [18, 19], the complete backreaction problem in rapidly rotating black holes of CS gravity has been approached in a numerical manner. Furthermore, investigations have also been conducted on static and rotating black string solutions within the dynamical CS modified gravity [20, 21].

The violation of energy conditions is closely linked to the emergence of scalar hair outside the black hole horizon. This violation is not only observed in the context of black hole scalarization but also characterizes CS-axion string-inspired running-vacuum cosmology [22, 23]. It signifies a fascinating interplay between curvature and matter, where the presence of matter interacting with curvature causes the deformation of the spacetime geometry around the black hole. This effect highlights the intricate relationship between matter and the curvature of spacetime.

This work is organized as follows: In Sect. 2 we briefly review the four-dimensional Chern-Simons gravitational theory and in Sect. 3 we present an exact slowly rotating hairy black hole solution in the aforementioned framework, emphasizing the corresponding violation of the energy conditions. In Sect. 4, we delve into the non-trivial effect of the angular momentum reversal in the near horizon regime. Finally, in Sect. 5 we present our conclusions.

2 Chern-Simons Gravitational Theory

We shall begin the presentation of our work by reviewing the Chern-Simons (CS) gravitational theory, thus allowing us to delve into the local solutions that the theory supports. The four-dimensional Chern-Simons gravity consists of an Einstein-Hilbert term, the kinetic term of a dynamical axion matter field and a linear coupling of the axion to the gravitational CS topologicalterm known as the Pontryagin density

[24, 25]. This term, henceforth denoted as R_{CS}, is constructed from the contraction of the Riemann tensor to its dual and reads

$$R_{CS} = \frac{1}{2} R^{\mu}{}_{\nu\rho\sigma} \widetilde{R}^{\nu}{}_{\mu}{}^{\rho\sigma}, \tag{1}$$

where the symbol $\widetilde{(\ldots)}$ denotes the dual of the Riemann tensor, defined as

$$\widetilde{R}_{\alpha\beta\gamma\delta} = \frac{1}{2} R_{\alpha\beta}{}^{\rho\sigma} \varepsilon_{\rho\sigma\gamma\delta}, \tag{2}$$

with $\varepsilon_{\rho\sigma\kappa\lambda} = \sqrt{-g(x)}\, \hat{\epsilon}_{\rho\sigma\kappa\lambda}$ the covariant Levi-Civita under the convention that the symbol $\hat{\epsilon}_{0123} = 1$, etc.

Therefore, the action under consideration for our work is given by

$$S = \int d^4x \sqrt{-g} \left[\frac{R}{2\kappa^2} - \frac{1}{2}(\partial_\mu b)(\partial^\mu b) - A\, b\, R_{CS} \right], \tag{3}$$

where $\kappa = M_{\rm Pl}^{-1}$ is the inverse of the reduced Planck mass, b is a pseudoscalar denoting the axion matter field and A is the coupling parameter of the axion field to the Chern-Simons term with dimension of length. We stress that the action (3) is shift-symmetric under the transformation $b \to b + c$, because we have not introduced any axionic potentials.

Variation of the action with respect to the metric and the axion field yields the following equations of motion

$$G_{\mu\nu} = \kappa^2\, T^b_{\mu\nu} + 4\kappa^2\, A C_{\mu\nu}, \tag{4}$$

$$\Box b = A\, R_{CS}, \tag{5}$$

where $T^b_{\mu\nu}$ is the stress energy-momentum tensor associated with the kinetic term of a matter field,

$$T^b_{\mu\nu} = \nabla_\mu b \nabla_\nu b - \frac{1}{2} g_{\mu\nu} (\nabla b)^2. \tag{6}$$

The quantity $C_{\mu\nu}$ is the Cotton tensor derived from the metric variation of $b R_{CS}$ and reads

$$C_{\mu\nu} = -\frac{1}{2} \nabla^\alpha \left[(\nabla^\beta b) \widetilde{R}_{\alpha\mu\beta\nu} + (\nabla^\beta b) \widetilde{R}_{\alpha\nu\beta\mu} \right]. \tag{7}$$

An important property of the Cotton tensor is that it satisfies a covariant non-conservation law [24, 25],

$$\nabla_\mu C^{\mu\nu} = -\frac{1}{4} (\nabla^\nu b) R_{CS}. \tag{8}$$

This result can be straightforwardly recovered using the equations of motion, by virtue of the Bianchi identity. This non-conservation law implies an exchange of energy between the axion field b with the gravitational anomaly for spacetime backgrounds for which the latter is non trivial, such as gravitational wave perturbations or rotating black holes [6–8], since spherical symmetry enforces a trivial vanishing of the Chern-Simons term and, in addition, the axion field is a pseudoscalar and can thus have no dynamics in a spherically symmetric framework.

3 The Search for Local Solutions: Slowly Rotating Kerr-Type Black Holes

Following the previous considerations, it can be easily deduced that our theory will always be reduced to GR in a spherically symmetric background. However, relaxing our spacetime symmetries down to axial symmetry, we may search for non-trivial rotating black holes. To get an exact rotating black hole solution, we will constrain ourselves to slowly rotating spacetimes. We shall perform our analysis to leading order in the angular momentum parameter a of the slowly-rotating Kerr-type spacetime, and suppress any contributions of order $\mathcal{O}(a^2)$.

The metric ansatz we are going to consider is the following

$$ds^2 = -H(r)dt^2 + F(r)dr^2 - 2r^2 a \sin^2\theta W(r)dtd\phi + r^2 d\Omega^2. \tag{9}$$

From the matter equation of motion (5), it can be immediately verified that the axion assumes the general solution of [1]

$$b = aAu(r)P_1(\cos\theta), \tag{10}$$

where P_1 denotes the Legendre polynomial of the first order. We note, however, that the axion field is of minimum order $\mathcal{O}(a)$. Focusing on the Einstein equations (4), we find that this implies that the tt and rr component of the gravitational equations of motion are satisfied in vacuum, since the corresponding stress energy tensor components are of order $\mathcal{O}(a^2)$, which we ignore in our approximation. Thus, up to first order in a, the temporal and radial components of the metric are the familiar Schwarzschild ones, i.e.

$$H(r) = \frac{1}{F(r)} = 1 - \frac{2M}{r}, \quad M \equiv G\mathcal{M} \tag{11}$$

with $G = 8\pi\,\kappa^{-2}$ the Newton's gravitational constant in (3+1)-dimensional spacetimes. Therefore, any backreaction on our spacetime will be encoded in the off-diagonal component of the metric. To this end, we consider the redefinition of

$$W(r) = \frac{2M}{r^3} + w(r)\,, \tag{12}$$

in order to extract the corrections, $w(r)$, on the recovered slowly rotating Hartle-Thorne spacetime (9), which are enforced by the axionic contribution. Using Eqs. (10), (11) and (12), we shift our focus on the non-trivial $t\phi$ component of the gravitational equations of motion (4). The corresponding differential equation can be easily integrated and it is found that the radial component of the axion with respect to the correction on the metric reads [1]

$$u(r) = -\frac{r^5 w'}{24\, A^2\kappa^2 M}. \tag{13}$$

Therefore, all the degrees of freedom of our solution are fixed except the corrections on the off-diagonal component w, which can be found via the axionic equation of motion, by virtue of equation (13). In particular, plugging our results on the axion equation (5) we find the following differential equation

$$\begin{aligned} &r^{11}(r-2M)w''' + 2r^{10}(6r-11M)w'' \\ &+ (28r^{10} - 50Mr^9 - 576\,A^2\kappa^2M^2r^4)w' + 3456\,A^2\kappa^2M^3 = 0. \end{aligned} \tag{14}$$

To solve the above equation we consider a series expansion on the correction function w. In particular, in order for the radial component of the axion field to vanish, $w'(r)$ needs to be at least of order $\mathcal{O}(r^{-5})$ and consequently $w(r)$ to be at least of order $\mathcal{O}(r^{-4})$, which in view of equation (12) implies that asymptotically the leading order of $g_{t\phi}$ is that of the Hartle-Thorne spacetime [3], i.e. asymptotically the spacetime coincides with the slowly rotating Kerr metric (9). We define w in a non-closed form as

$$w(r) = \sum_{n=4}^{\infty} \frac{d_n M^{n-2}}{r^n}, \tag{15}$$

where M^{n-2} is introduced in order to keep the coefficients d_n dimensionless. Hence, the problem of finding the geometric correction $w(r)$ has been reduced to the determination of the coefficients d_n. Making use of (15) into (14), we find after some algebra that the series coefficients can be extracted by the recursive relation

$$d_n = \frac{2(n-5)^2(n-1)}{n(n-6)(n-3)}d_{n-1} + \frac{576A^2\kappa^2}{n(n-3)M^4}d_{n-6}, \quad \text{for } n \geq 10. \tag{16}$$

with the following set of initial values

$$
\begin{aligned}
&d_4 = d_5 = 0, \\
&-28d_7 + 48d_6 = 0, \\
&-80d_8 + 126d_7 = 0, \qquad (17)\\
&256d_8 - 162d_9 = -3456\frac{A^2\kappa^2}{M^4}.
\end{aligned}
$$

Thus, we are left with four unknown coefficients, but with three equations. This means that we cannot completely determine the series coefficients from the differential equation (14).

A way out of this conundrum is to consider the weak field limit, i.e. the case where the axionic field does not backreact on the metric. This case is valid only in the asymptotic region of our spacetime, for which the axion equation of motion can be analytically solved [6–8]. Indeed, the correction terms related to the backreaction of the axion on the metric should be subdominant in the spatial asymptotics. Consequently, since the asymptotic behavior of the axion will determine the leading terms of its radial components, we may perform a simple matching of the coefficients extracted in the asymptotic region with the coefficients of the corresponding infinite series solution, which yield the initial values of [1]

$$
\begin{aligned}
&d_4 = d_5 = 0, \quad d_6 = -5\gamma^2, \quad d_7 = -\frac{60\gamma^2}{7}, \\
&d_8 = -\frac{27\gamma^2}{2}, \quad d_9 = 0, \quad \text{where} \quad \gamma^2 = \frac{A^2\kappa^2}{M^4}.
\end{aligned} \qquad (18)
$$

It should be stressed at this point that all of the series coefficients d_n are found to be non-positive and the convergence of the relevant series (15) for every real value of the constant $\gamma \in \mathbb{R}$ has been verified in [1].

We also mention that, although in this work we concentrated to lowest-order in an angular-momentum expansion a for the black hole solution, our solution goes beyond the truncation considered in [13] as far as the incorporation of terms inversely proportional to powers of the radial distance from the centre is concerned. We include arbitrary order of such terms in an analytic fashion, as follows from the above considerations, and thus we can approach the horizon of the black hole from the exterior region arbitrarily close.

3.1 Violation of Energy Conditions for Rotating Black Holes in Chern-Simons Gravity with Axion Hair

With the Chern-Simons black hole geometry at our disposal, it is important to comment on the violation of the energy conditions that the Chern-Simons black hole enforces. In particular, according to Bekenstein [26], minimally as well as non-minimally [27] coupled scalar fields that produce an energy-momentum tensor that

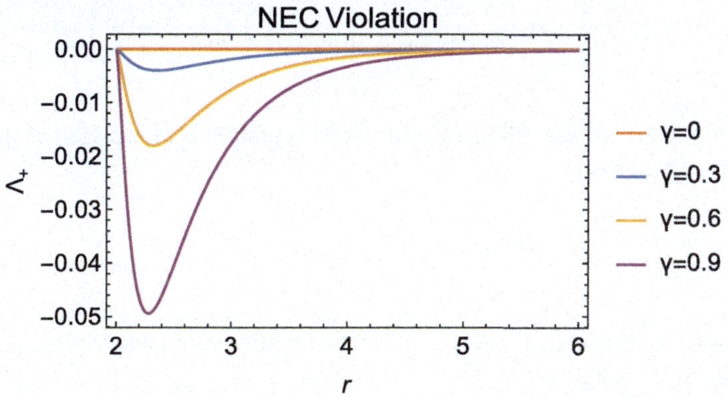

Fig. 1 Behavior of $\Lambda_+ = T_{\mu\nu} l^\mu l^\nu$ at the equitorial plane $\theta = \pi/2$ up to $\mathcal{O}(A^2)$ with respect to the radial coordinate. l^μ denotes a null vector

obeys the weak energy conditions (WEC) cannot be candidates for dressing a black hole with scalar hair. The case of the CS gravity is no exception, since the null energy conditions (NEC) are indeed violated (which implies the violation of all the energy conditions, see the discussion in Appendix of [1]).

This violation is attributed to the coupling of the axion with the Chern-Simons topological term, as this coupling is responsible for the appearance of the Cotton tensor in the gravitational equations of motion. Moreover, as the coupling constant takes larger values, the NEC violation intensifies and consequently the deformation of the spacetime becomes more important. In Fig. 1 we illustrate the precise way by means of which this violation occurs: near the horizon, the violation takes its maximum value, while far away is negligible. In addition, the NEC violation depends highly on the value of γ, which implies that axionic black holes of different sizes will in principle differ on the axionic hair they may support and their corresponding deformation.

4 Angular Momentum Reversal in the Near Horizon Regime

From the line element of the hairy black hole we extracted in CS gravity, it is immediately deduced that the back reaction effects appear solely on the off-diagonal component of the metric, at least for the first order in a approximation we adopted. To this end, we wish to analytically compute the angular momentum of the event horizon of our black hole solution and verify whether any novel non-trivial effects are present in our geometry.

In particular, making use of the polar isometry vector $\xi = \partial_\phi$, the total angular momentum of the spacetime is given by [2]

$$J = -\frac{1}{8\pi}\int\limits_{\mathcal{H}} d\Sigma_{\alpha\beta}\nabla^{\alpha}\xi^{\alpha} - \int\limits_{S} d\Sigma_{\alpha}\left(T^{\alpha}{}_{\beta} - \frac{1}{2}\delta^{\alpha}{}_{\beta}T\right)\xi^{\beta}, \tag{19}$$

where the first term of the right hand side of the equation is interpreted as the angular momentum of the horizon,

$$J_H = -\frac{1}{8\pi}\int\limits_{\mathcal{H}} d\Sigma_{\alpha\beta}\nabla^{\alpha}\xi^{\alpha}, \tag{20}$$

while the second one can be interpreted as the angular momentum of the matter field outside the event horizon:

$$J_M = -\int\limits_{S} d\Sigma_{\alpha}\left(T^{\alpha}{}_{\beta} - \frac{1}{2}\delta^{\alpha}{}_{\beta}T\right)\xi^{\beta}. \tag{21}$$

Obviously, if there is no matter ($T_{\mu\nu} = 0$) the only contribution comes from the black hole angular momentum and $J = J_H$; this is the case of the Kerr solution, which is a vacuum solution. Substituting the S^2 sphere of the horizon and integrating, we obtain that the black hole's angular momentum reads:

$$J_H = \left[1 + \frac{2rw(r) - r^2w'(r)}{6M}\right]_{r=2M} Ma. \tag{22}$$

On the other hand, the angular momentum of the axionic matter outside the horizon is easily found to read

$$J_M = -\left[\frac{2rw(r) - r^2w'(r)}{6M}\right]_{r=2M} Ma. \tag{23}$$

Thus, (22) and (23) yield the following result for the total angular momentum

$$J = J_H + J_M = Ma. \tag{24}$$

Hence, the total angular momentum is equal to Ma as in the Kerr spacetime, but in our case there is a non-trivial internal structure consisted of the black hole and axion angular momentum. The effect of the axion is to rotate the black hole, in the sense that it is necessary for the spacetime to be rotating in order for the axion to remain dynamical. In addition, as previously mentioned, the parameter γ, which appears in the correction function $w(r)$ is a measure of how strong the backreaction on the geometry is. Thus, as the backreaction gets stronger, i.e. for increasing γ, we expect the black hole's rotation to be affected in a non-trivial manner. In particular, the axion cloud around the black hole acquires an angular momentum in such a way that the total angular momentum remains constant and equal to Ma. As γ increases, however, the black hole's angular momentum decreases until a critical value, γ_c is reached, for

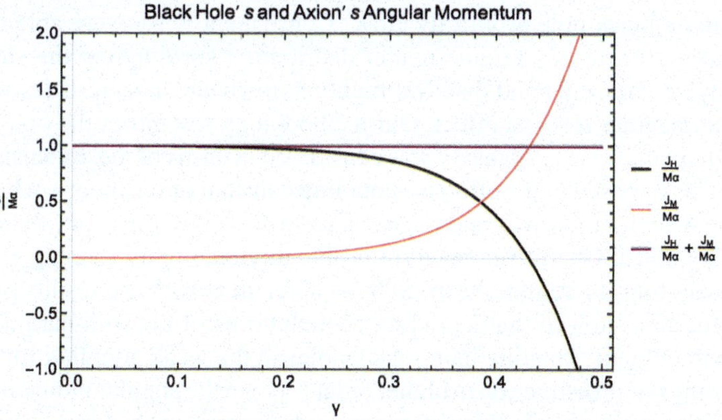

Fig. 2 The total angular momentum of the system of the Kerr-like black hole and the surrounding axionic matter. The monotonically decreasing line corresponds to the black hole's angular momentum at the event horizon, the monotonically increasing line respresents the angular momentum of the axion field, while the constant line represents the total angular momentum of the system

which the black hole event horizon has no angular momentum at all. After that value, the black hole starts to rotate in the opposite direction with increasing magnitude. Hence, the magnitude of each particular angular momentum can in principle reach large values, but the angular momentum of the spacetime as a whole, as measured by an observer at infinity, remains as if the background were slowly rotating.

One may understand this, by having in mind two competitive systems; the vacuum spacetime (vacuum solution of the Einstein's equations) and the axion, the pseudoscalar field which couples to the CS term. If we turn off the coupling, the system reduces to the Kerr metric which is a solution of GR in vacuum. Turning on the coupling, the axion becomes dynamical, acquires angular momentum and tends to rotate the black hole in the other way. When the coupling is strong enough, the spacetime around the black hole is counter rotating with regard to the measured Komar integral of the whole spacetime, as can be seen in Fig. 2. This is a manifestation of the exchange of energy between the axionic matter and the gravitational field. For more information on this phenomenon, the reader may refer to [2].

5 Conclusions and Outlook

Our work was focused on the scalarization of a slowly rotating black hole under the presence of an axion field coupled to the Chern-Simons term. We extracted an exact solution for a slowly rotating black hole taking into account the back-reaction of axionic hair. This solution is described in terms of the inverse powers of the radial distance from the center of the black hole and is valid for arbitrarily large values of the effective coupling constant. Such behavior enables us to approach the horizon of the axionic black hole arbitrarily close. Consequently, we were able to examine the characteristics of the axionic matter distribution surrounding the slowly rotating Kerr-type black hole and discover highly non-trivial characteristics, such as the angular momentum reversal effect. Our methodology resembles the one considered in [16], where the series expansion was considered in terms of the dimensionless spin parameter, as opposed to the effective coupling constant of the theory, which was our approach. Although the two approaches are substantially different, the conclusion that the dynamical Chern-Simons term is able to yield highly rotating effects, even in the slowly rotating regime, seem to be in agreement with the results presented in [16], where they showed that including enough terms of the spin-parameter, one is able to extract highly rotating hairy black holes in the weak coupling limit.

Extending our investigation to higher order $\mathcal{O}(a)$ in the angular momentum parameter would be intriguing. In this regime, we anticipate a much richer structure of the axionic Kerr-type black hole. This will provide the theoretical framework to study its angular momentum distribution and test whether the angular momentum reversal effect survives in higher rotating regimes.

References

1. N. Chatzifotis, P. Dorlis, N.E. Mavromatos, E. Papantonopoulos, Phys. Rev. D **105**(8), 084051 (2022). https://doi.org/10.1103/PhysRevD.105.084051, arXiv:2202.03496 [gr-qc]
2. N. Chatzifotis, P. Dorlis, N.E. Mavromatos, E. Papantonopoulos, Phys. Rev. D **106**(8), 084002 (2022). https://doi.org/10.1103/PhysRevD.106.084002, arXiv:2206.11734 [gr-qc]
3. C.W. Misner, K.S. Thorne, J.A. Wheeler, *Gravitation* (W. H. Freeman, San Francisco, 1973)
4. M.B. Green, J.H. Schwarz, E. Witten, *Superstring Theory Vols. 1 & 2: 25th Anniversary Edition* (Cambridge Monographs on Mathematical Physics, Cambridge University Press, 2012)
5. B.A. Campbell, N. Kaloper, K.A. Olive, Phys. Lett. B **285**, 199–205 (1992)
6. B.A. Campbell, M.J. Duncan, N. Kaloper, K.A. Olive, Phys. Lett. B **251**, 34–38 (1990)
7. B.A. Campbell, M.J. Duncan, N. Kaloper, K.A. Olive, Nucl. Phys. B **351**, 778–792 (1991)
8. M.J. Duncan, N. Kaloper, K.A. Olive, Nucl. Phys. B **387**, 215–235 (1992)
9. P. Svrcek, E. Witten, JHEP **06**, 051 (2006). arXiv:hep-th/0605206 [hep-th]
10. D.G. Boulware, S. Deser, Phys. Rev. Lett. **55**, 2656 (1985); J.T. Wheeler, Nucl. Phys. B **268**, 737 (1986)
11. P. Gonzalez, E. Papantonopoulos, J. Saavedra, JHEP **08**, 050 (2010). arXiv:1003.1381 [hep-th]
12. J. Zanelli, *Lecture Notes on Chern-Simons (Super-)Gravities*, 2nd edn. (2008). arXiv:hep-th/0502193 [hep-th]
13. N. Yunes, F. Pretorius, Phys. Rev. D **79**, 084043 (2009). arXiv:0902.4669 [gr-qc]
14. K. Yagi, N. Yunes, T. Tanaka, Phys. Rev. D **86**, 044037 (2012) [erratum: Phys. Rev. D **89**, 049902 (2014)]. arXiv:1206.6130 [gr-qc]
15. P. Pani, C.F.B. Macedo, L.C.B. Crispino, V. Cardoso, Phys. Rev. D **84**, 087501 (2011). https://doi.org/10.1103/PhysRevD.84.087501. arXiv:1109.3996 [gr-qc]
16. P.A. Cano, A. Ruipérez, JHEP **05**, 189 (2019) [erratum: JHEP **03**, 187 (2020)]. https://doi.org/10.1007/JHEP05(2019)189, arXiv:1901.01315 [gr-qc]
17. P.A. Cano, A. Ruipérez, Phys. Rev. D **105**(4), 044022 (2022). https://doi.org/10.1103/PhysRevD.105.044022, arXiv:2111.04750 [hep-th]
18. L.C. Stein, Phys. Rev. D **90**(4), 044061 (2014). https://doi.org/10.1103/PhysRevD.90.044061, arXiv:1407.2350 [gr-qc]

19. T. Delsate, C. Herdeiro, E. Radu, Phys. Lett. B **787**, 8–15 (2018). https://doi.org/10.1016/j.physletb.2018.09.060, arXiv:1806.06700 [gr-qc]
20. A. Cisterna, C. Corral, S. del Pino, Eur. Phys. J. C **79**(5), 400 (2019). arXiv:1809.02903 [gr-qc]
21. C. Corral, C. Erices, D. Flores-Alfonso, K. Lara, Phys. Rev. D **105**(2), 024050 (2022). arXiv:2111.00912 [hep-th]
22. S. Basilakos, N.E. Mavromatos, J. Solà Peracaula, Phys. Rev. D **101**(4), 045001 (2020). arXiv:1907.04890 [hep-ph]
23. N.E. Mavromatos, J. Solà Peracaula, Eur. Phys. J. Plus **136**(11), 1152 (2021). arXiv:2105.02659 [hep-th]
24. R. Jackiw, S.Y. Pi, Phys. Rev. D **68**, 104012 (2003). arXiv:gr-qc/0308071 [gr-qc]
25. S. Alexander, N. Yunes, Phys. Rep. **480**, 1–55 (2009). arXiv:0907.2562 [hep-th]
26. J.D. Bekenstein, Phys. Rev. D **51**(12), R6608 (1995)
27. A.E. Mayo, J.D. Bekenstein, Phys. Rev. D **54**, 5059 (1996). arXiv:gr-qc/9602057 [gr-qc]

Aspects of Thermal Stability for Black Holes

Panagiotis Dorlis

Abstract In this article some aspects of thermal stability for black holes is reviewed. We emphasize on how black hole solutions can be classified into equivalent topological classes, defined on the thermodynamic parameter space. Moreover, by taking into account the backreaction of a secondary hair onto the spacetime geometry, which is modified nontrivially, we derive, in a model independent way, the condition for the hairy black holes to reach a stable thermal equilibrium with the heat bath. Specifically, if the scalar hair, produces an inner horizon in the deformed geometry, a thermodynamically stable configuration will be reached. Such a mechanism might potentially produce black hole remnants from the early universe.

Keywords Black holes · Black hole thermodynamics · Hairy black holes · Black hole relics · Minimum horizon size

1 Introduction

In this work, I review the recent paper of mine and my colleagues [1], as a contribution to the 40th Conference on Recent Developments in High Energy Physics and Cosmology, Ioannina, Greece. With this review, a more coprehensive presentation of [1] is given, with an emphasis on the cosmological directions of the including ideas. Modified gravity theories, which are resulting from the presence of high curvature terms and from scalar fields coupled to gravity, known as scalar-tensor theories have been intensely studied. An important question in all such theories concerns their structure, the behaviour of black hole solutions and their stability. If the scalar field coupled to gravity backreacts on the background metric, then hairy black hole solutions can be generated in scalar-tensor theories [2]. Such a backreaction can drastically alter the causalstructure of black holes, with the appearance of an inner

P. Dorlis (✉)
Physics Department, School of Applied Mathematical and Physical Sciences, National Technical University of Athens, 15780 Zografou Campus, Athens, Greece
e-mail: psdolris0@gmail.com

J. Strologas and L. Perivolaropoulos (eds.), *HEP2023*, Springer Proceedings in Physics 333 https://doi.org/10.1007/978-3-031-76856-9_2

horizon being an example, with important consequencies to the stability issues, especially when combined with a regularized center for the black hole [3, 4]. In this work, we concern on the thermodynamical stability for the black holes, in which we find that the existence of an inner horizon is of crucial importance.

Recently in [5] the thermodynamical stability of black holes has been conjectured as a defect in thermodynamical parameter space. Specifically, by employing a generalized off-shell free energy, the authors of [5] constructed appropriate vector fields in the black hole thermodynamical parameter space, which vanish for the specific solutions of various gravitational models, thus leading to the interpretation of the black hole as a defect in the thermodynamic parametric space. Under this view, stability issues of the several black hole branches that appear for different temperatures are concerned with respect to the topological charges that can be defined for each defect. Moreover, a way to classify several black hole solutions into topologically equivalent classes is assigned. Specifically, every black hole solution can be ascribed with a global topological charge,[1] where the latter depends only on the thermodynamic asymptotic behavior of the black hole temperature at the small and large black hole limits. Such a procedure can be used for a better understanding of the black hole thermodynamics, and, as the authors of [5] conjecture, it may even shed light on such fundamental issues as the nature of quantum gravity, to which black holes are expected to play a fundamental role (e.g. in certain spacetime foam situations, microscopic black holes play a crucial rôle [6]).

In this work, we combine both the issues of the structure of the hairy black holes, with the issues concerning the thermodynamics of them. Specifically, under the view of black holes as defects to the abstract thermodynamic parametric space, we conclude that, if an inner horizon is produced due to the high curvature interactions, then the black hole is stabilized (thermodynamically) at the final stages of evaporation, leaving in that way a black hole remnant.

2 Black Holes in Thermal Equilibrium

In this Section we will review the thermodynamic stability of asymptotically flat black holes. To better understand, we consider a black hole in a heat bath, meaning that the total system is in thermal equilibrium at a temperature T. Thus, we have the following thermodynamic statement:

$$T_{BH} = T, \tag{1}$$

indicating that the black hole can be described in terms of the canonical ensemble. On the other hand, from the black hole physics it is known that the temperature of a black hole corresponds to the surface gravity, κ_g:

[1] Defined by the sum of every local topological charge ascribed to each branch.

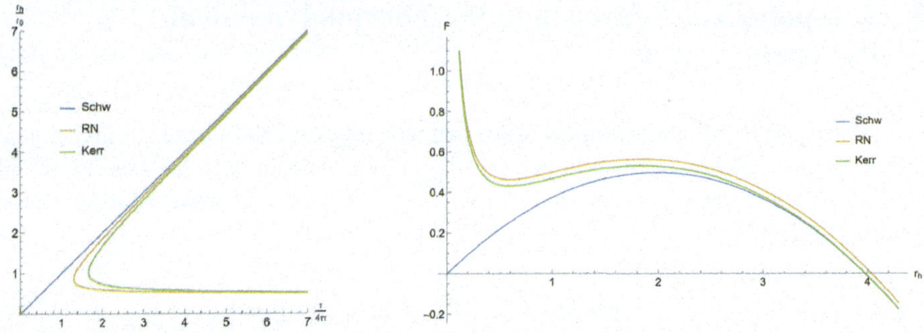

Fig. 1 Left: Branches of static, stationary and charged black holes in General Relativity. Right: The Free Energy for these black holes

$$T_{BH} = \frac{\kappa_g}{2\pi}. \tag{2}$$

Combining Eqs. (1) and (2), we get

$$\kappa_g = 2\pi T, \tag{3}$$

which combines a geometric quantity of the black hole (surface gravity) with the temperature of the heat bath T, which can vary freely and independently of the black hole. Considering the canonical ensemble, only the black hole's mass can be assumed not fixed and the surface gravity is expressed as a function of r_h, $\kappa_g = \kappa_g(r_h)$. Thus, Eq. (3) relates the size of the black hole and the temperature of the heat bath. Depending on the surface gravity dependence on r_h, more than one black hole configurations in thermal equilibrium with the heat bath can be achieved (see left panel of Fig. 1)[2]:

Stable and unstable thermal equilibrium is related to the free energy of the canonical ensemble, which is defined as follows

$$\mathcal{F} = E - \frac{\mathcal{S}}{\tau}, \tag{4}$$

where E is the total energy of the system, $\mathcal{S}$ is the entropy[3] and τ is the freely varying parameter introduced previously, $\tau = T^{-1}$. Since we consider the canonical ensemble, we consider E and S and consequently $\mathcal{F}$ as functions only of M. This is equivalent as to consider the above quantities as functions of the black hole radius r_h. Thus, the extremum points of the free energy are determined by its derivatives with respect to r_h, which can be confirmed in Fig. 1 (right panel).

[2] Where $\tau = T^{-1}$, with dimensions of time, Q is the charge of the black and α its angular momentum per unit mass.

[3] For a black hole, the total energy of the system corresponds to the ADM mass, while the entropy in an arbitrary diffeomorphism invariant theory is given by the well known Wald's formula [7].

3 A Topological Approach in the Thermodynamical Parameter Space

In a recent paper [5], a topological approach was introduced, in order to investigate the different branches of black holes in thermal equilibrium with the heat bath and classify solutions into *topologically equivalent classes*. A black hole solution can be mapped to the zero points of a vector field $\Phi(\vec{x})$ (defects), defined as follows:

$$\vec{\Phi} = \left(\Phi^{r_h}, \Phi^{\Theta}\right) = \left(\frac{\partial \mathcal{F}}{\partial r_h}, -\cot\Theta\csc\Theta\right) \tag{5}$$

where Θ is an auxillary parameter, taking values on the interval $0 \leq \Theta \leq \pi$.

The physical parameters (τ, r_h), together with the auxillary parameter Θ define a $3 - D$ parametric space, in which the zeros of the vector $\vec{\Phi}$ lie. The afforementioned mapping of the black hole solutions to the zeros of $\vec{\Phi}$ is guaranteed with this choice for Φ^{r_h}, since the condition of $\partial\mathcal{F}/\partial r_h = 0$ is equivalent with the condition $\tau = T_{BH}^{-1}$, which makes the free energy on-shell (solution of the Einstein's equations).

Considering τ as the "time" of the $3 - D$ parametric space, we can think of the equilibrium states ($\vec{\Phi} = 0$) like "point particles" moving on the $2 - D$ "space", constructed by the $\Theta - r_h$ plane. Thus, we define the following topological current on the 3-D parameter space (Duan's Φ—mapping) [8, 9]:

$$j^{\mu} = \frac{1}{2\pi}\epsilon^{\mu\nu\rho}\epsilon_{ab}\partial_{\nu}n^{a}\partial_{\rho}n^{b} \quad \text{where} \quad \mu, \nu, \rho = 0, 1, 2 \quad \text{and} \quad a, b = 1, 2, \tag{6}$$

with the notation $\partial_{\nu} = \partial/\partial x^{\nu}$ for $x^{\nu} = (\tau, r_h, \Theta)$, while ϵ corespond to the Levi-Civita symbols in the appropriate spaces. Moreover, the unit vector n is defined as $n^{\alpha} = \Phi^{\alpha}/||\Phi||$, where $\Phi^{1,2} = \Phi^{r_h,\Theta}$. Following [9], one can show that the topological current (6) reads

$$j^{\mu} = \delta(\vec{\Phi})\, J^{\mu}\left(\frac{\Phi}{x}\right), \quad J^{\mu}\left(\frac{\Phi}{x}\right) \equiv \frac{1}{2}\,\epsilon_{ab}\,\epsilon^{\mu\nu\rho}\partial_{\nu}\Phi^{a}\,\partial_{\rho}\Phi^{b} \ \text{(Jacobian Tensor)}, \tag{7}$$

where the presence of the delta function confirms the "point like" structure of the above current. Moreover, due to the Levi-Civita symbol the above current is identically conserved, $\partial_{\mu}j^{\mu} = 0$, meaning that we can define a conserved charge for each defect, by integration into some region, Σ, on the $\Theta - r_h$ plane (see Fig. 2). In Fig. 2, a black hole solution with a total number of N defects, for some $\tau = const$ surface, is represented.

Taking a region Σ_i that includes only one defect, say the $i - th$ defect, we can define the *local topological charge* corresponding to the $i - th$ defect, as follows:

$$q_i = \int_{\Sigma_i} d^2x j^0 \tag{8}$$

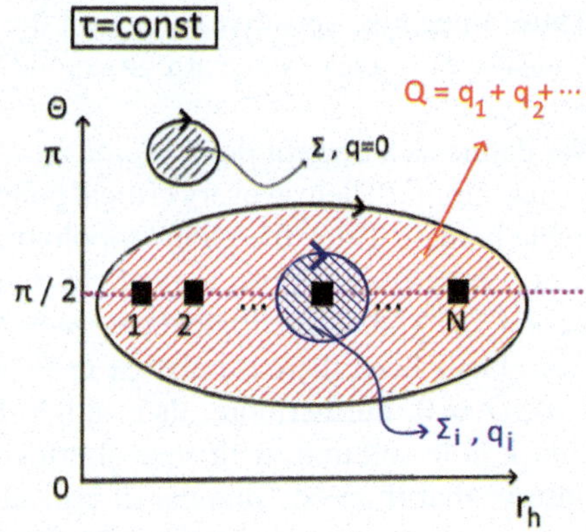

Fig. 2 Definition of the local and global topological charge on the $\Theta - r_h$ plane

Moreover, if the region Σ includes all of the defects, then the corresponding charge defines the *global topological charge* as the sum of all the local toppological charges:

$$Q = q_1 + q_2 + \cdots + q_N \tag{9}$$

The local topological charge characterizes each different branch of a given black hole solution separately. Specifically, in [5] was conjectured that the local topological charge is related to the thermal stability of the branch. In [1] the conjecture was confirmed, since on can prove that the local topological charge is related to the sign of the second dervative of the on-shell free energy:

$$q_i = sgn\left(\left.\frac{\partial^2 \mathcal{F}}{\partial r_h^2}\right|_i\right) \tag{10}$$

Thus, the local topological charge of each black hole branch concerns its thermodynamical stability, as follows:

$$q_i = 1 \rightarrow \text{Stable Branch}$$
$$q_i = -1 \rightarrow \text{Unstable Branch}$$

under the convention that we integrate clockwise on the $\Theta - r_h$ plane. Moreover, the thermodynamical stability of each branch can be related with the slope of the corresponding curve on the $r_h - \tau$ graph. Specifically, from the definition of the heat capacity as $C_Q = TdS/dT$ and using the chain rule for the derivative with respect to r_h, we get:

$$C_Q = T\frac{\partial r_h}{\partial T}\frac{\partial S}{\partial r_h}. \tag{11}$$

Thus, a negative (positive) slope on the $r_h = r_h(\tau)$ graph corresponds to a stable (unstable) branch, since it corresponds to a positive (negative) heat capacity, respectively.[4]

We define as a *critical point* on the $r_h = r_h(\tau)$ graph, the points where the slope becomes infinite. Then, at every critical point corresponds a pair of one stable and one unstable branch. Thus, the global topological charge depends on the number of the critical points. For example, to the Schwarzschild black hole, where there is no critical point, the global topological charge is $Q = -1$. For the Reissner-Nordstrom or the Kerr black hole, there is only one critical point and the global topological charge is zero, $Q = 0$. Furthermore, there might be more complicated graphs depending the black hole solution, with several critical points. However, it is straightforward to observe that if an odd number of critical points appear, then the global topological charge is certaintly zero, $Q = 0$, since stable ($q = 1$) and unstable ($q = -1$) branches appear in pairs. On the other hand, if the number of critical points is even, one branch (stable or unstable) is left without a pair. Thus, the global topological charge for an even number of critical points can take only two values $Q = \pm 1$. Moreover, assuming continuity of the $r_h = r_h(\tau)$ graph, whether there is an even or an odd number of critical points is determined by the asymptotic behavior of the graph, i.e. at the limits of large and small black hole's size, r_h. More precisely, if the slopes of the graph at large and small black holes is opposite then for sure $Q = 0$, since in this case an odd number of critical points appear. On the other hand, if the slopes are the same, the number of critical points is even and the global topological charge is non-zero, with its actual value, $Q = \pm 1$, determined by the sign of the slope on its asymptotic limits. Finally, we can conclude that the actual value of the global topological charge is determined by the asymptotic behavior of the graph and it can take only three values $Q = 0, \pm 1$, as shown in Fig. 3.

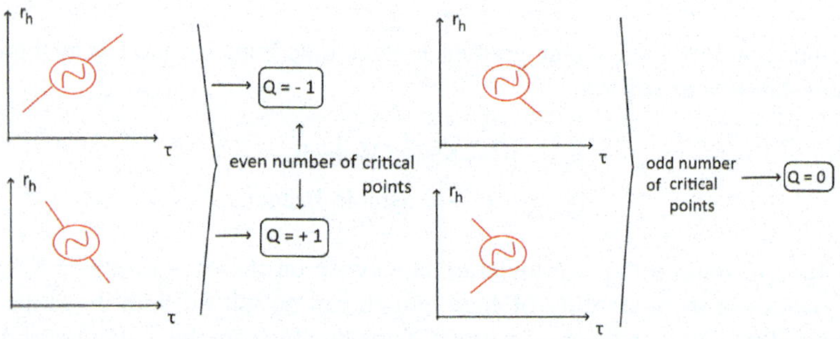

Fig. 3 The global topological charge for odd and even number of critical points

[4] Of course, this is true only for those solutions that the physical statement of $\partial S/\partial r_h > 0$ is valid.

4 Stabilizing Black Holes Though Secondary Hair: Black Hole Relics?

Primordial black holes can be formed at the early stages of the universe [10], when curvature fluctuations of the spacetime become large enough. Such black holes constitute a basic candidate in the literature for the dark matter observed into our universe (see [15] and references therein). However, one basic constraint[5] about their mass spectrum at the present epoch of the universe comes from the fact that black holes evaporate thermal radiation [11, 12]. Specifically, this thermal emission decreases the mass of the black hole and eventually leads to its disappearance. In this respect, any primordial black hole with mass $M < 10g$, would have dissapeared by now. An alternative scenario that overcomes such contraints, is that after evaporation a remnant is left over, as a result of the high curvature interactions introduced as effective lagrangians of quantum origin [13].

In order for the black hole relics to exist, the temperature—mass relation has to be modified, due to higher order interactions, in such a way that the black hole reaches a minimum horizon radius size, rather than totaly evaporate and completely disappear. Taking into account the black holes that tend to coincide to the Schwarzschild one at the limit of large black holes, the possible behaviors of the $r_h = r_h(\tau)$ graphs that induce a minimum horizon size are presented in Fig. 4. To the left panel of Fig. 4, the black hole reaches a minimum horizon size $r_{h,min}$, either with a positive or a negative heat capacity, C_Q, albeit with a non-vanishing temperature. Such a behavior, however, does not provide a well defined end of the evaporation process, since the black hole has still temperature and the evaporation is suddenly stopped. In [16] a numerical approach is considered in order to investigate the highly non-linear effects of the Gauss-Bonnet coupling in the string-inspired case of [17]. Such simulations reveal that indeed a stable branch appears at the final stages of evaporation, without, however, a zero temperature as an endpoint, which could be considered unphysical, since, if true, it would imply either a violation of the first law of thermodynamics or a fixed black-hole entropy, which is equally strange. We expect, although, that such results are misleading for the fate of the dilatonic black hole, due to the failure of the numerical method (see [1] for a more detailed discussion).

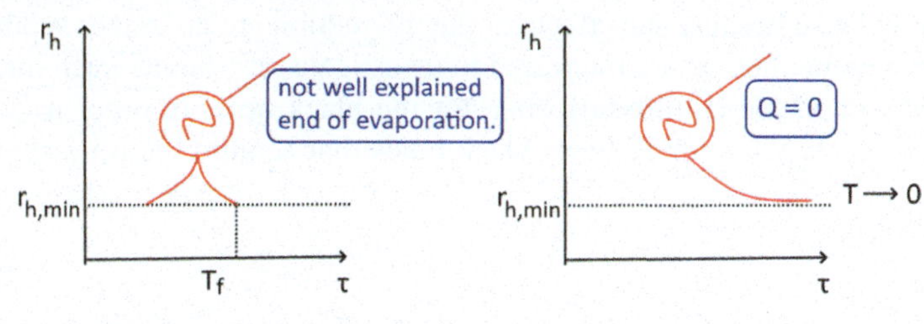

Fig. 4 Behavior of the $r_h = r_h(\tau)$ inducing a minimum horizon size

[5] For a recent update on the PBH constraints see Ref. [14].

To the right panel of Fig. 4 the behavior in which a stable branch appears at the end of evaporation, with the black hole tending to a vanishing temperature and consequently smoothly end its evaporation process, is presented. Such black holes are in the topological class of $Q = 0$ as defined in the previous section. Our goal is to investigate the conditions, under which such a behavior can be achieved as a result of an effective interaction between the matter and the gravitational field, leading to such remnants. Specifically, we will prove that such a behavior is achieved if the black hole tends to be extremal through a thermodynamically stable branch of positive local topological charge [1].

We shall be interested in interactions that have the form of a higher curvature coupling, with the strength of the interaction measured by a dimensionfull coupling constant A. We can write the general Lagrangian as

$$\mathcal{L} = \mathcal{L}_{EH} + \mathcal{L}_{matter} + A\,\mathcal{L}_{int}, \tag{12}$$

where $\mathcal{L}_{EH}$ is the Einstein-Hilbert action, $\mathcal{L}_{matter}$ the kinetic terms for the matter fields, while $\mathcal{L}_{int}$ accounts for the possible interactions between the matter fields and the gravitational field. Working in natural units,[6] all the physical quantities are reduced into units of length or energy. In this respect, we consider the dimensionfull coupling constant to be of length dimension, introducing in that way a new length scale into the system that characterizes the interaction of matter with gravity.[7] We claim that the impact of an interaction to a black hole can be realized by the dimensionless parameter γ

$$\gamma = \frac{A\kappa}{r_h^2}. \tag{13}$$

which is constructed by the three characteristc length scales of the system. Specifically, γ contains the coupling constant of the interaction A, the Planck length (through κ) and the size of the black hole, r_h. For a fixed coupling constant, γ becomes larger for smaller black holes. This means, that as the black hole shrinks (for example, through Hawking radiation), its interaction with the matter fields becomes more important and especially with an inverse square relation of its radius. Thus, in the large black hole regime, the interactions can be treated perturbatively, in the sense that we are able to keep terms up to first order in γ. However, this is not true for black holes of small size because the effects of the interaction are no longer subdominant. It is in this regime that we are interested to investigate on whether such interactions might lead to a stable equilibrium state for the black hole. In order to acomplish that, we focus our attention to hairy black holes that acquire the following general features:

---

[6] $\hbar = c = 1$.

[7] If the coupling constant is not itself of length dimension, the length scale can be defined by some power of it in order to acquire the appropriate dimension.

1. The theories support hairy black hole solutions acquiring a secondary scalar charge.
2. The corresponding local solutions have a smooth limit for a vanishing coupling constant, which denotes the absence of the interaction.
3. The dependence of the backreacting terms vanish asymptotically.
4. The coupling constant of the interaction appears only into the dimensionless factor γ.

Assuming spherical symmetry and staticity and in the case where the zero coupling limit $A \to 0$ corresponds to the Einstein-Hilbert action, the metric has to coincide with the Schwarzchild solution. Then, in such a case, the g_{tt} component of the metric would have the following form as a series expansion in γ:

$$g_{tt}(r) = -(1-x)\left(1 + \sum_n \gamma^n f_n(x)\right), \tag{14}$$

where $x = r_h/r$ and $f_n(x)$ some polynomial functions.

- Black hole's exterior: $r > r_h \to 0 < x < 1$; interior: $r < r_h \to x > 1$
- Event horizon: $r = r_h \to x = 1$; Asymptotic flatness: $f_n(0) = 0$

Moreover, at the limit of large black holes, the backreacting terms relax to zero ($\gamma \to 0$) and the metric is effectively reduced to the Schwarzschild black hole. At the exterior, the temporal component has to be negative, except for the horizon that has to be zero, which is guaranteed by the Schwarzschild term, $(1-x)$. This implies,

$$1 + \sum_n \gamma^n f_n(x) > 0, \quad \text{for } 0 < x < 1. \tag{15}$$

In the interior of the black hole, $x > 1$, an inner horizon might be present, if

$$1 + \sum_n \gamma^n f_n(x_0) = 0, \quad \text{for some } x_0 > 1. \tag{16}$$

Assuming the case where (16) has at most one physically acceptable root, there might be two asymptotic limits for the black hole in the $\tau - r_h$ plane. In agreement with the above considerations, we expect that one asymptotic limit corresponds to the case of the large black hole size, which is effectively the Schwarzschild black hole (plus a perturbation due to the interaction), which is the unstable branch. On the other hand, the stable black hole branch exists if there exists a finite (non-zero) r_h for which $\tau \to \infty$, since, if this is the case, a branch for which $\partial r_h/\partial T > 0$ in the $r_h - \tau$ plane has to be there, which implies a positive heat capacity and a corresponding positive topological charge for this branch. The surface gravity for a spherically symmetric metric is given by

$$\kappa_g = \left.\frac{|g'_{tt}|}{2\sqrt{-g_{tt}\, g_{rr}}}\right|_{r\to r_h}, \tag{17}$$

where prime denotes differentiation in respect to the variable r. For the metric determinant to be finite, $g_{tt}\, g_{rr}$ has to be finite. Thus, the surface gravity tends to zero, for those r_h for which $|g'_{tt}(r_h)| \to 0$. Then, form (14), we get

$$g'_{tt}(r_h) = -\frac{1}{r_h}\left(1 + \sum_n \gamma^n f_n(1)\right). \tag{18}$$

Thus, $g'_{tt}(r_h) \to 0$ for $r_h \to \infty$, which accounts for the Schwarzschild limit (free gravitational field) and for those finite r_h that satisfy:

$$1 + \sum_n \gamma^n f_n(1) = 0. \tag{19}$$

This condition describes the limit where the event horizon and the inner horizon match at $r = x_0 r_h$ and the black hole tends to be extremal. Thus, if the interaction produces an inner horizon for the black hole, the latter tends to a vanishing temperature through a stable branch. One can prove the inverse, too; if the black hole is allowed to become extremal due to the contribution of secondary hair, the final thermal stage of the black hole will always be reached through a stable branch [1].

As a final remark, we note that electric charge and rotation, also induce an inner horizon. However, electric charge and rotation probably do not survive at the final stages of evaporation [18]. This is not true in this case, since there is no extra charge; thus, the inner horizon, due to secondary hair, can survive until the final stages and stabilize (thermodynamically) the black hole.

References

1. N. Chatzifotis, P. Dorlis, N.E. Mavromatos, E. Papantonopoulos, Phys. Rev. D **107**(8), 084053 (2023). https://doi.org/10.1103/PhysRevD.107.084053, arXiv:2302.03980 [gr-qc]
2. Y. Fujii, K. Maeda, *The Scalar-tensor Theory of Gravitation* (Cambridge University Press, 2007)
3. A. Bonanno, A.P. Khosravi, F. Saueressig, Phys. Rev. D **103**(12), 124027 (2021). https://doi.org/10.1103/PhysRevD.103.124027, arXiv:2010.04226 [gr-qc]
4. R. Carballo-Rubio, F. Di Filippo, S. Liberati, C. Pacilio, M. Visser, JHEP **05**, 132 (2021) https://doi.org/10.1007/JHEP05(2021)132, arXiv:2101.05006 [gr-qc]
5. S.W. Wei, Y.X. Liu, R.B. Mann, Black hole solutions as topological thermodynamic defects. Phys. Rev. Lett. **129**(19), 191101 (2022). https://doi.org/10.1103/PhysRevLett.129.191101
6. J.A. Wheeler, Geons. Phys. Rev. **97**, 511–536 (1955)
7. R.M. Wald, Black hole entropy is the Noether charge. Phys. Rev. D **48**(8), R3427–R3431 (1993). arXiv:gr-qc/9307038 [gr-qc]
8. Y.S. Duan, M.L. Ge, SU(2) Gauge theory and electrodynamics with N magnetic monopoles. Sci. Sin. **9**(11), 1072 (1979)
9. YS. Duan, *The Structure of the Topological Current* (SLAC-PUB-3301, 1984)
10. B.J. Carr, S.W. Hawking, Mon. Not. R. Astron. Soc. **168**, 399–415 (1974). https://doi.org/10.1093/mnras/168.2.399
11. S.W. Hawking, Nature **248**, 30–31 (1974). https://doi.org/10.1038/248030a0

12. S.W. Hawking, Commun. Math. Phys. **43**, 199–220 (1975) [erratum: Commun. Math. Phys. **46**, 206 (1976)]. https://doi.org/10.1007/BF02345020
13. J.D. Barrow, E.J. Copeland, A.R. Liddle, Phys. Rev. D **46**, 645–657 (1992). https://doi.org/10.1103/PhysRevD.46.645
14. B. Carr, K. Kohri, Y. Sendouda, J. Yokoyama, Rep. Prog. Phys. **84**(11), 116902 (2021). https://doi.org/10.1088/1361-6633/ac1e31, arXiv:2002.12778 [astro-ph.CO]
15. I. Dalianis, PoS **CORFU2019**, 096 (2020). https://doi.org/10.22323/1.376.0096
16. F. Corelli, M. De Amicis, T. Ikeda, P. Pani, Nonperturbative gedanken experiments in Einstein-dilaton-Gauss-Bonnet gravity: nonlinear transitions and tests of the cosmic censorship beyond General Relativity. arXiv:2205.13007 [gr-qc]
17. P. Kanti, N.E. Mavromatos, J. Rizos, K. Tamvakis, E. Winstanley, Dilatonic black holes in higher curvature string gravity. Phys. Rev. D **54**, 5049–5058 (1996). arXiv:hep-th/9511071 [hep-th]
18. X. Calmet, B. Carr, E. Winstanley, Hawking radiation and black hole evaporation, in *Quantum Black Holes*. Springer Briefs in Physics. (Springer, Berlin, 2014)

Normal and Phantom Black Holes in Scalar-Tensor Theories

Athanasios Bakopoulos and Theodoros Nakas

Abstract In this work, we explore spherically symmetric black-hole solutions in a specific subclass of the Horndeski theories. By adopting a potential engineering approach, we derive exact analytic black-hole solutions endowed with scalar hair. We analyze the properties of these novel solutions and discover a spectrum that encompasses normal, ultra-compact, and ultra-sparse black holes. Each type of black hole emerges from a distinct range of scalar charge values. Rotating versions of these black holes are also obtained through a perturbation method, and their relative angular velocities are calculated. These rotation signatures may offer potential observational probes to test gravity. Our solutions showcase the richness of black hole geometries permitted in scalar-tensor theories beyond General Relativity. Further efforts to establish stability and astrophysical viability will be important in future directions.

Keywords Modified gravity · Horndeski theory · Scalar-tensor theories · Black holes · Compact objects

---

A. Bakopoulos (✉) · T. Nakas
Physics Division, School of Applied Mathematical and Physical Sciences, National Technical University of Athens, Zografou Campus, Zografou 15780, Greece
e-mail: atbakopoulos@gmail.com

T. Nakas
e-mail: theodoros.nakas@gmail.com

A. Bakopoulos
Division of Applied Analysis, Department of Mathematics, University of Patras, Rio Patras 26504, Greece

T. Nakas
Research Centre for Theoretical Physics and Astrophysics, Institute of Physics, Silesian University in Opava, Bezručovo nám. 13, CZ-746, 01, Opava, Czech Republic

J. Strologas and L. Perivolaropoulos (eds.), *HEP2023*, Springer Proceedings in Physics 333 https://doi.org/10.1007/978-3-031-76856-9_3

1 Introduction

General Relativity (GR) has been successful in explaining the vast majority of astrophysical phenomena, but it is still an effective theory that most likely breaks down under extreme conditions. Scalar-Tensor (ST) theories, by introducing only one additional scalar degree of freedom, provide a much broader framework for exploring alternative spacetime geometries. These theories have gained prominence due to the discovery of the Higgs boson [1], a discovery that irrefutably proved that scalar fields exist in nature.

Although many modified gravitational theories focus on cosmological implications, it is crucial to scrutinize their local implications, especially in relation to black holes. Despite early hurdles due to No-Scalar Hair theorems [2, 3], subsequent developments have led to the discovery of a diverse array of black hole solutions in ST theories [4–13]. These theories also allow for the existence of other intriguing objects, such as wormholes and solitons [14–16].

In this study, we explore a specific class of ST theories belonging to the Horndeski family and investigate the existence and the properties of analytic black-hole solutions and their slowly rotating extensions [17, 18]. By comparing the angular velocities of these solutions with those of their Schwarzschild counterparts, we gain insights into their astrophysical implications. In summary, investigating black-hole solutions within ST theories presents an exciting opportunity to extend our understanding of gravity beyond GR. By exploring the properties and implications of these solutions, we aim to uncover new insights into the nature of spacetime and astrophysical phenomena.

2 The Theoretical Framework

We start our analysis considering the following action functional

$$\mathcal{S} = \frac{1}{16\pi} \int d^4x \sqrt{-g} \left[R - \frac{1}{2}(\partial\Phi)^2 - V(\Phi) \right] . \tag{1}$$

The theory includes the Einstein-Hilbert term and a minimally coupled scalar field Φ with its potential and kinetic term. The variation of the action (1) with respect to the metric tensor $g^{\mu\nu}$ yields the tensorial equation

$$G^{\mu}{}_{\nu} = T^{(\Phi)\mu}{}_{\nu} . \tag{2}$$

This equation relates the Einstein tensor to the effective stress-energy tensor $T^{(\Phi)\mu}{}_{\nu}$, associated with the scalar field Φ, and defined as

$$T^{(\Phi)\mu}{}_{\nu} \equiv \frac{1}{2}\partial^{\mu}\Phi\partial_{\nu}\Phi - \frac{1}{2}\delta^{\mu}{}_{\nu}\left[\frac{(\partial\Phi)^2}{2} + V(\Phi)\right]. \tag{3}$$

Varying the action with respect to the scalar field Φ results in the equation of motion (4), describing the behavior of Φ.

$$\nabla^{\lambda}\nabla_{\lambda}\Phi - \partial_{\Phi}V = 0. \tag{4}$$

In this work, our focus lies on deriving asymptotically flat black-hole solutions with scalar hair. To achieve this, the following ansatz in Schwarzschild type coordinates for the line element is considered:

$$ds^2 = -e^{A(r)}\,B(r)\,dt^2 + \frac{dr^2}{B(r)} + r^2\left(d\theta^2 + \sin^2\theta\,d\varphi^2\right), \tag{5}$$

where the scalar field depends solely on the radial coordinate. The tensorial equation (2) gives rise to the independent equations (6), (7), and (8), which govern the metric functions and the potential

$$A'(r) = \frac{r}{2}\left[\Phi'(r)\right]^2, \tag{6}$$

$$B''(r) + \frac{3}{2}A'(r)B'(r) + \left\{A''(r) + \frac{A'(r)}{r} + \frac{\left[A'(r)\right]^2}{2} - \frac{2}{r^2}\right\}B(r) = -\frac{2}{r^2}, \tag{7}$$

$$V(\Phi) = \frac{2}{r^2} - \frac{2}{r}A'(r)B(r) - \frac{2B(r)}{r^2} + \frac{1}{2}\left[\Phi'(r)\right]^2 B(r) - \frac{2B'(r)}{r}. \tag{8}$$

In the analysis above, we use the convention of denoting differentiation with respect to the radial coordinate r, with the prime symbol. Note also that the scalar-field Eq. (4) can be derived from the three coupled differential equations (6)–(8), thus, it does not offer additional information [17, 18].

To solve the system of differential equations (6)–(8) with four unknown functions, we need to assume an expression for one of those functions. To this end, we adopt a potential engineering method based on the simplicity of Eq. (6) and choose a particular functional form for either the metric function A or the scalar field Φ. Consequently, we determine the remaining functions supporting a black-hole solution.

3 Black Hole Solutions and Their Properties

3.1 Solutions with a Coulombic Scalar Field

As a starting point of our analysis, we assume a Coulombic form for the scalar-field function, namely $\Phi(r) = q/r$, where q represents the scalar charge [17]. Subsequently, and upon solving the system of ordinary differential equations (ODE) (6)–(8), we obtain the following result:

$$A(r) = -\frac{q^2}{4r^2}, \tag{9}$$

$$B(r) = 1 - \frac{2m(r)}{r},$$

$$m(r) = \frac{r}{2} + \frac{4r^3}{q^2} + \frac{e^{\frac{q^2}{8r^2}} r^2}{q^2}\left[-12M + \sqrt{2\pi}\, q \,\mathrm{erf}\left(\frac{q}{2\sqrt{2}\,r}\right)\right]$$
$$- \frac{e^{\frac{q^2}{4r^2}} r^3}{q^3}\left\{4q - 12\sqrt{2\pi}\, M \,\mathrm{erf}\left(\frac{q}{2\sqrt{2}\,r}\right) + \pi q\left[\mathrm{erf}\left(\frac{q}{2\sqrt{2}\,r}\right)\right]^2\right\}, \tag{10}$$

$$V(\Phi) = \frac{2(24+\Phi^2)}{q^2} - \frac{12\,\Phi\, e^{\Phi^2/8}}{q^3}\left[12M - \sqrt{2\pi}\, q \,\mathrm{erf}\left(\frac{\Phi}{2\sqrt{2}}\right)\right]$$
$$+ \frac{(\Phi^2-12)e^{\Phi^2/4}}{q^3}\left\{4q - 12\sqrt{2\pi}\, M \,\mathrm{erf}\left(\frac{\Phi}{2\sqrt{2}}\right) + \pi q\left[\mathrm{erf}\left(\frac{\Phi}{2\sqrt{2}}\right)\right]^2\right\}. \tag{11}$$

In Eq. (10), the integration constants have been carefully selected to ensure a flat geometry at the boundary of the spacetime. Additionally, the equation $B(r_h) = 0$ determines the horizon radius r_h in terms of the mass parameter M and the scalar charge q. However, numerical methods are required to solve for r_h. To obtain a more meaningful solution though, we express the solution of $B(r_h) = 0$ in terms of the dimensionless quantity $r_h/(2M)$. Using Eq. (10) and algebraic manipulations, we find that

$$\frac{r_h}{2M} = \frac{6\,e^{\frac{q^2}{8r_h^2}}\left[\frac{q}{r_h} - \sqrt{2\pi}\, e^{\frac{q^2}{8r_h^2}} \,\mathrm{erf}\left(\frac{q}{2\sqrt{2}\,r_h}\right)\right]}{\frac{q}{r_h}\left\{4 - 4\,e^{\frac{q^2}{4r_h^2}} + \frac{q}{r_h}\sqrt{2\pi}\, e^{\frac{q^2}{8r_h^2}} \,\mathrm{erf}\left(\frac{q}{2\sqrt{2}\,r_h}\right) - \pi\, e^{\frac{q^2}{4r_h^2}}\left[\mathrm{erf}\left(\frac{q}{2\sqrt{2}\,r_h}\right)\right]^2\right\}}. \tag{12}$$

From Eq. (12), we observe that the value of $r_h/(2M)$ depends solely on the ratio q/r_h, indicating secondary type scalar hair. The graph of this relation is shown in Fig. 1a.

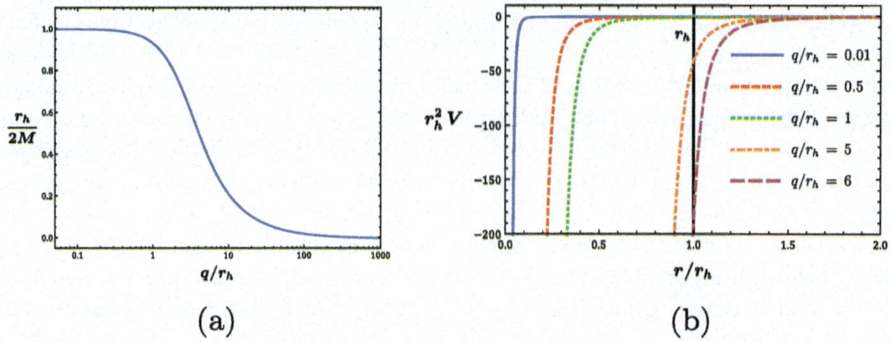

Fig. 1 **a** The graph of the fraction $r_h/(2M)$ in terms of q/r_h and **b** the potential $r_h^2 V(\Phi(r))$ in terms of the radial coordinate r for different values of the dimensionless parameter q/r_h

For small values of q/r_h, $r_h/(2M)$ approaches unity, yielding $r_h = 2M$ similar to the Schwarzschild geometry. Notice also that as q/r_h increases, $r_h/(2M)$ decreases. In the parameter space where $q > r_h$, the condition $B(r_h) = 0$ leads to an ultra-compact black hole with a smaller horizon radius than the corresponding Schwarzschild black hole of the same mass M. However, note that for $q/r_h > 10$, the horizon radius of the ultra-compact black hole becomes extremely small, indicating that this region may be nonphysical. Future observations from experiments like LIGO-VIRGO, the Event Horizon Telescope, and other astrophysical missions will provide valuable insights and constraints on the size and mass of (ultra-)compact objects.

Before discussing the behavior of the scalar potential $V(\Phi)$ given by Eq. (11), we address the No-Scalar Hair theorem [2], which extends the No-Hair theorems of General Relativity to include black holes with scalar hair. The validity of this theorem depends on the sign of the potential. To prove this, we multiply both sides of Eq. (4) by V and then integrate over the entire spacetime which lies outside the black hole. By doing so, one is led to

$$\int_{r_h}^{\infty} d^4x \sqrt{-g}\, V \left(\nabla_\mu \nabla^\mu \Phi - \partial_\Phi V \right) = 0. \tag{13}$$

In the above, we additionally assume asymptotic flatness and spherical symmetry of the scalar field, namely $\Phi = \Phi(r)$. Integrating now the first term by parts, we obtain

$$\int_{r_h}^{\infty} d^4x \sqrt{-g}\, \partial_\Phi V \left(\partial_\mu \Phi\, \partial^\mu \Phi + V \right) = 0. \tag{14}$$

The boundary term $\left[\sqrt{-g}\, V \partial^\mu \Phi\, \partial_\mu \Phi\right]_{r_h}^{\infty} = \left[\sqrt{-g}\, V B(r) (\partial_r \Phi)^2\right]_{r_h}^{\infty}$ vanishes at both boundaries. At the black hole horizon, it vanishes due to the factor $B(r)$ which becomes zero at $r = r_h$, while at infinity, $V(\Phi)$ vanishes due to asymptotic flatness. The first term inside the brackets of Eq. (14), $\partial_\mu \Phi \, \partial^\mu \Phi = g^{rr} \Phi'^2$, is positive throughout the exterior region of the black hole, hence, for Eq. (14) to be satisfied, $V(\Phi)$ must be negative. Thus, in theories with $V > 0$, non-trivial black hole solutions are impossible. Pertaining now to our solutions, Fig. 1b depicts the graphical form of the potential $r_h^2 \, V(\Phi(r))$ for different values of q/r_h. The potential is always negative and thus satisfying the evasion condition. For $q/r_h < 1$, the effect of the potential is significant inside the black hole but negligible at the horizon. On the other hand, ultra-compact black holes have a considerable value for the potential in the near-horizon region. The potential $V(\Phi(r))$ for all black-hole solutions are negative definite and monotonically increasing. It also vanishes at infinity, ensuring asymptotic flatness.

3.2 Normal and Phantom Black Holes

In contrast to what we did in the previous subsection, where we assumed a Coulombic form for the scalar field $\Phi(r)$, here we consider a particular expression for the metric function $A(r)$ [18]. Specifically, we adopt the following form:

$$A(r) = -\xi \, \ln\left(\frac{1 + r^2/q^2}{r^2/q^2}\right), \tag{15}$$

where ξ is a dimensionless constant and q is a positive physical parameter with length units. Upon using Eq. (6), we determine the functional expression of the scalar field $\Phi(r)$, namely it is

$$\Phi(r) = 2\sqrt{\xi} \, \ln\left(\frac{1 + \sqrt{1 + r^2/q^2}}{r/q}\right). \tag{16}$$

It is worth noting that for $\xi > 0$, the scalar field is real-valued, while for $\xi < 0$, it becomes purely imaginary. In the latter case, the scalar field is a phantom one, which upon a scale redefinition can be made real-valued as well, however, its kinetic term comes with an opposite sign in the action.

After obtaining the scalar field from Eq. (6), we now solve Eq. (7) to specify the metric function $B(r)$. In general, the solution for $B(r)$ is given in terms of the hypergeometric function. However, for simplicity, we will consider two specific cases: a normal solution with $\xi = 5$ and a phantom solution with $\xi = -2$. These choices yield the most simple analytic expressions for $B(r)$. Note that different values of ξ lead to different solutions. For more information see the appendix B in [18] where we present the complete solution. For $\xi = 5$, the metric function $B(r)$ is given by

$$B(r) = \left(1+\frac{r^2}{q^2}\right)^2\left[\frac{q^4}{r^4}+\frac{3q^6}{r^6}+\frac{17}{9}\frac{q^8}{r^8}-\frac{2M}{q}\frac{q^8}{r^8}\left(1+\frac{r^2}{q^2}\right)^{3/2}\right], \tag{17}$$

while for the phantom solution with $\xi = -2$, the expression for $B(r)$ is

$$B_{\mathfrak{p}}(r) = \left(1+\frac{r^2}{q^2}\right)^{-2}\left[\frac{r^4}{q^4}-\frac{2M}{q}\frac{r}{q}\left(\frac{3}{5}+\frac{r^2}{q^2}\right)-\frac{1}{3}\right]. \tag{18}$$

The subscript "$\mathfrak{p}$" indicates that we are referring to the phantom solution. Using the functions $A(r)$ and $B(r)$ that describe the spacetime geometry, along with the expression for the scalar field $\Phi(r)$, we can determine the form of the scalar potential $V(\Phi)$ via Eq. (8). For the regular field with $\xi = 5$, substituting Eqs. (15), (16), and (17) into (8), we obtain

$$V(\Phi) = \frac{\sinh^6\left(\frac{\Phi}{2\sqrt{5}}\right)}{18q^3}\Bigg\{121q - 54M\left|\coth\left(\frac{\Phi}{2\sqrt{5}}\right)\right| + \cosh\left(\frac{2\Phi}{\sqrt{5}}\right)\Bigg[17q - 18M\left|\coth\left(\frac{\Phi}{2\sqrt{5}}\right)\right|\Bigg] + 6\cosh\left(\frac{\Phi}{\sqrt{5}}\right)\Bigg[17q - 12M\left|\coth\left(\frac{\Phi}{2\sqrt{5}}\right)\right|\Bigg]\Bigg\}. \tag{19}$$

For the phantom field with $\xi = -2$, we first perform the redefinition $\Phi = i\tilde{\Phi}$ making the scalar field $\tilde{\Phi}$ real-valued and its Lagrangian density describing a phantom field. Then, from Eqs. (15), (16), and (18), one can compute that

$$V_{\mathfrak{p}}(\tilde{\Phi}) = \frac{\tanh^5\left(\frac{\tilde{\Phi}}{2\sqrt{2}}\right)}{15q^3\cosh\left(\frac{\tilde{\Phi}}{2\sqrt{2}}\right)}\left[48M + 45q\sinh\left(\frac{\tilde{\Phi}}{2\sqrt{2}}\right) + 5q\sinh\left(\frac{3\tilde{\Phi}}{2\sqrt{2}}\right)\right]. \tag{20}$$

Using equation $B(r_h) = 0$ and the expressions (17) and (18), we can establish the relations that connect the black hole mass M with its horizon radius r_h and scalar hair q. For the normal case (17), we find

$$\frac{r_h}{2M} = 9\left(1+\frac{q^2}{r_h^2}\right)^{3/2}\left(9+27\frac{q^2}{r_h^2}+17\frac{q^4}{r_h^4}\right)^{-1}, \tag{21}$$

while for the phantom case (18), we have

$$\left(\frac{r_h}{2M}\right)_{\mathfrak{p}} = \left(1-\frac{q^4}{3r_h^4}\right)^{-1}\left(1+\frac{3q^2}{5r_h^2}\right). \tag{22}$$

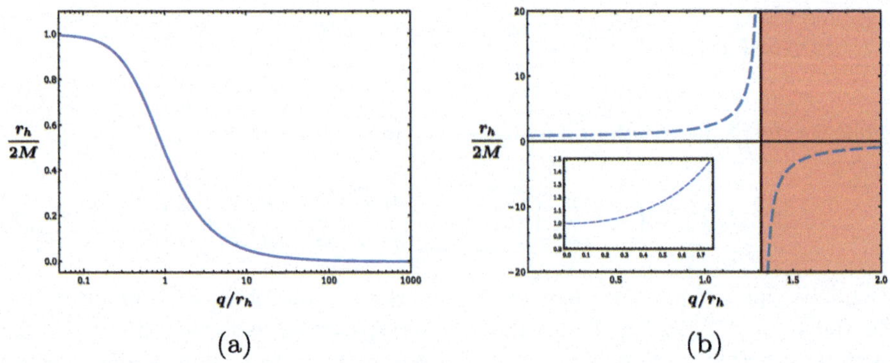

Fig. 2 The graphs of the ratio $r_h/(2M)$ in terms of q/r_h for **a** a normal scalar field and **b** a phantom one. The horizontal axis in (**a**) is logarithmic

Figure 2a, b illustrate the graphs of (21) and (22), respectively. In Fig. 2a, we notice that as q/r_h increases, $r_h/(2M)$ decreases, with $r_h/(2M)$ approaching unity for $q/r_h \to 0$. For $q/r_h \geq 1$, ultra-compact black holes with $r_h/(2M) < 0.5$ form. However, for $q/r_h > 10$, the horizon radii of ultra-compact black holes become extremely small, raising concerns about the validity of this region. In the phantom case shown in Fig. 2b, $r_h/(2M)$ initially increases with q/r_h but diverges and becomes negative at $q/r_h = \sqrt[4]{3}$. Therefore, the region where $q/r_h \geq \sqrt[4]{3}$ is physically invalid. In the region $0 < q/r_h < \sqrt[4]{3}$, the condition $r_h > 2M$ implies that black holes with phantom scalar fields are less dense compared to Schwarzschild black holes of the same mass M. These solutions, with $r_h/(2M)$ significantly exceeding unity, are referred to as "ultra-sparse black holes".

4 Slow-Rotating Solutions

In this section, we extend the static line-element (5) to include slowly rotating black-hole solutions. We treat the rotating solutions as axisymmetric perturbations on the static and spherically symmetric spacetime (5). The static solutions derived in the previous section will serve as background solutions. This approach follows the method proposed by Hartle [19] in General Relativity and later generalized by Pani and Cardoso for scalar-tensor theories [20]. The line element considered is:

$$ds^2 = -e^{A(r)} B(r)\, dt^2 + \frac{dr^2}{B(r)} + r^2 \left\{ d\theta^2 + \sin^2\theta \left[d\varphi - \varepsilon\, \omega(r)\, dt \right]^2 \right\}. \quad (23)$$

The expression involves the dimensionless auxiliary parameter ε for regulating the perturbations, while the function $\omega(r)$ is directly related to the angular velocity caused by rotational frame-dragging experienced by a locally stationary observer at distance r.

In the limit $\epsilon \to 0$, the unperturbed set of equations yield the general solutions described earlier. At first order in ε, the only independent field equation arises from the (t, φ) component, which can be brought to the following form

$$\left[r^4 e^{-\frac{A(r)}{2}} \omega'(r)\right]' = 0. \tag{24}$$

In our approximation, the angular velocity $\omega(r)$ depends only on the metric function $A(r)$ and is independent of the type of scalar field (regular or phantom) [17, 18]. By requiring that the asymptotic expansion of $\omega(r)$ matches the slowly rotating Schwarzschild solution's angular velocity, $\omega_{\mathrm{Sch}}(r) = 2J/r^3$, we can determine that

$$\omega(r) = \omega_0 - 6J \int_0^r \frac{1}{x^4} e^{\frac{A(x)}{2}} dx. \tag{25}$$

In the above, the integration constant ω_0 is specified by imposing the condition that the angular velocity becomes zero as r approaches infinity. Thus, we obtain

$$\omega_0 = 6J \int_0^\infty \frac{1}{r^4} e^{\frac{A(r)}{2}} dr. \tag{26}$$

Substituting the metric function A for the Coulombic black hole, we find

$$\omega_c(r) = -\frac{24J}{q^3}\left[\frac{q\, e^{-\frac{q^2}{8r^2}}}{r} - \sqrt{2\pi}\, \mathrm{erf}\left(\frac{q}{2\sqrt{2}\,r}\right)\right], \tag{27}$$

while for the normal and phantom black holes we find

$$\omega(r) = \frac{2J}{r^3}\left(1 + \frac{q^2}{r^2}\right)^{-3/2}, \quad \text{and} \quad \omega_p(r) = \frac{2J}{r^3}\left(1 + \frac{3q^2}{5\,r^2}\right). \tag{28}$$

Since the coulombic and the normal black holes have similar properties, from now on we will focus on the last two solutions Eq. (28). Specifically, we examine the relative angular velocities of a slowly rotating black hole with regular or phantom hair and a slowly rotating Schwarzschild black hole of equivalent mass.

Figure 3a, b depict the graphs of Eq. (28) for the angular velocities of slowly rotating black holes with regular and phantom hair compared to the slowly rotating Schwarzschild black hole of the same mass. When q/r_h is small, the angular velocities in both cases take values close to $\omega_{\mathrm{Sch}}(2M)$, while as $q \to 0$ the angular velocities become exactly equal to the angular velocity of the slowly rotating Schwarzschild

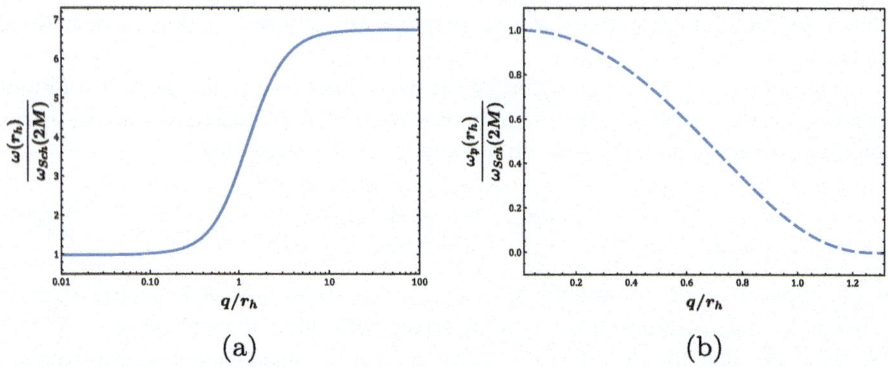

Fig. 3 Graphs of the ratio $\omega(r_h)/\omega_{Sch}(2M)$ in terms of q/r_h for **a** the normal and **b** the phantom solutions. The horizontal axis in (**a**) is logarithmic

black hole. In Fig. 3a, we observe that the ratio $\omega(r_h)/\omega_{\rm Sch}(2M)$ remains close to unity for $q/r_h \lessapprox 0.2$, while in the region from $q/r_h \approx 0.2$ to $q/r_h \approx 7$, the angular velocity $\omega(r_h)$ increases by a factor of seven compared to $\omega_{\rm Sch}(2M)$. Contrary to the normal solution, in Fig. 3b we see that the angular velocity of a phantom black hole is consistently smaller than that of a Schwarzschild black hole, and as q/r_h increases, the angular momentum decreases until it reaches zero at $q/r_h = \sqrt[4]{3}$. It should be noted that the considered phantom solutions describe ultra-sparse black holes, possessing an increasing horizon radius which blows up to infinity at $q/r_h = \sqrt[4]{3}$. The behavior of the angular velocity, for both normal and phantom black holes, can be intuitively understood using classical mechanics, where an increase in horizon radius corresponds to a larger moment of inertia that makes the rotation more difficult; on the other hand, a smaller horizon radius leads to a smaller moment of inertia and an easier rotation.

5 Conclusions

This research has investigated black hole solutions in scalar-tensor theories, focusing on a specific class of the Horndeski theories. By adopting a potential engineering approach and specifying the form of either the scalar field or a metric function, analytic black hole solutions were derived. The properties of these solutions were analyzed, revealing normal black holes which become ultra-compact for large values of the scalar charge, as well as phantom ones that become ultra-sparse.

The angular velocities of the slowly rotating versions of the aforementioned black holes were also examined. For normal black holes, we have found that as the value of the scalar charge increases, the angular velocity of the solution also increases. In contrast, phantom black holes exhibit smaller angular velocities that decrease to zero as the scalar charge approaches a critical value. These distinct rotational behaviors provide potential observational signatures to distinguish between normal and phantom black hole solutions.

Overall, this research demonstrates the richness of black hole solutions permitted within scalar-tensor theories beyond General Relativity. The existence of novel objects like ultra-compact and ultra-sparse black holes illustrates the value of exploring modified theories of gravity. Further work is still needed though to fully determine the astrophysical viability and implications of these solutions. A logical progression would involve the constraining of their allowed parameter space using astrophysical observations, along with conducting a comprehensive stability analysis.

To sum up, this study has uncovered new insights into possible black hole geometries and highlighted the flexibility of scalar-tensor theories to accommodate diverse spacetime solutions. Pursuing such solutions advances our conceptual understanding of gravity while providing potential observables to test gravity theories. Further developing the links between theory and observations will be key for maximizing the physics returns of current and planned astrophysical missions.

Acknowledgements The research project was supported by the Hellenic Foundation for Research and Innovation (H.F.R.I.) under the "3rd Call for H.F.R.I. Research Projects to support Post-Doctoral Researchers" (Project Number: 7212).

References

1. ATLAS Collaboration, G. Aad et al. Phys. Lett. B **716**, 1–29 (2012). https://doi.org/10.1016/j.physletb.2012.08.020, arXiv:1207.7214 [hep-ex]
2. J.D. Bekenstein, Phys. Rev. Lett **28**, 452–455 (1972). https://doi.org/10.1103/PhysRevLett.28.452
3. J.D. Bekenstein, Phys. Rev **D51**(12), R6608 (1995). https://doi.org/10.1103/PhysRevD.51.R6608
4. C. Herdeiro, E. Radu, Class. Quant. Grav. **32**(14), 144001 (2015). https://doi.org/10.1088/0264-9381/32/14/144001, arXiv:1501.04319 [gr-qc]
5. G. Antoniou, A. Bakopoulos, P. Kanti, Phys. Rev. Lett **120**(13), 131102 (2018). https://doi.org/10.1103/PhysRevLett.120.131102, arXiv:1711.03390 [hep-th]
6. D.D. Doneva, S.S. Yazadjiev, Phys. Rev. Lett **120**(13), 131103 (2018). http://dx.doi.org/10.1103/PhysRevLett.120.131103, arXiv:1711.01187 [gr-qc]
7. H.O. Silva, J. Sakstein, L. Gualtieri, T.P. Sotiriou, E. Berti Phys. Rev. Lett **120**(13), 131104 (2018). https://doi.org/10.1103/PhysRevLett.120.131104, arXiv:1711.02080 [gr-qc]
8. G. Antoniou, A. Bakopoulos, P. Kanti, Phys. Rev. D **97**(8), 084037 (2018). https://doi.org/10.1103/PhysRevD.97.084037, arXiv:1711.07431 [hep-th]
9. J. Ovalle, R. Casadio, E. Contreras, A. Sotomayor, Phys. Dark Univ. **31**, 100744 (2021). https://doi.org/10.1016/j.dark.2020.100744, arXiv:2006.06735 [gr-qc]
10. A. Bakopoulos, C. Charmousis, P. Kanti, N. Lecoeur, arXiv:2203.14595 [gr-qc]
11. T. Karakasis, E. Papantonopoulos, Z.-Y. Tang, B. Wang, Phys. Rev. D **103**(6), 064063 (2021). https://doi.org/10.1103/PhysRevD.103.064063, arXiv:2101.06410 [gr-qc]
12. N. Chatzifotis, P. Dorlis, N.E. Mavromatos, E. Papantonopoulos, Phys. Rev. D **106**(8), 084002 (2022). https://doi.org/10.1103/PhysRevD.106.084002, arXiv:2206.11734 [gr-qc]
13. P.G.S. Fernandes, Phys. Rev. D **103**(10), 104065 (2021). https://doi.org/10.1103/PhysRevD.103.104065, arXiv:2105.04687 [gr-qc]

14. G. Antoniou, A. Bakopoulos, P. Kanti, B. Kleihaus, J. Kunz, Phys. Rev. D **101**(2), 024033 (2020). https://doi.org/10.1103/PhysRevD.101.024033, arXiv:1904.13091 [hep-th]
15. A. Bakopoulos, C. Charmousis, P. Kanti, JCAP **05**(05), 022 (2022). https://doi.org/10.1088/1475-7516/2022/05/022, arXiv:2111.09857 [gr-qc]
16. A. Bakopoulos, N. Chatzifotis, C. Erices, E. Papantonopoulos, arXiv:2306.16768 [hep-th]
17. A. Bakopoulos, T. Nakas, JHEP **04**, 096 (2022). https://doi.org/10.1007/JHEP04(2022)096, arXiv:2107.05656 [gr-qc]
18. A. Bakopoulos, T. Nakas, Phys. Rev. D **107**(12), 124035 (2023). https://doi.org/10.1103/PhysRevD.107.124035, arXiv:2303.09116 [gr-qc]
19. J.B. Hartle, Astrophys. J **150**, 1005–1029 (1967). https://doi.org/10.1086/149400
20. P. Pani, V. Cardoso, Phys. Rev. D **79**, 084031 (2009). https://doi.org/10.1103/PhysRevD.79.084031, arXiv:0902.1569 [gr-qc]

Alleviating the Current-Era Cosmological Tensions in Quantum-Gravity Corrected String-Inspired Running-Vacuum Cosmology

Nick E. Mavromatos, Joan Solà Peracaula, and Adrià Gómez-Valent

Abstract In the context of the string-inspired running vacuum model (StRVM) of cosmology with anomalies and torsion-induced axion-like fields, we discuss one-loop quantum corrections (either from graviton loops or massive matter quantum fields) to the corresponding energy density. Such corrections are computed within the framework of weak quantum gravity in eras with approximately-de-Sitter-dark-energy dominance, during which the Hubble parameter $H(t)$ varies very slowly with the cosmic time t. The quantum-graviton corrections exhibit a logarithmic dependence on $H(t)$, of the form $H^n \ln(H^2)$, $n \in 2Z^+$. In modern eras, the $n = 2$ terns are dominant. It is then demonstrated that such corrections may contribute to the alleviation of both types of the currently observed cosmological tensions, namely H_0 and growth-of-structure tensions. We argue that such an effect is accomplished for a StRVM cosmology based on dynamically-broken primordial (pre-RVM inflation) supergravity (SUGRA). For this model, during the current de-Sitter era, the quantum graviton corrections dominate those due to matter fields, provided the energy scale of the dynamical SUGRA breaking lies a bit lower than the reduced Planck mass, which is a natural assumption, consistent with the transplanckian conjecture.

N. E. Mavromatos (✉)
National Technical University of Athens, School of Applied Mathematical and Physical Sciences, Department of Physics, 9 Iroon Polytechniou Street, Zografou Campus GR157 80, Athens, Greece
e-mail: mavroman@mail.ntua.gr

King's College London, Department of Physics, Strand, London WC2R 2LS, UK

J. S. Peracaula
Departament de Física Quàntica i Astrofísica, and Institute of Cosmos Sciences, Universitat de Barcelona, Av. Diagonal 647, 08028 Barcelona Catalonia, Spain

A. Gómez-Valent
INFN, Sezione di Roma 2, and Dipartimento di Fisica, Università di Roma Tor Vergata, via della Ricerca Scientifica 1, 00133 Roma, Italy

J. Strologas and L. Perivolaropoulos (eds.), *HEP2023*, Springer Proceedings in Physics 333 https://doi.org/10.1007/978-3-031-76856-9_4

1 Motivation and Summary

In this talk we shall review the work done in Ref. [1], which deals with a phenomenological study of quantum corrections of string-inspired running vacuum models of cosmology. Our main motivation is an attempt to resolve, by modifying the standard ΛCDM Cosmology, the cosmological tensions in the current-epoch data [2, 3]. These tensions concern either the measured value of the Hubble parameter today (H_0 tension), using nearby galaxies and comparing with Cosmic Microwave Basckground (CMB) data using ΛCDM fits [4], or growth-of-structure data. Although there could be mundane astrophysical or statistical explanations for these tensions [5], nonetheless, the fact that they persist triggered an enormous interest among the theoretical physics community [3], as they might be associated with truly novel deviations from the ΛCDM paradigm, thus pointing towards new fundamental physics explanations.

The running-vacuum-model (RVM) of cosmology offers the possibility to alleviate both types of tensions simultaneously. The RVM framework was initially motivated by semiqualitative arguments within the renormalization group [6–9], and it has been substantiated only recently by means of detailed studies in the context of string-theory inspired cosmologies [10, 11], and quantum field theoretical (QFT) calculations (renormalisation-group analyses) in curved spacetime [12–15]. According to the RVM framework, the entire evolution [16], and the thermodynamics, including entropy production [17], of the Universe can be accounted for by means of a smooth evolution of the vacuum energy density $\rho_{\rm RVM}(t)$ with the cosmic time t, from the inflationary to the modern eras. Due to *general covariance*, $\rho_{\rm RVM}(t)$ admits a perturbative expansion in *even* powers of $H(t)$:

$$\rho_{\rm RVM}(H(t)) = \frac{\Lambda(H^2(t))}{\kappa^2} = \frac{3}{\kappa^2}\Big(c_0 + \nu\, H^2(t) + \alpha\,\frac{H^4(t)}{H_I^2} + \cdots\Big). \tag{1}$$

In the above expression, $\kappa^2 = 8\pi {\rm G} = {\rm M}_{\rm Pl}^{-2}$ is the (3+1)-dimensional gravitational constant (with G the Newton's constant, and $M_{\rm Pl} = 2.435 \times 10^{18}$ GeV the reduced Planck mass), c_0 is a constant with dimensions of energy-squared, and ν, α are real dimensionless constants. All these constants assume positive values within the conventional RVM framework. The fixed quantity $H_I \sim 10^{-5}\, M_{\rm Pl}$ denotes the inflationary scale, as measured, e.g., by Planck Coll. [4]), and the ... denote higher even powers of $H(t)$.[1] The constant $3c_0\kappa^{-2}$ plays the rôle of a contribution to the cosmological constant, and the combination $3\kappa^{-2}(c_0 + \nu\, H_0^2)$ is set to the current value of the observed dark energy density [4], $\sim 10^{-122} M_{\rm Pl}^4$.

Phenomenologically, the entire history of the Universe can be sufficiently described by truncating the expansion (1) to the quartic terms H^4 [16]. It should

[1] Actually there are also terms $\dot{H}$, where the overdot denotes derivatve with repsect to the cosmic time. However, such terms can be expressed in terms of $H^2(t)$, as $\dot{H} = -(1+q)\, H^2$, where q is the deceleration parameter, which is assumed approximately constant during each cosmological epoch. Hence, the effects of the $\dot{H}$ terms can be absorbed (in each epoch) in the coefficient of H^2 in (1). In detailed phenomenological analyses of RVM, the effects of $\dot{H}$ can be calculated explicitly [18–20].

be remarked, however, that quantum field theory computations in RVM spacetimes, where massive matter fields are integrated out, yield H^2 and H^6 terms in an explicit way, but not H^4 [12, 14]. As discussed in [10, 11], the latter type of terms could arise in situations where condensates of primordial chiral-gravitational-wave tensor modes are formed in early pre-inflationary epochs of the string-inspired RVM models, called from now on stringy RVM (StRVM). The low-energy gravitational theory stemming from such string theories is characterised by torsion (provided by the field strength of the antisymmetric tensor field in the massless gravitational string multiplet) and Chern-Simons (CS) gravitational anomalies [21] (for a detailed discussion we refer the reader to [10, 11]). The StRVM cosmological evolution is characterised by different phases, in which the coefficient ν in (1) assumes different signs. In particular, it is negative during the inflationary era, and positive afterwards. Moreover, in the StRVM the constant $c_0 = 0$, given that a cosmological constant seems not compatible with either perturbative (scattering S-matrix) [22] or non-perturbative string theory (swampland criteria) [23]. In the StRVM context, approximately constant dark-energy density terms may arise either during the early RVM inflationary phase or at late eras. Such terms are *metastable*, and eventually disappear. They are the result of the formation of appropriate condensates [10, 11]. Such a metastability feature of the de Sitter phase seems also to characterise the local quantum field theory computations of Ref. [12, 14], under appropriate renormalization of the quantum vacuum in an expanding universe within the conventional RVM framework. The RVM vacuum fluid is characterised by a de-Sitter type equation of state (EoS) [6–9, 16]: $p_{\rm RVM} = -\rho_{\rm RVM}$, which is valid at early Universe epochs, dominated by the vacuum gravitational degrees of freedom. This is the case of StRVM at early epochs [11].[2] However, in post-inflationary epochs, such as radiation and matter-dominated eras, there are additional contributions to the energy density coming from the radiation or matter field excitations, ρ_m (where the index m = radiation, matter), which lie on top of the above-mentioned vacuum contributions. This in turn leads to a total energy density of the form $\rho_{\rm total} = \rho_{\rm RVM} + \rho_m$ in those eras. The EoS of such total contributions are approximately given by the equation of state of the dominant component in the respective epoch. Detailed quantum field theory computations in RVM background spacetimes [13], under appropriate vacuum-energy renormalization [12, 14], have confirmed such results.

At late eras of the cosmic evolution, the RVM framework is argued [18, 25] to be phenomenologically consistent with all the available cosmological data today [4], including cosmic-structure-formation data [19]. However, it also indicates observable deviations from the ΛCDM paradigm, due to the νH^2 term in (1), which dominates the current cosmological epoch. Fitting the current-era data with the RVM evolution framework leads to the estimate $0 < \nu = \mathcal{O}(10^{-3})$. Curiously enough, the same order of magnitude for ν is obtained by requiring consistency of the RVM framework with the Big-Bang-Nucleosynthesis data [26], under the assumption that the entirety

[2] We also note for completeness that similar situations, that is time-dependent vacuum energy satisfying a de Sitter equation of state, have appeared in the context of non-critical-string-theory models that involve brane spacetime defects [24].

of the observed [4] current-era dark energy is of RVM type. The H^4 (and higher order) terms in (1), which are dominant during the early eras of the cosmic evolution, are held responsible for driving an RVM-type inflation without the need for fundamental inflaton fields [9, 16]. Moreover, there are claims [20] that in a variant of RVM (type-II RVM), characterised by a mild phenomenological dependence of the gravitational constant on the cosmic time t, $\kappa \to \kappa(t)$, the currently observed H_0 and growth-of-structure tensions can be simultaneously alleviated [2]. As we shall argue in this talk, the StRVM may also provide such a simultaneous alleviation of both types of tensions, provided quantum corrections to the vacuum energy density are taken into account. Such corrections are due to quantum gravitons, in weak-gravity approximations, and they are of the form $H^2 \ln(H^2)$ during the modern eras [11, 27]. We shall show [1] that, under certain conditions, they could dominate the corrections induced by appropriately renormalised quantum matter fields in the spacetime background of an expanding universe [12, 14]. However, a more detailed study is required in order to ascertain whether the two types of RVM contributions (QFT versus stringy or quantum-gravity motivated effects) are comparable in magnitude and reinforce each other or point towards different directions.

The talk has the following structure: in the next Sect. 2, we formulate the StRVM cosmological model and discuss how one obtains inflation of RVM type, without external inflaton fields, upon the formation of condensates of chiral primordial GW modes. In Sect. 3, we discuss how, in the current era, the quantum-corrected StRVM could simultaneously alleviate the H_0 and galactic-growth-data tensions [1].

2 Stringy Running Vacuum Model of Cosmology

The fields appearing in the StRVM of [10, 11] belong to the bosonic part of the massless gravitational multiplet of the (closed sector of the) underlying microscopic superstring theory, which is also the ground state [28]. These are: the graviton (spin-2 symmetric tensor), the dilaton (scalar, spin-0 field) and the spin-1 antisymmetric tensor (or Kalb-Ramond (KR)) gauge field, $B_{\mu\nu} = -B_{\nu\mu}$. In (3+1)-dimensions, after string compactification, which we restrict ourselves for the purposes of this talk, the dual to the field strength of the KR field, $\mathcal{H}_{\mu\nu\rho} = \kappa^{-1}\, \partial_{[\mu}\, B_{\nu\rho]}$ (where the $[\dots]$ denotes total antisymmetrization of the respective indices), is an axion-like particle, the so-called string-model independent axion [29, 30]. The latter is essentially a Lagrange multiplier field implementing in the path integral a Bianchi identity constraint satisfied by $\mathcal{H}_{\mu\nu\rho}$ after its modification to include Green-Schwarz anomaly cancellation terms [31]:

$$\varepsilon_{abc}{}^{\mu}\, \mathcal{H}^{abc}{}_{;\mu} - \frac{\alpha'}{32\,\kappa}\, \sqrt{-g}\, R_{\mu\nu\rho\sigma}\, \widetilde{R}^{\mu\nu\rho\sigma} = 0, \tag{2}$$

where $\alpha' = M_s^{-2}$ is the Regge slope of the string, with M_s the string-mass scale. In general one has $\sqrt{\alpha'} \neq \kappa$, and essentially α' is a free, phenomenological parameter in string theory, at least as far as we understand it today. The semicolon ; denotes covariant derivative with respect to the standard Christoffel connection, $R_{\mu\nu\rho\sigma}$ denotes the Riemann tensor with respect to that connection,[3] with $\widetilde{R}_{\mu\nu\rho\sigma} = \frac{1}{2}\varepsilon_{\mu\nu\lambda\pi} R^{\lambda\pi}_{\quad\rho\sigma}$ the dual Riemann tensor, and $\varepsilon_{\mu\nu\rho\sigma} = \sqrt{-g}\, \epsilon_{\mu\nu\rho\sigma}, \varepsilon^{\mu\nu\rho\sigma} = \frac{\text{sgn}(g)}{\sqrt{-g}}\, \epsilon^{\mu\nu\rho\sigma}, \epsilon^{0123} = +1$, etc., the gravitationally covariant Levi-Civita tensor densities, totally antisymmetric in their indices. The second term in the left-hand side of (2) is proportional to the gravitational anomaly [32, 33], which is an exact one-loop result, due to circulation of chiral degrees of freedom in quantum loops.

In our discussion in [10], and in what follows, we ignore the dilaton, which can be set to zero in a self consistent manner. We also observe that, to quadratic order in a derivative expansion, $\mathcal{O}(\alpha')$, where we restrict ourselves here, the field $\mathcal{H}_{\mu\nu\rho}$ behaves as a non-propagating field. Then, upon path-integrating it out in the string effective action [28], one obtains the following action [29, 30]:

$$S_B^{\rm eff} = \int d^4x\sqrt{-g}\Big[-\frac{1}{2\kappa^2} R + \frac{1}{2}\, \partial_\mu b\, \partial^\mu b + \sqrt{\frac{2}{3}}\frac{\alpha'}{96\,\kappa} b(x) R_{\mu\nu\rho\sigma}\, \widetilde{R}^{\mu\nu\rho\sigma} + \cdots\Big], \tag{3}$$

where the dots … denote higher-derivative terms, not relevant for our discussion here. The field $b(x)$ is the pseudoscalar Lagrange multiplier which implements the Bianchi constraint (2) in the path integral of the string-inspired (low-energy) effective gravitational theory. As mentioned, this is the well-known gravitational, or KR axion, otherwise known as the string-model independent axion [30]. The latter name expresses the fact that this axion field characterises all string theories, independently of their types of compactified geometries. In string theory, of course, there is a plethora of other axions, which are the result of specific compactifications in the various string theories. Together with the KR axion, they may lead to a rich phenomenology [35] and cosmology [36]. Because to $\mathcal{O}(\alpha')$, one can view $\kappa^{-1}\,\mathcal{H}_{\mu\nu\rho}$ as a totally antisymmetric contorsion tensor, in the sense that the quadratic $\mathcal{H}_{\mu\nu\rho}\,\mathcal{H}^{\mu\nu\rho}$ in the effective low-energy gravitational lagrangian stemming from strings [28] can be absorbed in a generalised curvature with torsion [34], one may say that the KR axion in this model has a geometric origin [27]. The model (3) constitutes a CS modification of general relativity (GR) [21], and in general characterises theories in contorted geometrical backgrounds, such as Einstein-Cartan quantum electrodynamics [27, 29], in with the b field plays the rôle of the dual of the totally antisymmetric part of the (quantum) torsion, as in the string theory case.

In the cosmological model of [10, 11] it is further assumed that there was a very early phase in the primordial Universe evolution, after the Big Bang, during which the spin 3/2 supersymmetric partners of gravitons (which exist in the superstring case), the gravitinos, ψ_μ, could condense to scalar condensates $\sigma = \langle \overline{\psi}_\mu\, \psi^\mu \rangle \neq 0$, thus

[3] Our conventions and definitions here follow those of [10]: signature of metric $(+, -, -, -)$, Riemann Curvature tensor $R^\lambda_{\ \mu\nu\sigma} = \partial_\nu\, \Gamma^\lambda_{\mu\sigma} + \Gamma^\rho_{\mu\sigma}\, \Gamma^\lambda_{\rho\nu} - (\nu \leftrightarrow \sigma)$, Ricci tensor $R_{\mu\nu} = R^\lambda_{\ \mu\lambda\nu}$, and Ricci scalar $R = R_{\mu\nu} g^{\mu\nu}$.

acquiring masses close to the Planck scale, and therefore breaking local supersymmetry (supergravity) dynamically [37]. Such a scenario could lead to a primordial hill-top inflation [38], not necessarily of slow-roll type. This early inflationary phase does not have any observable consequences other than ensuring isotropy and homogeneity of the system of massless gravitons and KR axions, which is eventually characterised by a stiff-axion-matter dominated era. This can then be used to describe quantitatively the transition, via tunnelling [11], of the isotropic and homogeneous system of stiff KR axions to a second inflationary phase, of RVM type, that we now proceed to discuss. During the stiff axion era, *chiral* (left-right asymmetric) gravitational waves (GW) can form, as a consequence of non-spherically-symmetric domain-wall (DW) collapses and/or collisions. The DW owe their existence to, e.g. [11], a lifting of the degeneracy of the vacua of the double-well potential of the gravitino condensate due to dynamical percolation effects in the early Universe, which result in asymmetric occupation numbers of the two vacua [39]. The chiral GW can then condense, which leads in turn to the formation of condensates of the CS gravitational-anomaly terms in the effective action.

Assuming $\mathcal{N}(t)$ sources of such chiral primordial GW [40], we can show, following [41], that the GW induced CS condensate can be estimated as:

$$\langle R_{\mu\nu\rho\sigma}\, \widetilde{R}^{\mu\nu\rho\sigma}\rangle = \frac{\mathcal{N}(t)}{\sqrt{-g}}\,\frac{1.1}{\pi^2}\left(\frac{H}{M_{\rm Pl}}\right)^3 \mu^4\, \frac{\dot{b}(t)}{M_s^2} \equiv n_\star\, \frac{1.1}{\pi^2}\left(\frac{H}{M_{\rm Pl}}\right)^3 \mu^4\, \frac{\dot{b}(t)}{M_s^2}\,, \tag{4}$$

where H is approximately constant H during inflation. One can argue that this is a consistent solution of the pertinent cosmic evolution [10, 11]. The quantity $n_\star \equiv \mathcal{N}(t)/\sqrt{-g}$ is the proper density of sources of GW. For simplicity, the latter is assumed to be approximately constant during the RVM inflation. μ is the UV cutoff of the effective low-energy theory, which serves as an upper bound for the momenta of the quantum graviton modes. To arrive at (4), we have assumed isotropy and homogeneity of the string Universe, which can be guaranteed by the existence of the first hill-top inflation in the scenario of [11]. At this point it should be stressed that the estimate (4) is reliable only in the context of a local gravitational field theory. Unfortunately, as this estimate is sensitive to the UV cutoff μ, in the case of microscopic string theory models, towers of massive string states can in principle contribute to the value of the CS condensate. This makes an accurate estimation of the condensate (4) not possible at present, without invoking the full machinery of (non-perturbative) string theory. Nonetheless, the nature of the chiral GW ensures the non vanishing of the CS anomaly condensate.

The explicit computations of [10, 11, 40] have shown that the total vacuum energy density during inflation assumes a RVM form (1):

$$\rho^{\rm total}_{\rm vac} = -\frac{1}{2}\,\epsilon\, M_{\rm Pl}^2\, H^2 + 4.3\times 10^{10}\,\sqrt{\epsilon}\,\frac{|\overline{b}(0)|}{M_{\rm Pl}}\, H^4\,. \tag{5}$$

Above, we have used the following parametrisation of the approximately constant $\dot{b}$ axion background during inflation [10]:

$$b(t) = \overline{b}(t_0) + \sqrt{2\epsilon}\, H\, (t - t_0)\, M_{\rm Pl}\,, \quad \overline{b}(t_0) < 0\,, \tag{6}$$

where $0 < \epsilon < 1$ is a phenomenological parameter, and the time t_0 corresponds to the onset of the RVM inflation, implying that $\overline{b}(t_0)$ plays the rôle of a boundary condition for the KR axion. In arriving at the estimates of (5) we took into account the fact that the assumption on the approximate de Sitter (positive-cosmological-constant) nature of the condensate, so as to lead to an approximately constant $H \simeq H_I$ during the RVM inflation, requires [10, 11]: $|\overline{b}(t_0)| \ge N_e\, \sqrt{2\epsilon}\, M_{\rm Pl} \; = \mathcal{O}(10^2)\, \sqrt{\epsilon}\, M_{\rm Pl}$, with $N_e = \mathcal{O}(60 - 70)$ the number of e-foldings of the RVM inflation. The EoS of the string-inspired model during this inflationary phase has been explicitly computed in [11] and found to coincide with the RVM EoS [9, 16], which justifies *a posteriori* calling this model *Stringy RVM*. Because of the RVM form of the vacuum energy density (5), we observe that it is the fourth power H^4 of the Hubble parameter, being the dominant one in the early Universe, that drives inflation in this model, which thus is of RVM form, not requiring external inflaton fields for its realization. Notice that any even power of H beyond 2 would trigger inflation through that mechanism. However, in the works of [17], the structure (1) (or the generalized one with $H^{2n}, n > 2$) was assumed phenomenologically only, until such form got substantiated in the modern QFT formulation of the RVM presented in [12, 14], which encompasses the entire cosmic evolution. On the other hand, the explicit H^4 power in (5) is genuine in the stringy approach [10, 11], and can be accounted for in it. This justifies *a posteriori* our analysis on arriving at the RVM inflation from GW-induced anomaly condensates in the StRVM.

From (4) and the form of the action (3), the reader should observe that the CS condensate breaks the shift symmetry of the axion $b(x)$, leading to a linear potential for this field, of the form encountered in string theory, due to completely different reasons [42]. When string-world-sheet effects are taken into account, which lead to periodic modulations of the shift-symmetry-breaking potentials not only of the KR axion b, but also of the compactification axions [30], one can under appropriate circumstances, explained in [43], arrive at a situation in which the density of primordial black holes (pBH) produced during the RVM inflation in the StRVM is significantly enhanced, This affects the spectrum of GW in the radiation era, but also leads to significant fractions of such pBH playing the rôle of components of dark matter [44].

3 Modern Eras: Quantum Corrections in Stringy RVM and Cosmological Tensions

A detailed scenario for the post inflationary eras of the stringy RVM has been described in [10, 11], where we refer the interested readers. Here we simply mention that the decay of the metastable vacuum, at the end of the RVM inflation, leads to the generation of chiral fermionic matter and radiation. The chiral fermions generate their own gravitational CS terms, which *cancel* the primordial ones due to the Green-Schwarz mechanism, responsible for the RVM inflation. However, the chiral-fermion matter generate also global chiral anomalies in the gauge sector, which remain intact in post-RVM-inflation epochs. The axions, including the KR one, couple to such chiral-anomaly terms. It is possible then [10, 11], that during the cosmic epoch characterised by dominance of Quantum Chromodynamics (QCD) effects, instantons of the colour $SU_c(3)$ gauge group generate periodic potentials for the axions, and thus masses. In this way, such massive axions could play the rôle of components of dark matter.

In the StRVM context [11, 27], following methods developed for the prototype primordial supergravity model [37, 45], which characterises the early pre-RVM inflation epochs, and path-integrating out graviton quantum fluctuations around approximately (late-epoch) de Sitter cosmological space-times, one obtains logarithmic one-loop corrections to the energy density:

$$\delta\rho^{\rm QG}_{\rm RVM} \propto d_1\, H^2 \ln(H^2)\,, \tag{7}$$

where $H(t)$ is the mildly depending on the cosmic time t Hubble parameter in the approximately de Sitter current epoch, and the constant coefficients d_1:

$$d_1 \propto \kappa^2 \mathcal{E}_0, \ \text{ or } \ d_1 \propto \kappa^2 \mathcal{E}_0 \ln(\kappa^4 |\mathcal{E}_0|)\,. \tag{8}$$

Above, $|\mathcal{E}_0|$ denotes a bare (constant) vacuum energy density scale. In supergravity models $\mathcal{E}_0 < 0$, but in the absence of supersymmetry, which is the case of modern eras, $\mathcal{E}_0$ could be positive [1].

Corrections to the vacuum energy density with logarithmic dependence on $H(t)$ *also* appear as a result of integrating out massive fields in quantum field theories in Robertson-Walker background space times, under appropriate renormalization group analysis [12, 14]. Since in the present cosmological epoch there is appreciable matter content ($\Omega_m \simeq 0.26$) [4], such matter-generated effects are expected to be present. The question therefore arises as to whether such matter effects dominate over the quantum-gravity ones. The answer to this question is an open issue, and depends highly on the magnitude of $\mathcal{E}_0$, as a comparison with (8) shows. For instance, let us consider, for concreteness, the case of scalar real matter fields of mass m, non-minimally coupled to gravity with a parameter $\xi \in R$ [12, 14] (the conformal theory corresponds to $\xi = 1/6$). Path-ntegrating out these matter fields, one obtains the one-loop renormalised current-era energy density, in an expanding-universe background spacetime with Hubble parameter $H(t)$:

$$\begin{aligned} \rho^{\rm vac\ QFT}_{\rm RVM}(H) &= \rho^0_{\rm RVM} + \frac{3\,\nu_{\rm eff}(H)}{\kappa^2}\Big(H^2 - H_0^2\Big), \\ \nu_{\rm eff}(H) &\simeq \frac{1}{16\pi^2}\Big(\xi - \frac{1}{6}\Big)\,(\kappa^2 m^2)\, \ln\Big(\frac{m^2}{H^2}\Big) \end{aligned} \tag{9}$$

The quantity $\rho^0_{\rm RVM}$ is the standard current-era RVM energy density, which we associate with the measured value today of the cosmological constant through the relation $\Lambda = 8\pi G\rho^0_{\rm RVM}$ [15]. The result (9), is based on the assumption [12, 14] $|\ln(m^2/H^2)| \gg 1$. Qualitatively similar expressions characterise the fermion case [14]. We now notice that the corrections (9) differ from the quantum-graviton logarithmic corrections to the energy density. Indeed, $\delta\rho^{\rm QFT}_{\rm RVM} \propto (H^2 - H_0^2)\,(\kappa^2 m^2)\ln\left(\frac{m^2}{H^2}\right)$, which is small in the modern era, for which $H^2 \to H_0^2$, and cannot be cast in the form $R\ln(\kappa^2 R)$, with $R \sim 12H^2$ the de Sitter Ricci scalar, which is the case of the quantum-graviton corrections. Comparing $\delta\rho^{\rm QFT}_{\rm RVM}$ with (7) and (8), we therefore observe that, depending on the magnitude of the bare cosmological constant $\mathcal{E}_0$, matter effects could be suppressed compared to their quantum-graviton counterparts, since the factor $(\kappa^2 m^2)\,(H^2 - H_0^2) \ll (\kappa^2 m^2)\,H^2$ is very small at modern epochs. Thus, a non negligible possibility exists that quantum gravity corrections to the StRVM energy density dominate over quantum-matter-induced effects today. Making this assumption, we parametrise such corrections by the gravitational effective action [1, 27]:

$$S = -\int d^4x\,\sqrt{-g}\,\left[c_0 + R\,\left(c_1 + c_2 \ln\left(\frac{R}{R_0}\right)\right)\right] + S_m\,. \tag{10}$$

where $R_0 = 12\,H_0^2$ is the de Sitter curvature scalar today, and S_m is a matter/radiation action. The parameter $c_1 = 1/(2\kappa^2) + \delta c_1$, and δc_1 describes quantum corrections to the Newton's constant [1]. Unitarity in the gravity sector requires $c_1 > 0$ in our conventions.

A full phenomenological analysis of this class of StRVM models, including fits to all currently available cosmological data, specifically, CMB, supernovae, Baryon Acoustic Oscillations and cosmic chronometers, is given in Ref. [1], where we refer the interested reader. Attention is called to the fact that the parameter values needed for the H_0 tension alleviation are consistent with those obtained through BBN analysis in [26]. We also remark that, in StRVM, one can make the specific assumption [1] that the supergravity contributions, corresponding to a primordial supersymmetry breaking scale $\sqrt{|f|}$, which are responsible for a *negative* value of the bare cosmological constant $-f^2 < 0$ (as required by supersymmetry [37]), dominate all the quantum-graviton-induced logarithmic corrections to the energy density from the primordial to the current eras. In that case, the bare cosmological constant scale $\mathcal{E}_0$ appearing in (8) coincides with $-\sqrt{f} < 0$, and one obtains [1, 11, 27]:

$$\begin{aligned} c_1 - c_2 \ln(\kappa^2\,H_0^2) &= \frac{1}{2\kappa^2}\Big[1 + \frac{1}{2}\,\kappa^4\,f^2\,\Big(0.083 - 0.049\ln(3\kappa^4\,f^2)\Big)\Big]\,, \\ c_2 &= -0.0045\,\kappa^2\,f^2 < 0\,. \end{aligned} \tag{11}$$

The constants c_1, c_2 can then be constrained from data [1], if we require the simultaneous alleviation of the H_0 and growth tensions. The situation is summarised in Fig. 1, for the values of the dimensionless parameters $|\frac{c_2}{c_1+c_2}| = 9 \times 10^{-3}\,\kappa^4\,f^2 \lesssim \mathcal{O}(10^{-7})$

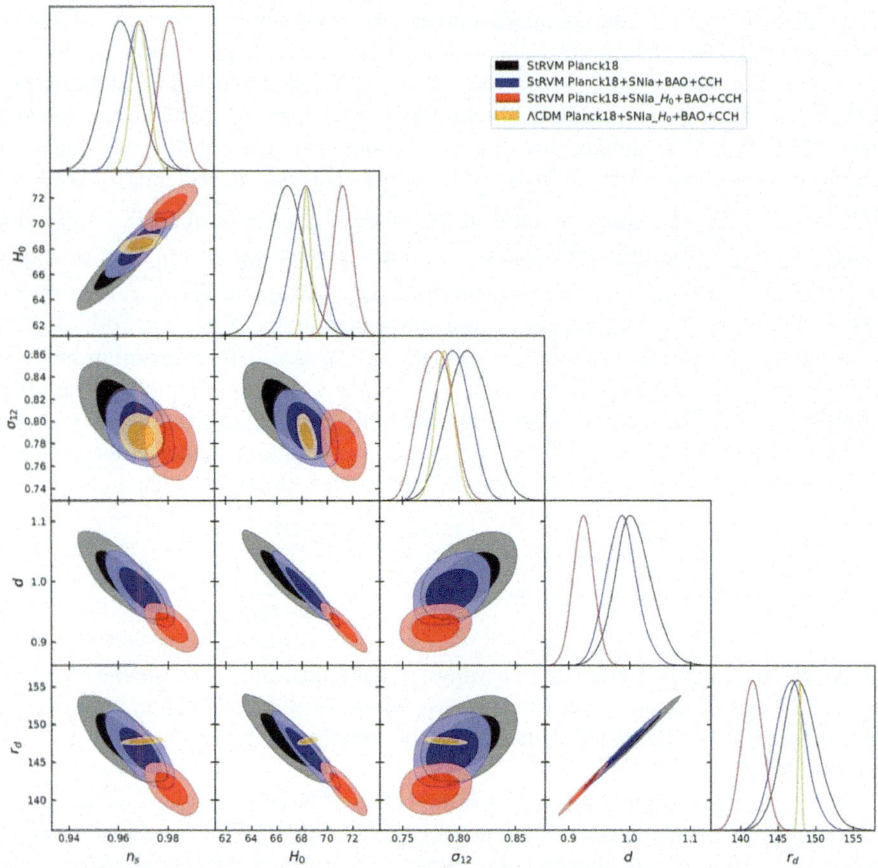

Fig. 1 The alleviation of the H_0 and growth tensions in StRVM, upon the inclusion of graviton quantum fluctuations at one-loop order in modern eras (see (10)). Here we quantify the amplitude of the matter power spectrum at linear scales by means of the root-mean-square mass fluctuations at length scales of 12 Mpc, that is, the parameter $\sigma_{12} = \sigma(r = 12\,\text{Mpc})$. The reasons why σ_{12} is a more appropriate parameter than $\sigma_8 = \sigma(r = 8h^{-1}\text{Mpc})$ for the phenomenology of the StRVM are explained in [1], which this figure is taken from. The ΛCDM predictions are depicted in orange colour

and $2\kappa^2\,(c_1 + c_2) = 0.924 \pm 0.017$. This yields the following estimate on the magnitude of $\sqrt{|f|}$, $\sqrt{|f|}\kappa \sim 10^{-5/4} \lesssim 1$, which in turn implies a sub-planckian scale for the primordial dynamical supergravity breaking, consistent with the transplanckian conjecture (that is, broadly speaking, the assumption that no physical scale exceeds the Planck scale). From (11), one also observes that such values are consistent with the perturbative deviations of c_1 from the (3+1)-dimensional gravitational constant $1/(2\kappa^2)$, despite the fact that $\kappa^2 H_0^2 = \mathcal{O}(10^{-122}) \ll 1$. One can also see [11, 27] that the quantum graviton corrections during the RVMinflationary eras are subleading, as compared to the H^4 terms in the vacuum energy density (5) induced by GW condensates. Hence, the presence of such quantum corrections does not affect the mechanism for inflation discussed in the previous section and in [10, 11].

Acknowledgements NEM would like to thank the organisers of the 40th Conference on Recent Developments in High Energy Physics and Cosmology (HEP2023, 5–7 April 2023) of the Hellenic High Energy Physics Society in the U. of Ioannina (Greece) for their kind invitation to give this plenary talk. The work of NEM is supported in part by the UK Science and Technology Facilities research Council (STFC) under the research grants ST/T000759/1 and ST/X000753/1. The work of JSP is funded in part by the projects PID2019-105614GB-C21, FPA2016-76005-C2-1-P (MINECO, Spain), 2021-SGR-249 (Generalitat de Catalunya) and CEX2019-000918-M (ICCUB, Barcelona). The work of AGV is funded by the Istituto Nazionale di Fisica Nucleare (INFN) through the project of the InDark INFN Special Initiative: "Dark Energy and Modified Gravity Models in the light of Low-Redshift Observations" (n. 22425/2020). N.E.M. and J.S.P also acknowledge participation in the COST Association Action CA18108 "*Quantum Gravity Phenomenology in the Multimessenger Approach (QG-MM)*". AGV and JSP take part in the COST Association Action CA21136 "Addressing observational tensions in cosmology with systematics and fundamental physics (CosmoVerse)".

References

1. A.Gómez-Valent, N.E. Mavromatos, J. Solà Peracaula, Class. Quant. Grav. **41**(1), 015026 (2024). https://doi.org/10.1088/1361-6382/ad0fb8, arXiv:2305.15774 [gr-qc]
2. L. Verde, T. Treu, A.G. Riess, Nature Astron. **3**, 891. arXiv:1907.10625 [astro-ph.CO]; L. Perivolaropoulos, F. Skara, New Astron. Rev. **95**, 101659 (2022)
3. E. Abdalla, et al., JHEAp **34**, 49–211 (2022); E. Di Valentino, A. Melchiorri, O. Mena, S. Vagnozzi, Phys. Rev. D **101**(6), 063502 (2020)
4. N. Aghanim, et al., [Planck], Astron. Astrophys. **641**, A6 (2020) [erratum: Astron. Astrophys. **652** (2021), C4]
5. W.L. Freedman, Nature Astron. **1**, 0121 (2017)
6. I.L. Shapiro, J. Solà, JHEP **02**, 006 (2002); Phys. Lett. B **682**, 105–113 (2009)
7. J. Solà, J. Phys. A **41**, 164066 (2008)
8. J. Solà, J. Phys. Conf. Ser. **453**, 012015 (2013); AIP Conf. Proc. 1606(1), 19–37 (2015)
9. J. Solà, A. Gómez-Valent, Int. J. Mod. Phys. D **24**, 1541003 (2015)
10. S. Basilakos, N.E. Mavromatos, J. Solà Peracaula, Phys. Rev. D **101**(4), 045001 (2020); Phys. Lett. B **803**, 135342 (2020)
11. N.E. Mavromatos, J. Solà Peracaula, Eur. Phys. J. Plus **136**(11), 1152 (2021); Eur. Phys. J. ST **230**, 9, 2077–2110 (2021)
12. C. Moreno-Pulido, J. Solà Peracaula, Eur. Phys. J. C **82**(6), 551 (2022); Eur. Phys. J. C **80**(8), 692 (2020)
13. C. Moreno-Pulido, J. Solà Peracaula, Eur. Phys. J. C **82**(12), 1137 (2022)
14. C. Moreno-Pulido, J. Solà Peracaula, S. Cheraghchi, Eur. Phys. J. C **83**, 637 (2023)
15. J. Solà Peracaula, Philos. Trans. R. Soc. Lond. A **380**, 20210182 (2022)
16. E.L.D. Perico, J.A.S. Lima, S. Basilakos, J. Solà, Phys. Rev. D **88**, 063531 (2013); J.A.S. Lima, S. Basilakos, J. Solà, Mon. Not. R. Astron. Soc. **431**, 923–929 (2013)
17. J.A.S. Lima, S. Basilakos, J. Solà, Eur. Phys. J. C **76**(4), 228 (2016); Gen. Rel. Grav. **47**, 40 (2015); J. Solà Peracaula, H. Yu, Gen. Rel. Grav. **52**(2), 17 (2020)
18. A. Gómez-Valent, J. Solà, S. Basilakos, JCAP **01**, 004 (2015)
19. A. Gómez-Valent, J. Solà, Mon. Not. R. Astron. Soc. **448**, 2810–2821 (2015); A. Gómez-Valent, E. Karimkhani, J. Solà, JCAP **12**, 048 (2015)

20. J. Solà Peracaula, A. Gómez-Valent, J. de Cruz Perez, C. Moreno-Pulido, EPL **134**(1), 19001 (2021); Universe **9**(6), 262 (2023)
21. R. Jackiw, S.Y. Pi, Phys. Rev. D **68**, 104012 (2003); S. Alexander, N. Yunes, Phys. Rep. **480**, 1–55 (2009)
22. S. Hellerman, N. Kaloper, L. Susskind, JHEP **06**, 003 (2001); W. Fischler, A. Kashani-Poor, R. McNees, S. Paban, JHEP **07**, 003 (2001)
23. H. Ooguri, C. Vafa, Nucl. Phys. B **766**, 21–33 (2007); G. Obied, H. Ooguri, L. Spodyneiko, C. Vafa, arXiv:1806.08362 [hep-th]; S.K. Garg, C. Krishnan, JHEP **11** (2019), 075; H. Ooguri, E. Palti, G. Shiu, C. Vafa, Phys. Lett. B **788**, 180–184 (2019); For reviews see: E. Palti, Fortsch. Phys. **67**(6), 1900037 (2019), and references therein
24. J.R. Ellis, N.E. Mavromatos, D.V. Nanopoulos, Gen. Rel. Grav. **32**, 943–958 (2000)
25. G. Papagiannopoulos, P. Tsiapi, S. Basilakos, A. Paliathanasis, Eur. Phys. J. C **80**(1), 55 (2020); P. Tsiapi, S. Basilakos, Mon. Not. R. Astron. Soc. **485**(2), 2505–2510 (2019)
26. P. Asimakis, S. Basilakos, N.E. Mavromatos, E.N. Saridakis, Phys. Rev. D **105**(8), 8 (2022)
27. N.E. Mavromatos, Universe **7**(12), 480 (2021); Philos. Trans. A. Math. Phys. Eng. Sci. **380**(2222), 20210188 (2022)
28. M.B. Green, J.H. Schwarz, E. Witten, *Superstring Theory Vols. 1 & 2: 25th Anniversary Edition* (Cambridge University Press, 2012) ISBN 978-1-139-53477-2, 978-1-107-02911-8; ISBN 978-1-139-53478-9, 978-1-107-02913-2
29. M.J. Duncan, N. Kaloper, K.A. Olive, Nucl. Phys. B **387**, 215–235 (1992)
30. P. Svrcek, E. Witten, JHEP **06**, 051 (2006)
31. M.B. Green, J.H. Schwarz, Phys. Lett. B **149**, 117–122 (1984)
32. T. Eguchi, P.B. Gilkey, A.J. Hanson, Phys. Rep. **66**, 213 (1980)
33. L. Alvarez-Gaume, E. Witten, Nucl. Phys. B **234**, 269 (1984)
34. F.W. Hehl, P. Von Der Heyde, G.D. Kerlick, J.M. Nester, Rev. Mod. Phys. **48**, 393–416 (1976); I.L. Shapiro, Phys. Rep. **357**, 113 (2002)
35. A. Arvanitaki, S. Dimopoulos, S. Dubovsky, N. Kaloper, J. March-Russell, Phys. Rev. D **81**, 123530 (2010)
36. D.J.E. Marsh, Phys. Rep. **643**, 1–79 (2016)
37. J. Alexandre, N. Houston, N.E. Mavromatos, Phys. Rev. D **88**, 125017 (2013); Int. J. Mod. Phys. D **24**(04), 1541004 (2015)
38. J. Ellis, N.E. Mavromatos, Phys. Rev. D **88**(8), 085029 (2013)
39. Z. Lalak, B.A. Ovrut, S. Thomas, Phys. Rev. D **51**, 5456–5474 (1995); Z. Lalak, S. Lola, B.A. Ovrut, G.G. Ross, Nucl. Phys. B **434**, 675–696 (1995); D. Coulson, Z. Lalak, B.A. Ovrut, Phys. Rev. D **53**, 4237–4246 (1996)
40. N. E. Mavromatos, Lorentz Symmetry Violation in String-Inspired Effective Modified Gravity Theories, Lect. Notes Phys. **1017**, 3–48 (2023). https://doi.org/10.1007/978-3-031-31520-6_1, arXiv:2205.07044 [hep-th]
41. S.H.S. Alexander, M.E. Peskin, M.M. Sheikh-Jabbari, Phys. Rev. Lett. **96**, 081301 (2006); D.H. Lyth, C. Quimbay, Y. Rodriguez, JHEP **03**, 016 (2005)
42. L. McAllister, E. Silverstein, A. Westphal, Phys. Rev. D **82**, 046003 (2010)
43. N.E. Mavromatos, V.C. Spanos, I.D. Stamou, Phys. Rev. D **106**(6), 063532 (2022)
44. B. Carr, F. Kuhnel, M. Sandstad, Phys. Rev. D **94**(8), 083504 (2016); S. Clesse, J. García-Bellido, Phys. Dark Univ. **15**, 142–147 (2017)
45. S. Basilakos, N.E. Mavromatos, J. Solà, Universe **2**(3), 14 (2016)

Higher-Order Corrections at the LHC: Current Status and Prospects

C. G. Papadopoulos

Abstract Precise theoretical predictions are necessary for revealing the nature of elementary particle interactions at the highest energies available at the LHC and the future collider FCC. I will review the scattering amplitudes computation starting with the leading-order (LO) and the next-to-leading-order (NLO) approximations. The recent progress at the two-loop frontier will be discussed along with next-to-next-to-leading-order (NNLO) results for $2 \rightarrow 3$ processes.

Keywords Scattering amplitudes · Higher-order corrections · Two-loop feynman integrals · Amplitude reduction at the integrand level

1 Introduction

During the last decade, we have witnessed three of the most significant discoveries of all time in the field of fundamental interactions, which can be summarised in the detection of the Higgs particle [1, 2] at the LHC, the detection of the gravitational waves (GW) [3] and the event-horizon-scale image of the supermassive black hole (BH) candidate M87* [4]. These discoveries have been made possible due to very sophisticated and technically advanced facilities such as the LHC accelerator, the ATLAS and CMS detectors, the Laser Interferometer Gravitational-Wave detectors, and the Event Horizon Telescope. On the other hand, these discoveries would have been impossible without the use of very advanced calculations and simulations regarding scattering processes in the case of Higgs particle, the description of merging BH in the case of GW detection, and the accretion process in the case of BH imaging, exploiting our best possible understanding of quantum field theory, especially QCD, general relativity (GR) and BH physics. It is fair to acknowledge that we are now better positioned to improve our understanding of fundamental interactions, including electroweak, strong, and gravitational interactions.

C. G. Papadopoulos (✉)
INPP, NCSR "Demokritos", 15310 Athens, Greece
e-mail: costas.papadopoulos@inp.demokritos.gr
URL: http://www.inp.demokritos.gr/staff-members/costas-papadopoulos/

J. Strologas and L. Perivolaropoulos (eds.), *HEP2023*, Springer Proceedings in Physics 333 https://doi.org/10.1007/978-3-031-76856-9_5

In this presentation, I will focus on the precise theoretical predictions necessary to describe scattering processes at high-energy colliders.

2 LO

Leading-order calculations refer to tree-order Feynman graphs. In the 80's the CALKUL collaboration [5] initiated a systematic approach to multi-particle processes. With the introduction of spinor techniques [6] for helicity amplitude calculations, analytic as well as numerical computation has been significantly advanced. `MadGraph` was one of the first attempts to automatise within a computer programme the calculation of tree-order scattering amplitudes [7].

Nevertheless, calculations based on Feynman graphs suffered from the $n!$ growth of the number of individual graphs as a function of the number of particles, n, involved in the process. The first attempt to go beyond Feynman graphs was the work of Berends and Giele [8]. By setting up a recursive equation they were able to prove the surprisingly compact result of Parke and Taylor for n-gluon amplitude [9]. Later on, the recursive equations have been extended to deal with any quantum field theory [10–12]. The so-called Dyson-Schwinger recursive equations refer to an arbitrary $1 \to (n-1)$ process and to all orders in perturbation theory [13, 14], as shown below.

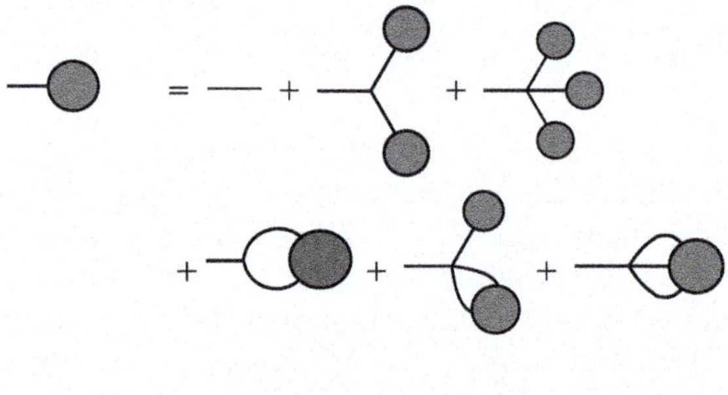

The first line refers to tree-order calculation (no loop involved), and it has been implemented in a numerical code, `HELAC` [12, 15]. `HELAC` is the first computational programme able to compute arbitrary $n-$particle scattering at LO. Nowadays the use of recursive equations to deal with tree-order amplitudes has become the standard procedure [16–18].

In Tables 1 and 2 one can see the gain in complexity achieved with the recursive equations.

Table 1 Number of Feynman graphs contributing to $n-$gluon process at tree order

$gg \to ng$	2	3	4	5	6	7	8	9
# FG	4	25	220	2,485	34,300	559,405	10,525,900	224,449,225

Table 2 Number of Berends-Giele currents needed to calculate $n-$gluon colour-stripped amplitude at tree order

$gg \rightarrow ng$	2	3	4	5	6	7	8	9
#	5	15	35	70	126	210	330	495

3 NLO

At the NLO, calculations become not only more complex but more insightful from the physics point of view. It is the first time to encounter, the renormalisation of the so-call ultraviolet (UV) divergencies [19] and the application of the factorisation of long- and short-distance physics [20].

Schematically for a process at the parton level, the NLO cross section is given by

$$\begin{aligned}\sigma_{NLO} = & \int_m d\Phi_m^{D=4}(|M_m^{(0)}|^2 + 2Re(M_m^{(0)*}M_m^{(CT)}(\epsilon_{UV})))J_m(\Phi) \qquad \leftarrow LO \\ & + \int_m d\Phi_m^{D=4} 2Re(M_m^{(0)*}M_m^{(1)}(\epsilon_{UV}, \epsilon_{IR}))J_m(\Phi) \qquad \leftarrow Virtual \\ & + \int_{m+1} d\Phi_{m+1}^{D=4-2\epsilon_{IR}}|M_{m+1}^{(0)}|^2 J_{m+1}(\Phi) \qquad \leftarrow Real\end{aligned}$$

$J_m(\Phi)$ is the jet function satisfying $J_{m+1} \rightarrow J_m$, guaranteeing infrared safeness. IR and UV refer to the infrared and ultraviolet divergencies respectively. UV divergencies are cancelled by the Lagrangian counter-terms (CT) introducing the scale dependence on the renormalisation scale μ_R. IR divergencies are cancelled by collinear counter-terms that are absorbed in the parton distribution functions introducing the QCD factorization scale μ_F.

The general form of the one-loop amplitude can be written as follows:

$$\mathcal{A} \quad = \quad \sum_{I \subset \{0,1,\cdots,m-1\}} \int \frac{\mu^{(4-d)} d^d q}{(2\pi)^d} \frac{\bar{N}_I(\bar{q})}{\prod_{i \in I} \bar{D}_i(\bar{q})}$$

where $\bar{D}_i(\bar{q})$ represent inverse Feynman propagators and $\bar{N}_I(\bar{q})$ is the numerator function, with all momenta in $d-$dimensions.

The amplitude can be written in terms of known Feynman Integrals [21] using either tensor-reduction [22] or unitarity methods [23].

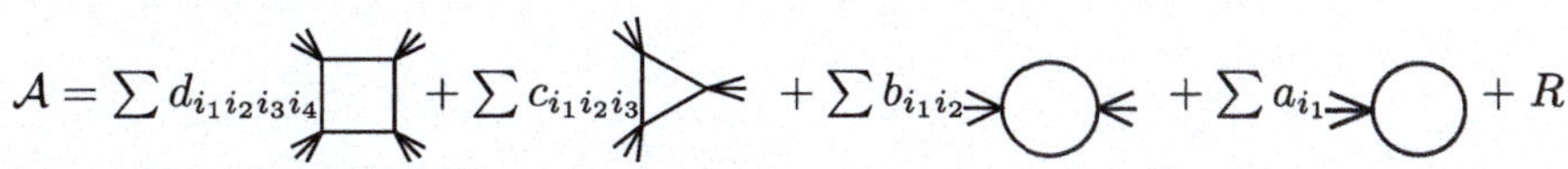

where R denotes collectively the so-called *rational terms*.

In contrast with the above-mentioned methods that work at the *integral level*, OPP [24] for the first time introduced the reduction at the *integrand level*, as shown in the following decomposition

$$
\begin{aligned}
N(q) = & \sum_{i_0<i_1<i_2<i_3}^{m-1} \left[d(i_0 i_1 i_2 i_3) + \tilde{d}(q; i_0 i_1 i_2 i_3)\right] \prod_{i \neq i_0, i_1, i_2, i_3}^{m-1} D_i \\
& + \sum_{i_0<i_1<i_2}^{m-1} \left[c(i_0 i_1 i_2) + \tilde{c}(q; i_0 i_1 i_2)\right] \prod_{i \neq i_0, i_1, i_2}^{m-1} D_i \\
& + \sum_{i_0<i_1}^{m-1} \left[b(i_0 i_1) + \tilde{b}(q; i_0 i_1)\right] \prod_{i \neq i_0, i_1}^{m-1} D_i \\
& + \sum_{i_0}^{m-1} \left[a(i_0) + \tilde{a}(q; i_0)\right] \prod_{i \neq i_0}^{m-1} D_i
\end{aligned}
$$

by introducing the so-called spurious terms, $\tilde{d}$, $\tilde{c}$, $\tilde{b}$, $\tilde{a}$, that vanish upon integration. OPP has been implemented in `CutTools` [25]. On the other hand, using Dyson-Schwinger equations applied to every one-loop topology, `HELAC-1LOOP` [26], based on the OPP method, was the first computational framework to numerically evaluate the one-loop amplitude for an arbitrary scattering process. Combined with the `HELAC-DIPOLES` [27] that addresses the one-parton unresolved contribution (so-called real corrections) to the NLO cross-section, the `HELAC-NLO` [28] was the first computational framework to numerically evaluate NLO cross sections for an arbitrary scattering process. It is fair to say that the so-called NLO-revolution [29] provided the necessary tools [18, 30–33] for the physics analysis at the LHC. It is worthwhile to mention that calculations with eight particles attached to the loop have been recently completed [34], based on `HELAC-NLO`.

4 NNLO

NNLO corrections for $2 \to 3$ processes is the current frontier to provide precise theoretical calculations to the LHC experiments [35]. The NNLO cross section at the parton level can be written as follows:

$$
\begin{aligned}
\sigma_{NNLO} = \ & \sigma_{NLO} \\
& + \int_m d\Phi_m \left(2Re(M_m^{(0)*} M_m^{(2)}) + \left|M_m^{(1)}\right|^2\right) J_m(\Phi) \qquad VV \\
& + \int_{m+1} d\Phi_{m+1} \left(2Re\left(M_{m+1}^{(0)*} M_{m+1}^{(1)}\right)\right) J_{m+1}(\Phi) \qquad RV \\
& + \int_{m+2} d\Phi_{m+2} \left|M_{m+2}^{(0)}\right|^2 J_{m+2}(\Phi) \qquad RR
\end{aligned}
$$

The VV (virtual-virtual) part includes the two-loop amplitude $M_m^{(2)}$. $RV + RR$ terms (real-virtual and real-real), address the one-loop one-particle-unresolved and the tree-order two-particle-unresolved contributions, respectively. A plethora of methods and computer programmes have been developed to deal with $RV + RR$ terms, including `antenna-S` [36, 37], `colorfull-NNLO` [38], `sector-improved residue subtraction` [39], `nested soft-collinear` [40], `local analytic sector subtraction` [41], `projection to born` [42], q_T-subtraction [43] and N-jetiness [44].

Besides non-trivial performance issues regarding the immense computer resources necessary for the numerical evaluation of $RV + RR$ terms [45], it is fair to say that the main bottleneck for NNLO calculations is the computation of two-loop amplitudes (VV terms). In fact, due to the advancement of two-loop amplitude and Feynman Integrals (FI) computations over the last few years, NNLO QCD calculations for a few $2 \to 3$ processes have been completed, including $pp \to \gamma\gamma\gamma + X$ [46, 47], $pp \to 3$ jets $+ X$ [48, 49] and $pp \to Wb\bar{b} + X$ [50], with the last one being the only NNLO calculation with one external massive leg.

In contrast to the one-loop case, the computation of the two-loop amplitude necessitates Feynman Integrals which are not known analytically yet whereas the automation of the integral-level amplitude reduction is limited for the moment to $2 \to 2$ and $2 \to 3$ processes with gluons and quarks [51].

4.1 Two-Loop Feynman Integrals

Beyond one loop, integration-by-part-identities (IBPI) [52, 53] are indispensable for the calculation of Feynman Integrals. Multi-loop FI can be defined as

$$
F[a_1, \ldots, a_N] = C_L \int \frac{1}{D_1^{a_1} \ldots D_N^{a_N}} \prod_{i=1}^{L} [d^d k_i]
$$

with a_i being zero, positive or negative integers and C_L a proper normalisation. In writing the above equation we assume that the family of FIs depend on m independent external momenta and therefore the number of independent scalar products between external and loop-momenta is given by $N = L(L+1)/2 + Lm$. Inverse propagators, $D_1 \ldots D_N$, $D_i = (\{k_1, k_2\} + p_i)^2 - M_i^2$ form a basis that allows to express all N scalar products.

IBP identities express the fact that the total derivative of a regulated FI vanishes, which in the case of two loops can be written as,

$$\int d^d k d^d l \;\; \frac{\partial}{\partial \{k^\mu, l^\mu\}} \left(\frac{\{k^\mu, l^\mu, \upsilon^\mu\}}{D_1^{a_1} \ldots D_N^{a_N}} \right) = 0$$

IBPI equations reduce *all* FI to a finite subset of Master Integrals (MI), G,

$$F[a_1, \ldots, a_N] = \sum_i R_i(\{p\}, d)\, G_i\left[a'_1, \ldots, a'_N\right]$$

where $R_i(\{p\}, d)$ are rational coefficients of the kinematical invariants formed by the external momenta and the dimension d. Now the problem has been reduced in calculating a finite set of MI.

MI can be evaluated by Feynman parameters using the sector decomposition method [54–56], or Mellin-Barnes [57] representation. Nevertheless, the Differential Equations (DE) method [58] proved the most efficient one in obtaining analytic or semi-analytic results for multi-loop FI. Since the MI is a function of external momenta, one can set up differential equations by differentiating and using IBPI to bring FI to MI.

$$p_j^\mu \frac{\partial}{\partial p_i^\mu} G[a_1, \ldots, a_n] \to \sum C_{b_1,\ldots,b_n} F[b_1, \ldots, b_n] \to \sum C_{a'_1,\ldots,a'_n} G[a'_1, \ldots, a'_n]$$

By using a proper basis of MI, one can bring the system of equations in a so-called canonical form [59],

$$\begin{aligned} \partial_m f(\varepsilon, \{x_i\}) &= \varepsilon A_m(\{x_i\})\, f(\varepsilon, \{x_i\}) \\ \partial_m A_n - \partial_n A_m &= 0 \quad [A_m, A_n] = 0 \end{aligned}$$

This allows to express the result in terms of iterated integrals [60] and in many cases of interest in terms of a particular class of functions, the Goncharov Polylogarithms (GP) [61]

$$\mathcal{G}(a_n, \ldots, a_1; x) = \int_0^x dt \frac{1}{t - a_n} \mathcal{G}(a_{n-1}, \ldots, a_1, t)$$

Boundary conditions for the system of DE can be obtained either by the expansion by regions method [62] or regularity conditions.

Fully numerical approaches to solve the system of DE have also been developed [63, 64] over the last few years.

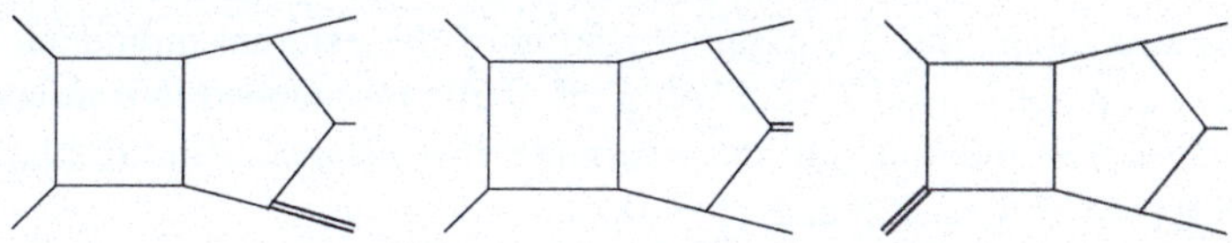

Fig. 1 The three planar pentaboxes of the families P_1 (left), P_2 (middle) and P_3 (right) with one external massive leg

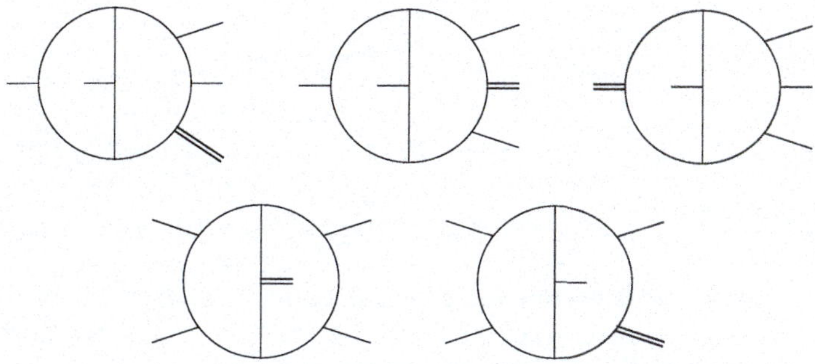

Fig. 2 The five non-planar families with one external massive leg. In the upper line the hexabox families N_1, N_2, N_3 are shown. In the lower line the double-pentagon families N_4, N_5 are shown

Five-point two-loop integrals with one off-shell leg are the current frontier. In Figs. 1 and 2, the top-sector representative integral is presented.

The P_1 family has been computed analytically in terms of GP already in 2015 [65]. The full set of planar families has been computed analytically [66] and numerically [67] based on generalised power series expansion [68], in 2020. The non-planar family N_1 has been computed in terms of GP whereas N_2, N_3 in terms of GP including one-dimensional integral representations [69]. N_1, N_2, N_3 have also been computed numerically [70] based on generalised power series expansion.

In 2014, I introduced the so-called Simplified Differential Equations approach (SDE) [71]. Based on SDE we have been able to compute all MI up to five-point two-loop integrals with one off-shell leg. To describe briefly the SDE approach, let us denote by $\mathbf{g}$ the set of MI, W_a the so-called letters in the canonical form which are functions of the external momenta and $\tilde{M}_a$ the matrices whose elements are rational numbers.

$$d\mathbf{g} = \epsilon \sum_a d \log (W_a)\, \tilde{M}_a \mathbf{g}$$

The canonical DE above is still a multi-variate equation with the W_a functions being algebraic functions of the kinematical invariants in general, and as such it is not straightforward to express the result in terms of GPs.

The SDE approach provides a factorisation of the canonical form, in terms of a univariate DE that is straightforwardly solvable in terms of GPs, in most cases. More specifically it accounts for the parametrisation of the external momenta, so that n momenta, with at least one of them being off-shell, are mapped to n momenta with one less off-shell momentum plus the x variable. For instance in the P_1 planar family with $q_i^2 = 0$ for $i = 1, 2, 4, 5$ and $q_3^2 \neq 0$,

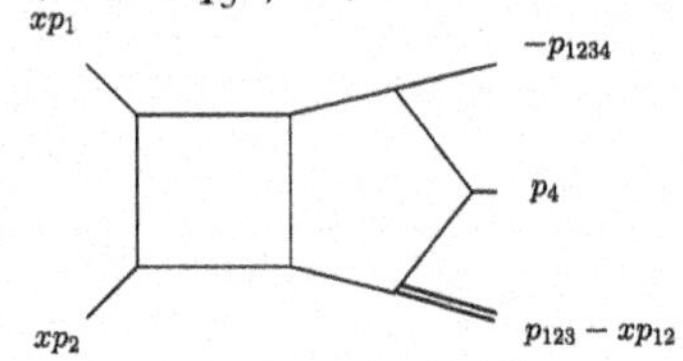

the mapping $\{q_1, q_2, q_3, q_4, q_5\} \to \{p_1, p_2, p_3, p_4, p_5, x\}$ defined by (all $p_i^2 = 0$)

$$q_1 \to xp_1,\; q_2 \to xp_2,\; q_3 \to p_1 + p_2 + p_3 - x(p_1 + p_2),\; q_4 \to p_4,\; q_5 \to p_5$$

explicitly separates or factorises the dependence on x. In fact, by expressing the initial invariants $q_i \cdot q_j$ in terms of the new invariants $p_i \cdot p_j$ and the variable x, and then differentiating with respect to x the DE takes the form

$$\frac{d\mathbf{g}}{dx} = \epsilon \sum_b \frac{1}{x - \ell_b} M_b \mathbf{g}$$

where ℓ_b depends on the new invariants $p_i \cdot p_j$, but are independent of x and M_b are just matrices of rational numbers. The fact that the DE has now simple poles in the variable x allows for a straightforward expression of its solution in terms of GP

$$\begin{aligned}
\mathbf{g} = {} & \epsilon^0 \mathbf{b}_0^{(0)} + \epsilon \Big(\sum \mathcal{G}_a \mathbf{M}_a \mathbf{b}_0^{(0)} + \mathbf{b}_0^{(1)} \Big) \\
& + \epsilon^2 \Big(\sum \mathcal{G}_{ab} \mathbf{M}_a \mathbf{M}_b \mathbf{b}_0^{(0)} + \sum \mathcal{G}_a \mathbf{M}_a \mathbf{b}_0^{(1)} + \mathbf{b}_0^{(2)} \Big) \\
& + \epsilon^3 \Big(\sum \mathcal{G}_{abc} \mathbf{M}_a \mathbf{M}_b \mathbf{M}_c \mathbf{b}_0^{(0)} + \sum \mathcal{G}_{ab} \mathbf{M}_a \mathbf{M}_b \mathbf{b}_0^{(1)} + \sum \mathcal{G}_a \mathbf{M}_a \mathbf{b}_0^{(2)} + \mathbf{b}_0^{(3)} \Big) \\
& + \epsilon^4 \Big(\sum \mathcal{G}_{abcd} \mathbf{M}_a \mathbf{M}_b \mathbf{M}_c \mathbf{M}_d \mathbf{b}_0^{(0)} + \sum \mathcal{G}_{abc} \mathbf{M}_a \mathbf{M}_b \mathbf{M}_c \mathbf{b}_0^{(1)} \\
& + \sum \mathcal{G}_{ab} \mathbf{M}_a \mathbf{M}_b \mathbf{b}_0^{(2)} + \sum \mathcal{G}_a \mathbf{M}_a \mathbf{b}_0^{(3)} + \mathbf{b}_0^{(4)} \Big) + \ldots \\
\mathcal{G}_{ab\ldots} := {} & \mathcal{G}(\ell_a, \ell_b, \ldots; x)
\end{aligned}$$

where $\mathbf{b}_0^{(i)}$ are the boundary constants[1].

---

[1] See references [66, 69] for details.

Table 3 Skeleton data for several processes (***Process***), including the flavour of the particles encountered in the loop (***Loop-Flavors***), the treatment of the colour (***Color***), the disk size of the skeleton (***Size***), the time for its creation (on a single laptop) (***Time***) and the number of numerators computed (***Num/s***)

Process	#	*Loop-Flavors*	*Color*	*Size*	*Time*	*Num/s*
$gg \to gg$	2	$\{g, c, \bar{c}\}$	Lead.	8.9 MB	15.017 s	4560
$gg \to gg$	2	$\{g, q, \bar{q}, c, \bar{c}\}$	Full	110.6 MB	6 m 54.574 s	89392
$gg \to q\bar{q}$	2	$\{g, q, \bar{q}, c, \bar{c}\}$	Full	16.1 MB	3 m 14.509 s	13856
$gg \to ggg$	2	$\{g, c, \bar{c}\}$	Lead.	300.0 MB	21 m 42.609 s	81480

4.2 *Two-Loop Amplitude Reduction*

A generic two-loop amplitude can be decomposed following an OPP-type approach

$$\frac{N\left(l_1, l_2; \{p_i\}\right)}{D_1 D_2 \ldots D_n} = \sum_{m=1}^{\min(n,n_p)} \sum_{S_{m;n}} \frac{\Delta_{i_1 i_2 \ldots i_m}\left(l_1, l_2; \{p_i\}\right)}{D_{i_1} D_{i_2} \ldots D_{i_m}} \tag{1}$$

where $S_{m;n}$ denotes the subsets of m out of n elements, and $n_p = 8$ or $n_p = 11$ depending if the reduction is performed in 4 or in d dimensions. The residues $\Delta_{i_1 i_2 \ldots i_m}\left(l_1, l_2; \{p_i\}\right)$ are composed by terms corresponding to irreducible integrals and spurious terms that vanish upon integration. As usual, the determination of the residue terms proceeds through the cut equations $D_{i_1} = D_{i_2} = \ldots = D_{i_m} = 0$. Irreducible integrals are expressed in terms of MI through IBPI.

The left-hand side of Eq. (1) is evaluated by `HELAC-2LOOP`. This is achieved by expressing the $n-$particle $2-$loop amplitude in terms on $(n+2)-$particle 1-loop amplitude and then follow an appropriately adapted `HELAC-1LOOP` construction. As in the one-loop case, a skeleton is constructed that includes all numerators for a given process. For details please refer to [72].

In the Table 3 below we present the first results on the construction of the skeleton for several processes, in both leading-colour and full-colour approximations.

5 Summary and Outlook

In the last few years, significant progress on the two-loop frontier has been achieved. In this presentation, I have reviewed most of the major accomplishments of this progress. For two-loop MI the current frontier is five-point two-loop integral families with up to one massive leg. Based on their analytic representation numerical tools have been developed [73–75]. With the completion of the HELAC-2LOOP framework, we expect that the numerical evaluation of two-loop amplitudes with up to five external particles and with up to one external off-shell momentum will be available. The successful application of the SDE approach to MI families with internal masses [76, 77] paves the road for new results, significantly widening the range of two-loop amplitudes to be computed. Based on all these developments, NNLO QCD results are expected to be copiously produced in the near future.

Acknowledgements The research project was supported by the Hellenic Foundation for Research and Innovation (H.F.R.I.) under the "2nd Call for H.F.R.I. Research Projects to support Faculty Members and Researchers" (Project Number: 02674).

References

1. G. Aad et al., Phys. Lett. B **716**, 1 (2012)
2. S. Chatrchyan et al., Phys. Lett. B **716**, 30 (2012)
3. B.P. Abbott et al., Phys. Rev. Lett. **116**(6), 061102 (2016)
4. K. Akiyama et al., Astrophys. J. Lett. **875**, L1 (2019)
5. P. De Causmaecker, R. Gastmans, W. Troost, T.T. Wu, Nuclear Phys. B **206**(1), 53 (1982)
6. R. Kleiss, W.J. Stirling, Nucl. Phys. B **262**, 235 (1985)
7. T. Stelzer, W.F. Long, Comput. Phys. Commun. **81**, 357 (1994)
8. F.A. Berends, W.T. Giele, Nucl. Phys. B **306**, 759 (1988)
9. S.J. Parke, T.R. Taylor, Phys. Rev. Lett. **56**, 2459 (1986)
10. F. Caravaglios, M. Moretti, Phys. Lett. B **358**, 332 (1995)
11. P. Draggiotis, R.H.P. Kleiss, C.G. Papadopoulos, Phys. Lett. B **439**, 157 (1998)
12. A. Kanaki, C.G. Papadopoulos, Comput. Phys. Commun. **132**, 306 (2000)
13. E.N. Argyres, R.H.P. Kleiss, C.G. Papadopoulos, Nucl. Phys. B **391**, 42 (1993)
14. E.N. Argyres, R.H.P. Kleiss, C.G. Papadopoulos, Phys. Lett. B **308**, 292 (1993). [Addendum: Phys.Lett.B 319, 544 (1993)]
15. A. Cafarella, C.G. Papadopoulos, M. Worek, Comput. Phys. Commun. **180**, 1941 (2009)
16. M.L. Mangano, M. Moretti, F. Piccinini, R. Pittau, A.D. Polosa, JHEP **07**, 001 (2003)
17. T. Gleisberg, S. Hoeche, JHEP **12**, 039 (2008)
18. S. Actis, A. Denner, L. Hofer, J.N. Lang, A. Scharf, S. Uccirati, Comput. Phys. Commun. **214**, 140 (2017)
19. G. 't Hooft, Nucl. Phys. B **61**, 455 (1973)
20. J.C. Collins, D.E. Soper, G.F. Sterman, Adv. Ser. Direct. High Energy Phys. **5**, 1 (1989)
21. G. 't Hooft, M.J.G. Veltman, Nucl. Phys. B **153**, 365 (1979)
22. G. Passarino, M.J.G. Veltman, Nucl. Phys. B **160**, 151 (1979)
23. Z. Bern, L.J. Dixon, D.C. Dunbar, D.A. Kosower, Nucl. Phys. B **425**, 217 (1994)
24. G. Ossola, C.G. Papadopoulos, R. Pittau, Nucl. Phys. B **763**, 147 (2007)
25. G. Ossola, C.G. Papadopoulos, R. Pittau, JHEP **03**, 042 (2008)
26. A. van Hameren, C.G. Papadopoulos, R. Pittau, JHEP **09**, 106 (2009)
27. M. Czakon, C.G. Papadopoulos, M. Worek, JHEP **08**, 085 (2009)
28. G. Bevilacqua, M. Czakon, M.V. Garzelli, A. van Hameren, A. Kardos, C.G. Papadopoulos, R. Pittau, M. Worek, Comput. Phys. Commun. **184**, 986 (2013)
29. G.P. Salam, PoS **ICHEP2010**, 556 (2010)
30. C.F. Berger, Z. Bern, L.J. Dixon, F. Febres Cordero, D. Forde, H. Ita, D.A. Kosower, D. Maitre, Phys. Rev. D **78**, 036003 (2008)

31. J. Alwall, R. Frederix, S. Frixione, V. Hirschi, F. Maltoni, O. Mattelaer, H.S. Shao, T. Stelzer, P. Torrielli, M. Zaro, JHEP **07**, 079 (2014)
32. F. Cascioli, P. Maierhofer, S. Pozzorini, Phys. Rev. Lett. **108**, 111601 (2012)
33. F. Buccioni, J.N. Lang, J.M. Lindert, P. Maierhöfer, S. Pozzorini, H. Zhang, M.F. Zoller, Eur. Phys. J. C **79**(10), 866 (2019)
34. G. Bevilacqua, M. Lupattelli, D. Stremmer, M. Worek, Phys. Rev. D **107**(11), 114027 (2023)
35. A. Huss, J. Huston, S. Jones, M. Pellen, J. Phys. G **50**(4), 043001 (2023)
36. A. Gehrmann-De Ridder, T. Gehrmann, M. Ritzmann, JHEP **10**, 047 (2012)
37. T. Gehrmann, P.F. Monni, JHEP **12**, 049 (2011)
38. G. Somogyi, Z. Trocsanyi, JHEP **08**, 042 (2008)
39. M. Czakon, D. Heymes, Nucl. Phys. B **890**, 152 (2014)
40. F. Caola, K. Melnikov, R. Röntsch, Eur. Phys. J. C **77**(4), 248 (2017)
41. L. Magnea, E. Maina, G. Pelliccioli, C. Signorile-Signorile, P. Torrielli, S. Uccirati, JHEP **12**, 107 (2018). [Erratum: JHEP 06, 013 (2019)]
42. M. Cacciari, F.A. Dreyer, A. Karlberg, G.P. Salam, G. Zanderighi, Phys. Rev. Lett. **115**(8), 082002 (2015). [Erratum: Phys.Rev.Lett. 120, 139901 (2018)]
43. S. Catani, M. Grazzini, Phys. Rev. Lett. **98**, 222002 (2007)
44. R. Boughezal, C. Focke, X. Liu, F. Petriello, Phys. Rev. Lett. **115**(6), 062002 (2015)
45. M. Czakon, Z. Kassabov, A. Mitov, R. Poncelet, A. Popescu, (2023). arXiv:2304.05993 [hep-ph]
46. H.A. Chawdhry, M.L. Czakon, A. Mitov, R. Poncelet, JHEP **02**, 057 (2020)
47. S. Kallweit, V. Sotnikov, M. Wiesemann, Phys. Lett. B **812**, 136013 (2021)
48. M. Czakon, A. Mitov, R. Poncelet, Phys. Rev. Lett. **127**(15), 152001 (2021). [Erratum: Phys.Rev.Lett. 129, 119901 (2022), Erratum: Phys.Rev.Lett. 129, 119901 (2022)]
49. X. Chen, T. Gehrmann, E.W.N. Glover, A. Huss, M. Marcoli, JHEP **10**, 099 (2022)
50. H.B. Hartanto, R. Poncelet, A. Popescu, S. Zoia, Phys. Rev. D **106**(7), 074016 (2022)
51. S. Abreu, J. Dormans, F. Febres Cordero, H. Ita, M. Kraus, B. Page, E. Pascual, M.S. Ruf, V. Sotnikov, Comput. Phys. Commun. **267**, 108069 (2021)
52. F.V. Tkachov, Phys. Lett. B **100**, 65 (1981)
53. K.G. Chetyrkin, F.V. Tkachov, Nucl. Phys. B **192**, 159 (1981)
54. T. Binoth, G. Heinrich, Nucl. Phys. B **585**, 741 (2000)
55. A.V. Smirnov, M.N. Tentyukov, Comput. Phys. Commun. **180**, 735 (2009)
56. J. Carter, G. Heinrich, Comput. Phys. Commun. **182**, 1566 (2011)
57. M. Czakon, Comput. Phys. Commun. **175**, 559 (2006)
58. A.V. Kotikov, Phys. Lett. B **254**, 158 (1991)
59. J.M. Henn, Phys. Rev. Lett. **110**, 251601 (2013)
60. K.T. Chen, Bull. Am. Math. Soc. **83**, 831 (1977)
61. A.B. Goncharov, M. Spradlin, C. Vergu, A. Volovich, Phys. Rev. Lett. **105**, 151605 (2010)
62. B. Jantzen, A.V. Smirnov, V.A. Smirnov, Eur. Phys. J. C **72**, 2139 (2012)
63. M. Hidding, Comput. Phys. Commun. **269**, 108125 (2021)
64. X. Liu, Y.Q. Ma, Comput. Phys. Commun. **283**, 108565 (2023)
65. C.G. Papadopoulos, D. Tommasini, C. Wever, JHEP **04**, 078 (2016)
66. D.D. Canko, C.G. Papadopoulos, N. Syrrakos, JHEP **01**, 199 (2021)
67. S. Abreu, H. Ita, F. Moriello, B. Page, W. Tschernow, M. Zeng, JHEP **11**, 117 (2020)
68. F. Moriello, JHEP **01**, 150 (2020)
69. A. Kardos, C.G. Papadopoulos, A.V. Smirnov, N. Syrrakos, C. Wever, JHEP **05**, 033 (2022)
70. S. Abreu, H. Ita, B. Page, W. Tschernow, JHEP **03**, 182 (2022)
71. C.G. Papadopoulos, JHEP **07**, 088 (2014)
72. G. Bevilacqua, D.D. Canko, A. Kardos, C.G. Papadopoulos, J. Phys. Conf. Ser. **2105**(5), 012010 (2021)
73. D. Chicherin, V. Sotnikov, JHEP **20**, 167 (2020)
74. D. Chicherin, V. Sotnikov, S. Zoia, JHEP **01**, 096 (2022)
75. S. Abreu, D. Chicherin, H. Ita, B. Page, V. Sotnikov, W. Tschernow, S. Zoia (2023). arXiv:2306.15431 [hep-ph]
76. N. Syrrakos, JHEP **10**, 041 (2021)
77. N. Syrrakos, JHEP **05**, 131 (2023)

A Field Theory Approach to Dark Energy and Its Observational Implications

Anupam Singh

Abstract From the perspective of Particle Physics and Field Theory, Dark Energy is a low energy phenomenon. However, in the context of the Standard Model of Particle Physics obtaining fields with sufficiently low masses that can both be protected technically and be relevant for Dark Energy physics can be a challenge. Pseudo Nambu Goldstone Bosons tied to non-zero Neutrino masses provide an approach to the Dark Energy problem that appears to be very promising. Not only does such an approach solve the traditional problems that Dark Energy was invented to solve but it provides avenues to explore the connected areas of physics, cosmology and astrophysics. As some examples of this, we will discuss the gravitational collapse of dark energy field configurations to form SMBHs (Super Massive Black Holes) with masses comparable to the masses at the centers of galaxies. Further, we will discuss the gravitational waves produced by dark energy fields and show that these gravitational waves can explain the periodicity of the Ice Ages through the amplitude and frequency of the ellipticity variation of earth's orbit created by such Dark Energy Gravitational Waves. Finally, we discuss phase transitions in these Dark Energy models, the resulting Black Hole formation and the prediction of an earlier generation of Quasars and structure formation at redshifts $z \sim 18$.

Keywords Dark energy · Field theory · Neutrino masses · Phase transitions · Black holes · Quasars · Cosmology · Gravitational waves · Observations

1 Introduction

In the last century, a large fraction of fundamental physics was uncovered as a result of precision experiments carried out at high energy accelarators on earth. During the same period, humankind also started focusing on information on fundamental physics that could be obtained from astronomical observations. There is now a wide range of sophisticated satellites gathering data from the sky which is giving us fundamental

A. Singh (✉)
Physics Department, LNMIIT, Jaipur, India
e-mail: singh@lnmiit.ac.in

J. Strologas and L. Perivolaropoulos (eds.), *HEP2023*, Springer Proceedings in Physics 333 https://doi.org/10.1007/978-3-031-76856-9_6

new insights and opening unexpected doors to new physics. A prime example of this is the accurate determination of the Hubble constant and the independent determination of the age of the universe which forced us to critically re-examine the simplest cosmological model—a universe with a zero cosmological constant [1, 2]. These observations led us [3] to examine the idea of a non-vanishing vacuum energy due to fields as playing an important role in the current Universe. Since then, there has been a growing body of work that has confirmed the existence of what we now call dark energy. For an exhaustive exposition on Dark Energy, please see [4].

Here we wish to focus on field theory models of dark energy and its observational implications. Field theory models for dark energy motivated by particle physics were discussed earlier [3]. The fields involved must have very light masses so as to be cosmologically relevant in the current epoch. Investigations of particle physics models with naturally light masses have been done by several authors [5, 6]. Thus, it was noted that we can get interesting effects in models of neutrino masses with Pseudo Nambu Goldstone Bosons (PNGB's). The mass scales can then be related to the neutrino masses. Finally, any small parameters required can be protected from radiative corrections by the symmetry underlying the Nambu-Goldstone bosons [7].

PNGBs have a celebrated history because at the late times relevant for observational cosmology any dynamical fields must have small masses compared to the usual masses encountered in the standard model. PNGBs provide a natural mechanism from obtaining light fields. Of the PNGBs, the most celebrated is the axion (please see e.g. [8–12] for more details). Finally, of all the particles we have currently detected, the particles with the smallest non-vanishing masses are the neutrinos. Thus, it should not be surprising that PNGBs linked to neutrino masses can be important for understading the cosmological dynamics at the current time.

2 Dark Energy Collapse and Super Massive Black Hole Formation

Here we report on our study of the gravitational dynamics of Dark Energy. The dynamics of fields in cosmological space-times has been discussed previously (see e.g. Kolb and Turner [14]). Similarly, gravitational collapse has also been discussed previously (see e.g. Weinberg [15]). We build on this to write down and study the evolution equations for the Dark Energy field in our model.

To study gravitational collapse in a cosmological setting we can start with the metric:

$$ds^2 = dt^2 - U(r,t)dr^2 - V(r,t)\left[d\theta^2 + \sin^2\theta d\phi^2\right] \tag{1}$$

where, (t, r, θ, ϕ) are the comoving coordinates describing a space time point. This metric is a generalization of the FRW metric used to study homogeneous cosmological space-times. Note that the functions $U(r, t)$ and $V(r, t)$ are functions of both space and time and can capture inhomogeneous gravitational collapse.

Let us keep the initial discussion as general as possible. Thus, we will consider a field Φ with a Lagrangian L given by

$$L = \frac{1}{2}\partial^{\mu}\Phi\partial_{\mu}\Phi - \mathcal{V}(\Phi) \tag{2}$$

where $\mathcal{V}$ is the potential for the field Φ which for now can be an arbitrary function. After we obtain the general equations for any potential, we will specialize to consider the physically motivated PNGB models.

By using standard techniques we can write the evolution equations in the form:

$$\ddot{V} = 2\left[-1 + \frac{V''}{2U} - \frac{V'U'}{4U^2} - \frac{\dot{V}\dot{U}}{4U} + 8\pi G V\left(\frac{\rho}{2} - \frac{P}{2} - \frac{(\Phi')^2}{3U}\right)\right] \tag{3}$$

$$\ddot{U} = 2U\left[-\frac{\ddot{V}}{V} + \frac{\dot{U}^2}{4U^2} + \frac{\dot{V}^2}{2V^2} - 4\pi G(\rho + 3P)\right] \tag{4}$$

$$\ddot{\Phi} = \frac{\Phi''}{U} - \dot{\Phi}\left[\frac{\dot{V}}{V} + \frac{\dot{U}}{2U}\right] + \frac{\Phi'}{U}\left[\frac{V'}{V} - \frac{U'}{2U}\right] - \frac{\partial\mathcal{V}(\Phi)}{\partial\Phi} \tag{5}$$

where a dot represents a partial derivative w.r.t. t and a prime represents a partial derivative w.r.t. r. Further,

$$\rho = \frac{1}{2}\dot{\Phi}^2 + \mathcal{V}(\Phi) + \frac{(\Phi')^2}{2U} \tag{6}$$

and

$$P = \frac{1}{2}\dot{\Phi}^2 - \mathcal{V}(\Phi) + \frac{(\Phi')^2}{6U}. \tag{7}$$

The above equations are true for any general potential $\mathcal{V}(\Phi)$. The simplest potential one can write down for the physically motivated PNGB fields [5] can be written in the form:

$$\mathcal{V}(\Phi) = m^4\left[K - \cos(\frac{\Phi}{f})\right] \tag{8}$$

As discussed in [5] m is of order the neutrino mass and K is of order 1. We will consider such a potential for studying the dynamics in the next section.

The evolution equations described above can be solved numerically to study the gravitational collapse of the field configurations. The key issue to understand and address is the timescale on which the gravitational collapse happens and see is this timescale is shorter than the age of the Universe. From our numerical study we see that the collapse of the field configuration occurs on timescales $\leq 10^6$ years. Indeed, we followed this up with a more careful numerical analysis using the 3+1 BSSN formalism of numerical relativity [16] where we showed the formation of Black Holes as a result of the collapse of Dark Energy field configurations. Moreover, we found that the masses of these Black Holes were consistent with the masses of Black Holes found at the centers of galaxies.

3 Gravitational Waves from Dark Energy and Ice Age Periodicity

The production of Gravitational Waves from Dark Energy and its impact on the orbital eccentricity of the earth and Ice Age Periodicty has been discussed by us previously [21]. Here, we wish to highlight that this is an important observational consequence of dark energy. Further, as we will see the resonance mechanism which can be used to understand the production of the gravitational waves gives a unique signature and a fingerprint of the nature of the dark energy field and thus provides a direct link between the identity of the Dark Energy field and its specific observational consequence.

Thus, as we will see the mass of the Dark Energy field determines the frequency of the gravitational wave which in turn determines the periodicity of the Ice Ages which is measured observationally (see e.g. [17, 18]). Further, the amplitude [19] is determined by the Dark Energy density—as we will show in our analysis below. Thus, the observed values for the amplitude and the values which follow from our analysis of the Dark Energy dynamics and resonance again match.

To quantitatively understand these issues, the natural place to start is the equation of the gravitational wave field $h_{\mu\nu}$. By using the fact that the primary mechanism for the transfer of energy from the Dark Energy field to the gravitational wave field is the resonance phenomenon, we will arrive at both the frequency and amplitude of the gravitational wave field.

The gravitational waves in vacuum obey:

$$\Box h_{\mu\nu} = 0 \tag{9}$$

this is a consequence of the Einstein Equation:

$$R_{\mu\nu} - \frac{1}{2} g_{\mu\nu} R = \kappa T_{\mu\nu} \tag{10}$$

with the source term zero i.e. when $T_{\mu\nu} = 0$ (please see for example [22]).

When $T_{\mu\nu} \neq 0$, the gravitational wave components $h_{\mu\nu}$ obey:

$$\Box h_{\mu\nu} = S_{\mu\nu} \tag{11}$$

The source term $S_{\mu\nu}$ will be non-zero if you have non-trivial scalar field configurations present. This will happen generically in any realistic collapse which happens in nature.

For a scalar field having a frequency ω (determined by the mass m of the scalar field), the source term becomes:

$$S_{\mu\nu} = S^{(0)}_{\mu\nu} e^{i\omega t} \tag{12}$$

The gravitational waves now becomes:

$$\Box h_{\mu\nu} = S^{(0)}_{\mu\nu} e^{i\omega t} \tag{13}$$

Looking at the equation for the gravitational wave components in Fourier space, we get:

$$\frac{d^2 h_{\mu\nu}}{dt^2} + k^2 h_{\mu\nu} = S^{(0)}_{\mu\nu} e^{i\omega t} \tag{14}$$

This is the equation for of a driven harmonic oscillator.

As is well known from the solution of the driven harmonic oscillator, this implies two things:

(1) The resonance enhances the amplitude when the driving frequency and driven frequency match. In our case, the time period of the Gravitational Waves produced matches the time period of the scalar field.

(2) The amplitude grows at resonance.

The solution to the driven oscillator at resonance implies that energetics permitting, the gravity wave amplitude grows to large values given by:

$$h^{(0)}_{\mu\nu} = \frac{S^{(0)}_{\mu\nu}}{(k^2 - \omega^2)} \tag{15}$$

implying as $k \rightarrow \omega$ the gravity wave amplitude is limited only by energetics.

Energetics will limit the amplitude from growing indefinitely and will limit the amplitude to finite values.

We now use the energetics to arrive at an estimate of the amplitude of the gravitational waves produced. We note that the system has two important sub-components: dark energy and gravitational waves. These two sub-components interact and so can exchange energy. Thus, the system's long term state will be dynamically driven towards the point where the energy density in gravitational waves ρ_{gw} equals the energy density in the dark energy field ρ_{DE}. Hence, we will use the condition (which we henceforth refer to as the equipartition condition):

$$\rho_{gw} \simeq \rho_{DE} \tag{16}$$

to arrive at the long term amplitude of the gravitational waves. The following analysis is based on the assumption that the equipartition condition holds. We note that the equipartition condition can be justified from a Statistical Mechanics perspective. In order to see this, consider the equation for the gravitational wave field $h_{\mu\nu}$ in Fourier space. For a given momentum, the gravitational wave field is a harmonic oscillator coupled to the Dark Energy field. Thus, we are dealing with a system of oscillators, the statistical properties of which can be determined by a canonical ensemble. It then follows, that by using standard statistical mechanics techniques one can arrive at the equipartition of energy—please see for example [23] and references therein for additional details on this and related issues.

We now estimate the amplitude of the gravitational waves. Using the equipartition condition.

At this point, we need to consider observational data. We will start with the Dark Energy density. We know that the Dark Energy density is the dominant component of the energy density. For arriving at the gravitational wave amplitude inside our galaxy we need to first get the dark energy density within our galaxy. This can be estimated based on available data (see [24]). We summarize this data by noting that $\sim 10^{12} M_{\odot}$ is confined to a volume with dimensions ~10 kpc. Thus, we estimate that on galaxy length scales (within our galaxy) $\rho_{DE} \sim 10^{-1}$ erg/cm^3.

We can now use the equipartition condition to arrive at the the energy density in gravitational waves. This can then be translated into the amplitude of the gravitational waves by using the standard expression relating these two quantities (See [22].)

For a gravitational wave with angular frequency ω and amplitude h_0, the energy density ρ_{gw} is:

$$\rho_{gw} \simeq \frac{c^2 \omega^2 h_0^2}{(32\pi G)} \tag{17}$$

Of course, the angular frequency is related to the time period T by $\omega = 2\pi/T$. From the resonance condition as well as from the observational data on the time period T we get that $T \sim 10^5 years$. Substituting into the equation for the gravitational wave energy density and using the equipartition condition we get $h_0 \sim 10^{-2}$ as the amplitude for the dark energy gravitational waves.

It is emphasized at this point that the gravitational wave amplitude as calculated above matches the data on the periodic eccentricity variation of earth's orbit [19].

Thus, the resonance analysis, allowed us not only to gain insights into the underlying physics, but also to quantitatively understand the correct magnitude for both the frequency and the amplitude of the gravitational waves produced by the dark energy dynamics.

In summary, we see that the gravitational waves produced by dark energy dynamics can produce the periodic variations in the Earth's orbital eccentricity resulting in the Ice Age periodicty.

Further, we note that, the gravitational waves with the computed amplitude and low frequencies are compatible with the observational constraints on gravitational waves (see for example [29] for a discussion on the constraints).

We note that the observation of such waves (other than through the variation of earth's orbit leading to the ice age periodicity) will be very challenging. The primary reason this will be challenging is because of the low frequency $\sim 10^{-13}$ Hz. It should be noted that this corresponds to a time period $T \sim 10^5 years$ which is larger than a human lifetime by a factor of 1000.

An exciting development in the area of measuring low frequency gravitational waves is the NANOGrav effort—please see e.g. [27] for a discussion of the 15-year results published this year. However, it should be noted given the 15-year timespan of the data collected so far, the lowest frequency gravitational waves that NANOGrav can observe or constrain is limited to be in the nanohertz or 10^{-9} Hz region. This is again many orders of magnitude higher in frequency than the extremely low frequency signal $\sim 10^{-13}$ Hz for the dark energy gravitational waves being discussed here.

Further, with regard to CMB anisotropies we note (as discussed in detail e.g. in [25]) that it is impossible to reliably measure CMB anisotropies below ~ 20 GHz and this needs to be contrasted with the extremely low frequency $\sim 10^{-13}$ Hz for the dark energy gravitational waves being discussed here. Furthermore, it should be noted that even the "low frequency band" of the LiteBIRD Mission [26] is designed to look at frequencies in the 40 GHz to 235 GHz range—about 20 orders of magnitude away in frequency space compared to the frequencies of gravitational waves produced by dark energy.

Thus, while we have every hope that with ingenuity, eventually, a method which is different than using the geological data and ice age periodicity will be found to do novel and perhaps more detailed observations for the extremely low frequency signal at $\sim 10^{-13}$ Hz, for now, we have to admit that it is a challenge to find another technique to do this.

4 Dark Energy Phase Transitions and Intermediate Mass Black Holes at High Redshifts

Extensive discusssions of the dynamics of fields and phase transitions in cosmological settings has been done previously (see e.g. Kolb and Turner [14]). We have built on these techniques to study and understand the physics of phase transitions in a general cosmological setting (please see e.g. [36–41]) and applied it to the phase transitions in dark energy and the dynamics of dark energy fields.

The finite temperature behavior of the see-saw model of neutrino masses has been previously studied [5, 6]. Phase transitions were found which result in the formation of topological defects. The critical temperature in the see-saw model is naturally linked to the neutrino masses.

In earlier papers, it has been shown that Dark Energy can play a significant role in structure formation and the physics of the galaxy [16, 33]. In particular, the dynamical length scale associated with the Dark Energy is comparable to the galactic length scale. Further, Dark Energy collapse can result in the formation of black holes.

In many of our investigations of Dark Energy, we focus mostly on the gravitational interactions of the Dark Energy. It is the dominant interaction of Dark Energy. In this context, the gravitational waves from Dark Energy constitute a new and interesting signal. Thus, we studied the gravitational waves arising from Dark Energy dynamics [20]. A result from that paper is useful is that the time period of the gravitational waves emitted is $\sim 10^5$ years.

As already discussed, the gravitational waves emitted by Dark Energy can result in the periodicity of the ice ages through the ellipticity variations of the orbit of the earth [21].

It is clear that Dark Energy has many important consequences for our Universe. We now wish to focus on phase transitions in Dark Energy and its implications for our Universe. To do this we will build on the work of Holman and Singh [6]. They investigated the finite temperature see-saw model of neutrino masses and demonstrated the existence of interesting phase transitions. The critical temperature is determined by the neutrino masses. Further the phase transition results in the formation of topological defects. Finally, the gravitational collapse of these field configurtions results in black hole formation [32, 33]. It has been shown that these Black Holes are Supermassive Black Holes (SMBHs). In fact, their masses are comparable to the observed masses of Black Holes at the centers of galaxies. Knowing the link between neutrino masses and the critical temperture of the phase transition [6], we could infer when to expect the peak in black hole formation to take place. This was done in [33] and indeed the observations matched the theoretical expectation. It should be noted that at the time of writing of [33] the only data available on neutrino masses was the Solar Neutrino data [35] from which one could arrive at an estimate of m_2—the mass of the second family of neutrinos. Of course, neutrinos have a family structure like every other known elementary particle and we will label the three families of light neutrinos as 1, 2 and 3 in increasing order of mass. This can be used to arrive at a prediction for the redshift at which we expect an earlier generation of Quasars [42].

We now wish to use this to relate the epochs at which the Phase Transitions occur resulting in the peaks in the associated Black Hole formation. We have already labeled the three families of light neutrinos as 1, 2 and 3 in increasing order of mass, so they will have masses given by m_1, m_2 and m_3. Let us now label the critical temperatures associated with the phase transitionsas T_1, T_2, T_3 and scale factors associated with these phase transitions as R_1, R_2, R_3 respectively. The scale factors can themselves be associated with the corresponding redshifts z_1, z_2, z_3 respectively.

Using the physics of the phase transitions and the evolution of temperature, scale fator and redshift in an expanding universe [6, 14, 15, 33] we note that:

$$\frac{T_3}{T_2} = \frac{m_3}{m_2} \tag{18}$$

$$\frac{T_3}{T_2} = \frac{R_2}{R_3} \tag{19}$$

and

$$\frac{R_2}{R_3} = \frac{1+z_3}{1+z_2}. \tag{20}$$

From observations of the Quasar distribution with redshift, see for example [34], we get:

$$z_2 \sim 2.5. \tag{21}$$

The value of z_3 has a dependence on the lightest neutrino mass m_1 and so the result is displayed as a plot of z_3 versus m_1 in the figure below.

Figure 1 displays the loaction of the Peak in redshift (z) as function of the mass of the lightest neutrino (m_1 in meV).

If the lightest neutrino mass is insignificant compared to the masses of the heavier neutrinos (i.e. $m_1 << m_2, m_3$), then we get:

$$z_3 \sim 18. \tag{22}$$

It is not possible to extract the value of m_1 from neutrino oscillation data alone, but we can actually evaluate the peak redshift as a function of the mass of lightest neutrino mass. Indeed, the location of the peak at high redshifts may enable us to extract the value of the lightest neutrino mass which has so far eluded us.

The reason for this is that while the neutrino oscillations are determined by the mass differences, the temperatures T_2, T_3 are actually determined by the masses m_2, m_3. This in turn determines the location of the peaks in redshift space located at z_2, z_3. If $m_1 << m_2, m_3$, then m_2, m_3 are approximately equal to the mass differences $\Delta m_{21}, \Delta m_{31}$ which can be determined from the neutrino oscillations data [31]. However, if the value of m_1 is not negligible, then the values for m_2, m_3 depend on

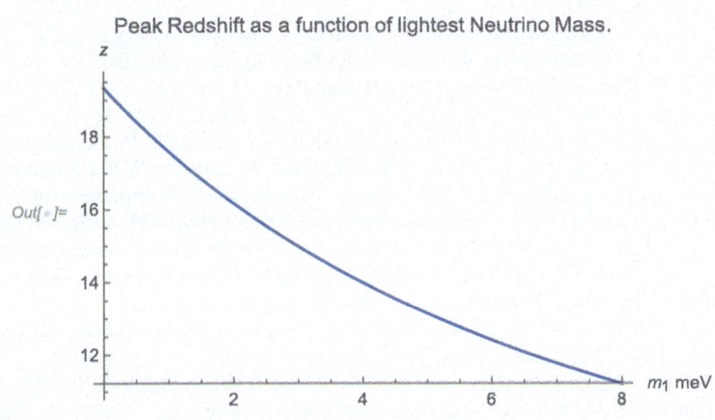

Fig. 1 Peak Redshift as a funcion of m_1

m_1 in addition to the mass differences $\Delta m_{21}, \Delta m_{31}$ which are determined from the neutrino oscillations data [31]. Thus, by determining the exact location of the peaks which depend on m_2, m_3 and using the existing data on the mass differences $\Delta m_{21}, \Delta m_{31}$ determined from the neutrino oscillations data [31] we can extract the value of the lightest neutrino mass m_1 which it is impossible to extract from neutrino oscillations data by itself.

In summary, we note that the field theory approach to dark energy and its comparison to observational data as described here has the potential to both enrich and expand our knowledge of our Universe.

References

1. M.J. Pierce, D.L. Welch, R.D. McClure, S. van den Bergh, R. Racine, P.B. Stetson, Nature **371**, 385–389 (1994)
2. W.L. Freedman et al., Nature **371**, 757–762 (1994)
3. A. Singh, Phys. Rev. D **52**, 6700–6707 (1995)
4. A. Silvestri, M. Trodden, Rep. Prog. Phys. **72**(9), 096901
5. A.K. Gupta, C.T. Hill, R. Holman, E.W. Kolb, Phys. Rev. D **45**, 441 (1992)
6. R. Holman, A. Singh, Phys. Rev. D **47**, 421 (1993)
7. G.'T. Hooft, Naturalness, chiral symmetry, and spontaneous chiral symmetry breaking, in *Recent Developments in Gauge Theories*, ed. by G. 't Hooft, et al. (Plenum Press, New York, 1980)
8. J. Preskill, M. Wise, F. Wilczek, Phys. Lett. B **120**, 127 (1983)
9. L. Abbott, P. Sikivie, Phys. Lett. B **120**, 133 (1983)
10. M. Dine, W. Fischler, Phys. Lett. B **120**, 137 (1983)
11. R. Peccei, H. Quinn, Phys. Rev. Lett. **38**, 1440 (1977); S. Weinberg, Phys. Rev. Lett. **40**, 223 (1978); F. Wilczek, Phys. Rev. Lett. **40**, 279 (1978); M. Dine, W. Fischler, M. Srednicki, Phys. Lett. B **104**, 199 (1981)
12. E.W. Kolb, A. Singh, M. Srednicki, Phys. Rev. D **58**, 105004 (1998)
13. M. Gell-Mann, P. Ramond, R. Slansky, Supergravity. North Holland Publication (1979); T. Yanagida, in *Proceedings of the Workshop on Unified Theories and Baryon Number in the Universe*, KEK, Japan (1979)
14. E.W. Kolb, M.S. Turner, *The Early Universe*, vol. 73 (Addison-Wesley Pub. Co., 1990)
15. S. Weinberg, *Gravitation and Cosmology* (John Wiley & Sons, Inc., 1972)
16. V. Jhalani, H. Kharkwal, A. Singh, J. Exp. Theor. Phys. **123**(5), 827–831 (2016). ISSN: 1063-7761
17. J.R. Petit, J. Jouzel, D. Raynaud, N.I. Barkov, J.-M. Barnola, I. Basile, M. Bender, J. Chappellaz, M. Davis, G. Delaygue, M. Delmotte, V.M. Kotlyakov, M. Legrand, V.Y. Lipenkov, C. Lorius, L. Pépin, C. Ritz, E. Saltzman, M. Stievenard, Climate and atmospheric history of the past 420,000 years from the Vostok ice core, Antarctica. Nature **399**(6735), 429–436 (1999)
18. J.D. Hays, J. Imbrie, N.J. Shackleton, Variations in the Earth's orbit: pacemaker of the ice ages. Science **194**(4270), 1121–113 (1976). https://doi.org/10.1126/science.194.4270.1121
19. L.A. Hinov, Annu. Rev. Earth Planet. Sci **28**, 430 (2000)
20. V. Jhalani, A. Mishra, A. Singh, J. Exp. Theor. Phys. **125**(4), 638–643 (2017). ISSN: 1063-7761
21. A. Singh, Dark energy gravitational wave observations and ice age periodicity. Phys. Lett. B **802**, 135226 (2020). ISSN: 0370-269. https://doi.org/10.1016/j.physletb.2020.135226
22. L.D. Landau, E.M. Lifshitz, *The Classical Theory of Fields*, vol. 2, 4th edn. eBook ISBN: 9780080503493; Paperback ISBN: 9780750627689. Imprint: Butterworth-Heinemann, Published Date: 1st January 1987

23. F. Reif, *Fundamentals of Statistical and Thermal Physics*, International Edition. McGraw-Hill Book Co. (1965). ISBN: 0-07-Y85615-X
24. P.J. McMillan, Mon. Not. R. Astron. Soc. **414**, 2446–2457 (2011)
25. M. Bucher, Physics of the cosmic microwave background anisotropy. Int. J. Mod. Phys. D **24**(02), 1530004 (2015)
26. A. Suzuki et.al., The litebird satellite mission—Sub-kelvin instrument (2018). arXiv:1801.06987
27. The NANOGrav 15 yr data set: evidence for a gravitational-wave background. Astrophys. J. Lett. **951**(1), id.L8, 24 (2023)
28. B.F. Schutz, F. Ricci, Gravitational waves, sources and detectors, in *Gravitational waves*, ed. by I. Ciufolini et al. (Institute of Physics, Bristol, 2001). arXiv:1005.4735
29. L. Barack et al., Class. Quantum Grav. **36**(14), (2019)
30. J.M. Ezquiaga, M. Zumalacárregui, Front. Astron. Space Sci. **5**, 44 (2018). https://doi.org/10.3389/fspas.2018.00044
31. Neutrino masses, mixing, and oscillations by M.C. Gonzalez-Garcia and M. Yokoyama in P.A. Zyla et al. (Particle Data Group). Prog. Theor. Exp. Phys. **2020**, 083C01 (2020)
32. V. Jhalani, H. Kharkwal, A. Singh, Gravitational collapse of dark energy field configurations and supermassive black hole formation. J. Exp. Theor. Phys. **123**(5), 827–831 (2016). ISSN: 1063–7761
33. A. Singh, Quasar production: topological defect formation due to a phase transition linked with massive neutrinos. Phys. Rev. D **50**(2), 671 (1994)
34. I. Paris et al., The Sloan Digital Sky Survey quasar catalog: fourteenth data release. Astron. Astrophys. **613**, A51 (2018)
35. X. Shi, D.N. Schramm, J.N. Bahcall, Phys. Rev. Lett. **69**, 717 (1992)
36. D. Boyanovsky, D.-S. Lee, A. Singh, Phys. Rev. D **48**, 800 (1993)
37. D. Boyanovsky, R. Holman, D.S. Lee, J.P. Silva, A. Singh, Nucl. Phys. B **441**(3), 595–608 (1995)
38. D. Boyanovsky, H.J. de Vega, R. Holman, D.-S. Lee, A. Singh, Phys. Rev. D **51**, 4419 (1995)
39. D. Boyanovsky, D. Cormier, H.J. De Vega, R. Holman, A. Singh, M. Srednicki, Phys. Rev. D **56**(4), 1939 (1997)
40. A. Chodos, F. Cooper, W. Mao, H. Minakata, A. Singh, Phys. Rev. D **61**(4), 045011 (2000)
41. A. Chodos, F. Cooper, W. Mao, A Singh, Phys. Rev. D **63**(9), 096010 (2001)
42. A. Singh, Dark energy black holes with intermediate masses at high redshifts: an earlier generation of quasars and observations. Mod. Phys. Lett. A **38**, 2350135 (2023). arXiv:2207.02143. https://doi.org/10.1142/S0217732323501353

T-Model Higgs Inflation in Supergravity

Constantinos Pallis

Abstract We focus on a simple, natural and predictive version of T-model inflation in Supergravity employing as inflaton the Higgs field which leads to the spontaneous breaking of a $U(1)_{B-L}$ symmetry at the SUSY GUT scale. We use a renormalizable superpotential, fixed by a $U(1)$ R symmetry, and a Kähler potential which parameterizes the Kähler manifold $SU(2,1)/(SU(2) \times U(1)) \times (SU(2)/U(1))$ with scalar curvature $\mathcal{R}_K = -6/N + 2/N_0$ where $0 < N_0 < 6$. The spectral index n_s turns out to be close to its present central observational value and the tensor-to-scalar ratio r increases with $N \lesssim 36$. The model can be nicely linked to MSSM offering an explanation for the magnitude of the μ parameter consistently with the phenomenological data. It also allows for baryogenesis via non-thermal leptogenesis with gravitino as light as 1 TeV.

Keywords Cosmology · Inflation · Supergravity models

1 From Minimal to T-Model Higgs Inflation

The identification of the inflaton—i.e., the scalar field driving inflation—remains an important open issue in the are(n)a of inflationary model building [1]. We aspire to identify ϕ with the radial component of a Higgs field, $\Phi = \phi e^{i\theta}$, within a *Grand Unified Theory* (GUT). We adopt the term *Higgs inflation* (HI) for the relevant models [2] and we clarify that it is not limited to those which employ as higgsflaton [3] the *Standard Model* (SM) Higgs field [4]. We find it technically convenient to exemplify HI in our talk taking as reference theory an "elementary" GUT based on the gauge group $G_{B-L} = G_{\rm SM} \times U(1)_{B-L}$—where $G_{\rm SM}$ is the gauge group of the SM and B, L denote baryon and lepton number respectively. Despite its simplicity, G_{B-L} is strongly motivated by neutrino physics and leptogenesis mechanisms—see e.g.

C. Pallis (✉)
Laboratory of Physics, Faculty of Engineering, Aristotle University of Thessaloniki, Thessaloniki, Greece
e-mail: kpallis@gen.auth.gr

J. Strologas and L. Perivolaropoulos (eds.), *HEP2023*, Springer Proceedings in Physics 333 https://doi.org/10.1007/978-3-031-76856-9_7

Refs. [5, 6]. Therefore, we choose as inflationary potential the one employed for the realization of the Higgs mechanism,

$$V_{\rm HI}(\phi) = \lambda^2(\phi^2 - M^2)^2/16 \simeq \lambda^2\phi^4/16 \;\; \text{for} \;\; M \ll m_{\rm P} = 1. \tag{1}$$

If ϕ is canonically normalized, the theoretically derived values $n_{\rm s} \simeq 0.947$ and $r \simeq 0.28$ are not compatible with the observational ones, since the combined BICEP2/*Keck Array* and *Planck* results require [7, 8]

$$n_{\rm s} = 0.965 \pm 0.009 \;\; \text{and} \;\; r \lesssim 0.032 \;\; \text{at}\, 95\%\, \text{c.l.} \tag{2}$$

On the contrary, observationally friendly are models called α-attractors which employ chaotic potentials and so can be activated with $V_{\rm HI}$ in Eq. (1). These are based on the specific relation established between the initial, ϕ, and the canonically normalized inflaton $\widehat{\phi}$ and can be classified into *E-Model Inflation* (EMI) [9, 10] (or α-Starobinsky model [11]) and *T-Model Inflation* (TMI) [12]. I.e.

$$\phi = \begin{cases} 1 - {\rm Exp}\left(-\sqrt{2/N}\widehat{\phi}\right) & \text{for EMI,} \\ \tanh\left(\widehat{\phi}/\sqrt{2N}\right) & \text{for TMI,} \end{cases} \quad \text{with} \;\; N > 0. \tag{3}$$

Such relations between ϕ and $\widehat{\phi}$ can be achieved in the presence of a pole in the inflaton kinetic term. TMI is "taylor made" for HI, since it arises from a kinetic pole of order two which includes the GUT-invariant quantity $|\Phi|^2 := \Phi^\dagger\Phi$. In particular, the lagrangian $\mathcal{L}$ of $\phi = \phi(t)$ reads

$$\mathcal{L} = \sqrt{-\mathfrak{g}}\left(N\dot{\phi}^2/f_K^2 - V_{\rm HI}(\phi)\right) \;\; \text{with} \;\; \dot{} = d/dt, \;\; f_K = 1 - \phi^2 \;\; \text{and} \;\; N > 0. \tag{4}$$

Here t is the cosmic time, $\mathfrak{g}$ is the determinant of the space-time metric $g_{\mu\nu}$ with signature $(+, -, -, -)$. If we extract $\widehat{\phi}$, we obtain

$$\frac{d\widehat{\phi}}{d\phi} = J = \frac{\sqrt{2N}}{f_K} \;\; \Rightarrow \;\; \phi = \tanh\frac{\widehat{\phi}}{\sqrt{2N}} \;\; \text{and} \;\; V_{\rm HI}(\widehat{\phi}) \simeq \frac{\lambda^2}{16}\tanh^4\frac{\widehat{\phi}}{\sqrt{2N}}, \tag{5}$$

by virtue of Eq. (1). $V_{\rm HI}$ expressed as a function of $\widehat{\phi}$ develops a plateau for $\widehat{\phi} > 1$ [12, 13] which renders it convenient for the realization of an observationally viable *T-Model HI* (THI) with the parameter N accordingly constrained.

Since the natural framework of a GUT is *Supersymmetry* (SUSY)—and its topical extension, *Supergravity* (SUGRA)—where the gauge hierarchy problem is set under control we consider worthwhile to embed THI in SUGRA. Section 2 below is devoted to this aim. Then, in Sect. 3, we propose a possible post-inflationary completion which assures generation of the μ term of *Minimal SUSY SM* (MSSM) and the *baryon number of the universe* (BAU) via *non-thermal leptogenesis* (nTL).

2 SUGRA Embedding

The part of the SUGRA lagrangian including the (complex) scalar fields z^α can be written as

$$\mathcal{L} = \sqrt{-\mathfrak{g}}\left(K_{\alpha\bar{\beta}}\partial_\mu z^\alpha \partial^\mu z^{*\bar{\beta}} - V_{\rm F} + \cdots\right), \tag{6}$$

where the kinetic mixing is controlled by the Kähler potential, K, and the relevant metric is defined as

$$K_{\alpha\bar{\beta}} = K_{,z^\alpha z^{*\bar{\beta}}} > 0 \quad \text{with} \quad K^{\bar{\beta}\alpha}K_{\alpha\bar{\gamma}} = \delta^{\bar{\beta}}_{\bar{\gamma}}. \tag{7}$$

We focus on the F-term part of the SUGRA scalar potential, $V_{\rm F}$, which is given in terms of K, and the superpotential, W, by

$$V_{\rm F} = e^K\left(K^{\alpha\bar{\beta}}D_\alpha W D^*_{\bar{\beta}}W^* - 3|W|^2\right) \quad \text{with} \quad D_\alpha W = W_{,z^\alpha} + K_{,z^\alpha}W \tag{8}$$

the Kähler covariant derivative *with respect to* (w.r.t) the field z^α—the symbol, Z as subscript denotes derivation w.r.t Z. The non-SUSY THI introduced in Sect. 1 can be reproduced if we adopt (for alternative choices see Ref. [13])

$$W = \lambda S\left(\bar{\Phi}\Phi - M^2/2\right)/2, \tag{9a}$$

$$K = -N\ln\frac{1-|\bar{\Phi}|^2-|\Phi|^2}{\sqrt{(1-2\bar{\Phi}\Phi)(1-2\bar{\Phi}^*\Phi^*)}} + N_0\ln\left(1+\frac{|S|^2}{N_0}\right), \tag{9b}$$

which are consistent with the $U(1)_{B-L}$ and an R symmetry acting on the various superfields as follows

$$(B-L)(S,\bar{\Phi},\Phi) = (0,-2,2) \quad \text{and} \quad R(S,\bar{\Phi},\Phi) = (1,0,0). \tag{10}$$

W leads to a $B-L$ phase transition since the vacuum of the SUSY limit of $V_{\rm F}$ lies at the direction

$$\langle S\rangle \simeq 0 \quad \text{and} \quad |\langle\Phi\rangle| = |\langle\bar{\Phi}\rangle| \simeq M/\sqrt{2} \quad \text{for} \quad M \ll 1, \tag{11}$$

i.e., $U(1)_{B-L}$ is spontaneously broken via the vacuum expectation values $\langle\Phi\rangle$ and $\langle\bar{\Phi}\rangle$. On the other hand, K exhibits a number of features: For $0 < N_0 < 6$ it assures a stabilization of the $S = 0$ direction [15] during THI without invoking higher order terms [16]; it causes no contribution to the exponent in $V_{\rm F}$ of Eq. (8)—cf. Ref. [14]—and it parameterizes the Kähler manifold [13]

$$(SU(2,1)/(SU(2)\times U(1))_{\bar{\Phi}\Phi} \times (SU(2)/U(1))_S \tag{12}$$

with constant scalar curvature $\mathcal{R}_K = -6/N + 2/N_0$—the indices above indicate the moduli which parameterize the respective Kähler submanifolds.

2.1 Inflationary Potential and Observables

Adopting the following parameterizations for the relevant fields

$$\Phi = \phi e^{i\theta}\cos\theta_\Phi, \quad \bar{\Phi} = \phi e^{i\bar{\theta}}\sin\theta_\Phi \quad \text{with} \quad 0 \leq \theta_\Phi \leq \pi/2 \quad \text{and} \quad S = \frac{s + i\bar{s}}{\sqrt{2}}, \tag{13}$$

we can easily verify that there is a D-flat direction

$$\langle\theta\rangle_{\rm I} = \langle\bar{\theta}\rangle_{\rm I} = 0, \; \langle\theta_\Phi\rangle_{\rm I} = \pi/4 \quad \text{and} \quad \langle S\rangle_{\rm I} = 0, \tag{14}$$

which can be qualified as inflationary path—the subscript I denotes values during THI. Substituting it and Eqs. (9a) and (9b) into Eq. (8), $V_{\rm HI}$ in Eq. (1) can be reproduced from the just one surviving term, $\langle |W_{,S}|^2\rangle_{\rm I}$. To achieve TMI, though, we have to establish the correct non-minimal kinetic mixing shown in Eq. (5). To this end, we compute $K_{\alpha\bar{\beta}}$ along the path in Eq. (14) which takes the form

$$\langle K_{\alpha\bar{\beta}}\rangle_{\rm I} = \mathsf{diag}\left(\langle M_{\bar{\Phi}\Phi}\rangle_{\rm I}, \langle K_{SS^*}\rangle_{\rm I}\right) \;\; \text{with} \;\; \langle M_{\bar{\Phi}\Phi}\rangle_{\rm I} = \frac{\kappa\phi^2}{2}\begin{bmatrix} 2/\phi^2 - 1 & 1 \\ 1 & 2/\phi^2 - 1 \end{bmatrix}, \tag{15}$$

$\kappa = N/f_K^2$ and $K_{SS^*} = 1$. Then we diagonalize $\langle M_{\bar{\Phi}\Phi}\rangle_{\rm I}$ via a similarity transformation as follows:

$$U_{\bar{\Phi}\Phi}\langle M_{\bar{\Phi}\Phi}\rangle_{\rm I} U_{\bar{\Phi}\Phi}^{\sf T} = \mathsf{diag}\left(\kappa_+, \kappa_-\right) \quad \text{where} \quad U_{\bar{\Phi}\Phi} = \frac{1}{\sqrt{2}}\begin{bmatrix} 1 & 1 \\ -1 & 1 \end{bmatrix}, \tag{16}$$

$\kappa_+ = \kappa$ and $\kappa_- = \kappa f_K$. Canonically normalizing the various fields, we obtain the desired form of J in Eq. (5). We can also verify that the direction of Eq. (14) is stable w.r.t the fluctuations of the non-inflaton fields and determine M demanding that the GUT scale $M_{\rm GUT} \simeq 2/2.433 \times 10^{-2}$ is identified with the mass of the $B - L$ gauge boson at the vacuum of Eq. (11), i.e. [13],

$$\langle M_{BL}\rangle = \sqrt{2N}gM/\langle f_K\rangle = M_{\rm GUT} \;\; \Rightarrow \;\; M \simeq M_{\rm GUT}/g\sqrt{2N} \;\; \text{with} \;\; g \simeq 0.7. \tag{17}$$

The allowed parameter space of our model can be delineated enforcing the basic inflationary prerequisites [17], i.e.,

$$N_\star \simeq 55 \quad \text{and} \quad A_{\rm s} \simeq 4.588 \cdot 10^{-5}, \tag{18}$$

where $N_\star$ is the number that the pivot scale $k_\star = 0.05/\text{Mpc}$ suffers during THI and A_s is the amplitude of the curvature perturbation generated by the inflaton when $k_\star$ crosses outside the inflationary horizon for $\phi = \phi_\star$. Enforcing Eq. (18) we constrain $\phi_\star$ and λ and obtain the allowed curve in the $n_s - r$ plane—see Fig. 1—by varying N; note that we use $N_0 = 1$ throughout. A comparison with the observational data [7, 8] is also displayed there together with some typical values of the parameters for $N = 12$. We observe that n_s and r increase with N. More specifically, we obtain

$$0.963 \lesssim n_s \lesssim 0.964, \quad 0.1 \lesssim N \lesssim 36 \quad \text{and} \quad 0.0005 \lesssim r \lesssim 0.039, \tag{19}$$

where $w_{\rm rh} \simeq 0.3$ is the barotropic index during the inflaton-decaying domination and $N_\star \simeq 56$. The proximity of $\phi_\star$ to 1 signals a mild tuning in the initial conditions since $\Delta_\star = 1 - \phi_\star \simeq (0.2 - 7)\%$ increasing with N.

3 Post-inflationary Completion

THI can be embedded in a $B - L$ extension of MSSM promoting to gauge the pre-existing global $U(1)_{B-L}$. The terms of the total super- and Kähler potential which control the coexistence of the inflationary and the MSSM sectors are

$$\Delta W = \lambda_\mu S H_u H_d + \lambda_{ij\nu} \bar{\Phi} N_i^c N_j^c, \tag{20a}$$

$$\Delta K = \sum_{a=1}^{5} N_0 \ln\left(1 + |X_\alpha|^2/N_0\right) \quad \text{with} \quad X_\alpha = N_i^c, H_u, H_d \tag{20b}$$

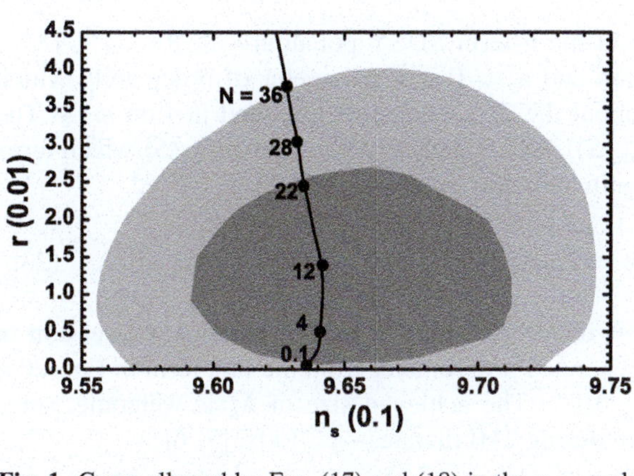

N	12
$\phi_\star/0.1$	9.75
$\Delta_\star(\%)$	2.5
$\phi_f/0.1$	3.9
$w_{\rm rh}$	0.266
$N_\star$	56.4
$\lambda/10^{-5}$	8.6
$n_s/0.1$	9.64
$-\alpha_s/10^{-4}$	6.4
$r/10^{-2}$	1.4

Fig. 1 Curve allowed by Eqs. (17) and (18) in the $n_s - r$ plane for various N's indicated along it. The marginalized joint 68% [95%] c.l. regions [7, 8] from PR4, BK18, BAO and lensing data-sets are depicted by the dark [light] shaded contours. The relevant field values, parameters and observables corresponding to $N = 12$ are listed in the Table

where H_u and H_d are the electroweak Higgs superfields & N_i^c the ith generation right-handed neutrinos with $i = 1, \ldots, 3$. These fields are supposed to be stabilized at zero during THI, i.e. $\langle X^\alpha \rangle_{\rm I} = 0$, and their charge assignments are

$$(B-L)(N_i^c, H_u, H_d) = (1, 0, 0) \quad \text{and} \quad R(N_i^c, H_u, H_d) = (1, 0, 0). \tag{21}$$

The phenomenological consequences of the terms in Eqs. (20a) and (20b) include the generation of the μ term and the implementation of nTL explored in Sects. 3.1 and 3.2 below. Hereafter we restore units, i.e., we take $m_{\rm P} = 2.433 \cdot 10^{18}$ GeV.

3.1 μ Term and MSSM Phenomenology

The contributions from the soft SUSY-breaking terms, although negligible during THI, may shift slightly $\langle S \rangle$ from zero in Eq. (11). In fact, the relevant potential terms are

$$V_{\rm soft} = \left(\lambda A_\lambda S \bar{\Phi} \Phi - {\rm a}_S S \lambda M^2/4 + {\rm h.c.}\right) + m_\gamma^2 \left|X^\gamma\right|^2, \tag{22}$$

where $X^\gamma = S, \Phi, \bar{\Phi}, H_u$ and H_d, $m_\gamma \ll M$, A_λ and ${\rm a}_S$ are soft SUSY-breaking mass parameters and we assume $H_u = H_d \simeq 0$ during and after THI. Rotating S in the real axis by an appropriate R-transformation, choosing conveniently the phases of A_λ and ${\rm a}_S$—so as the total low energy potential $V_{\rm tot} = V_{\rm SUSY} + V_{\rm soft}$ to be minimized—and substituting in $V_{\rm soft}$ the Φ and $\bar{\Phi}$ values from Eq. (11), we get

$$\langle V_{\rm tot}(S) \rangle = \lambda^2 \frac{M^2 S^2}{4N} - \lambda M^2 S {\rm a}_\mu m_{3/2} \;\; \text{with} \;\; {\rm a}_\mu = \frac{|A_\lambda| + |{\rm a}_S|}{2 m_{3/2}}, \tag{23a}$$

where the first term comes from the F-term SUSY potential—cf. Refs. [5, 18]—, $m_{3/2}$ is the gravitino ($\widetilde{G}$) mass and ${\rm a}_\mu > 0$ is a parameter of order unity which parameterizes our ignorance about the dependence of $|A_\lambda|$ and $|{\rm a}_S|$ on $m_{3/2}$. The extermination condition of $\langle V_{\rm tot}(S) \rangle$ w.r.t S leads to a non vanishing $\langle S \rangle$ which turns out to be several orders of magnitude below $m_{3/2}$—cf. Ref. [18]. Indeed,

$$d\langle V_{\rm tot}(S) \rangle / dS = 0 \;\Rightarrow\; \langle S \rangle = 2N {\rm a}_\mu m_{3/2}/\lambda \simeq \sqrt{N} N_\star {\rm a}_\mu m_{3/2} / 2\pi \sqrt{6 A_{\rm s}}, \tag{23b}$$

where we employed the $\lambda - A_{\rm s}$ condition implied by the rightmost expression in Eq. (18)—see Ref. [13]. The generated μ parameter from the first term in Eq. (20a) is $\mu = \lambda_\mu \langle S \rangle \sim m_{3/2}$ for $\lambda_\mu \sim 10^{-6}$. The achieved size of λ_μ is welcome, since stability of the $H_u - H_d$ system during THI dictates—cf. Refs. [5, 6]:

$$\lambda_\mu < \lambda (1 + N_0) \phi_{\rm f}^2 / 4 N_0 m_{\rm P}^2 \sim 10^{-5} \;\; \text{for} \;\; N_0 = 1. \tag{24}$$

Table 1 λ_μ values compatible with the CMSSM best-fit points

CMSSM REGION	$\lvert A_0\rvert$ (TeV)	m_0 (TeV)	$\lvert\mu\rvert$ (TeV)	a_μ	λ_μ (10^{-6})	
					$N=1$	$N=36$
(I) A/H Funnel	9.9244	9.136	1.409	1.086	1.81	0.3
(II) $\tilde{\tau}_1-\tilde{\chi}$ Coannihilation	1.2271	1.476	2.62	0.831	*14.48*	5
(III) $\tilde{t}_1-\tilde{\chi}$ Coannihilation	9.965	4.269	4.073	2.33	5.2	0.9
(IV) $\tilde{\chi}_1^{\pm}-\tilde{\chi}$ Coannihilation	9.2061	9.000	0.983	1.023	1.35	0.2

Thanks to the presence of $\sqrt{A_s}\sim 10^{-5}$ in the denominator of the expression in Eq. (23b), almost any $\mu/m_{3/2}<1$ value is accessible for $\lambda_\mu<10^{-5}$ without causing any ugly hierarchy between $m_{3/2}$ and μ.

Adopting the *Constrained MSSM* (CMSSM) [19], as a low-energy framework, we can check its compatibility with the proposed inflationary model doing some simple estimates. Namely, CMSSM is relied on the following 4 and 1/2 free parameters

$$\mathrm{sign}\mu,\quad \tan\beta=\langle H_u\rangle/\langle H_d\rangle,\quad M_{1/2},\quad m_0\ \text{and}\ A_0, \tag{25}$$

where the three last mass parameters denote the common gaugino mass, scalar mass and trilinear coupling constant, respectively, defined (normally) at $M_{\rm GUT}$. Imposing a number of cosmo-phenomenological constraints the best-fit values of $|A_0|$, m_0 and $|\mu|$ can be determined as in Ref. [19]. Some of their results are listed in the first four lines of Table 1. We see that there are four allowed regions characterized by the specific mechanism for suppressing the relic density of the lightest neutralino ($\tilde{\chi}$)—which turns out to be the lightest SUSY particle—to the acceptable level in Ref. [17], i.e., $\Omega_{\tilde{\chi}}h^2\lesssim 0.12$—note that $\tilde{\tau}_1$, $\tilde{t}_1$ and $\tilde{\chi}_1^{\pm}$ stand for the lightest stau, stop and chargino eigenstate whereas A and H are the CP-odd and the heavier CP-even Higgs bosons of MSSM respectively. The proposed regions pass all the currently available LHC bounds [20] on the masses of the various sparticles and assure for the mass of the lighter CP-even Higgs boson $m_h\simeq 125$ GeV.

Enforcing the conditions for the electroweak symmetry breaking, a value for the parameter $|\mu|$ can be achieved in each of the regions in Table 1. Taking this $|\mu|$ value as input we can extract the λ_μ values, if we first derive a_μ setting, e.g.,

$$m_0=m_{3/2}\ \text{and}\ |A_0|=|A_\lambda|=|a_S|. \tag{26}$$

The outputs of our computation are listed in the two rightmost columns of Table 1 in bold for $N = 1$ and $N = 36$. From these we infer that the required λ_μ values—decreasing with N increasing—, in all cases besides the one written in italics, are comfortably compatible with Eq. (24). As regards nTL,—see Sect. 3.2 below—the $\widetilde{G}$ constraint does not favor any specific case since the unstable $\widetilde{G}$ with $m_{3/2} \gtrsim 1$ TeV turns out to be cosmologically safe within our framework.

3.2 Non-thermal Leptogenesis and Neutrino Masses

Soon after the end of THI, the (canonically normalized) inflaton

$$\widehat{\delta\phi} = \langle J\rangle\delta\phi \;\; \text{with} \;\; \delta\phi = \phi - M \;\; \text{and} \;\; \langle J\rangle = \sqrt{2N} \tag{27}$$

acquires mass—reducing with increasing N in its domain of Eq. (19)—given by

$$\widehat{m}_{\delta\phi} = \left\langle V_{\rm HI,\widehat{\phi}\widehat{\phi}}\right\rangle^{1/2} = \left\langle V_{\rm HI,\phi\phi}/J^2\right\rangle^{1/2} = \frac{\lambda M}{2\sqrt{N}} \simeq (4.4 - 26)\cdot 10\ \text{EeV}, \tag{28}$$

where 1 EeV $= 10^9$ GeV. This value is lower than that encountered in models of induced-gravity HI [6] and close to that obtained in several versions of non-minimal HI [5]. Meantime, $\widehat{\delta\phi}$ settles into a phase of damped oscillations abound the minimum in Eq. (11) reheating the universe at a temperature [6]

$$T_{\rm rh} = \left(72/5\pi^2 g_*\right)^{1/4}\left(\widehat{\Gamma}_{\delta\phi} m_{\rm P}\right)^{1/2} \;\; \text{with} \;\; \widehat{\Gamma}_{\delta\phi} = \widehat{\Gamma}_{\delta\phi\to N_i^c} + \widehat{\Gamma}_{\delta\phi\to H}. \tag{29}$$

Here $g_* = 228.75$ counts the MSSM effective number of relativistic degrees of freedom and we take into account the following decay widths

$$\widehat{\Gamma}_{\delta\phi\to N_i^c} = \frac{g_{iN^c}^2}{16\pi}\widehat{m}_{\delta\phi}\left(1 - \frac{4M_{iN^c}^2}{\widehat{m}_{\delta\phi}^2}\right)^{3/2} \;\; \text{with} \;\; g_{iN^c} = \frac{\lambda_{iN^c}}{\langle J\rangle}, \tag{30a}$$

$$\widehat{\Gamma}_{\delta\phi\to H} = \frac{2}{8\pi} g_H^2 \widehat{m}_{\delta\phi} \;\; \text{with} \;\; g_H = \frac{\lambda_\mu}{\sqrt{2}}. \tag{30b}$$

For $T_{\rm rh} < M_i N^c$, the out-of-equilibrium decay of N_i^c generates a lepton-number-asymmetry yield which is partially converted through sphaleron effects into a yield of the observed BAU

$$Y_B = -0.35\cdot\frac{5}{2}\frac{T_{\rm rh}}{\widehat{m}_{\delta\phi}}\sum_i \frac{\widehat{\Gamma}_{\delta\phi\to N_i^c}}{\widehat{\Gamma}_{\delta\phi}}\varepsilon_i. \tag{31}$$

Here $\langle H_u \rangle \simeq 174$ GeV, for large $\tan\beta$, ε_i is the lepton asymmetry generated per N_i^c decay and $m_{\rm D}$ is the Dirac mass matrix of neutrinos, ν_i. The validity of Eq. (31) requires the decay of $\widehat{\delta\phi}$ into at least one pair of N_i^c and the avoidance of any erasure of the produced lepton asymmetry yield. These conditions can be fulfilled if we demand

$$\widehat{m}_{\delta\phi} \geq 2M_{1N^c} \;\text{ and }\; M_{1N^c} \gtrsim 10T_{\rm rh} \;\text{ with }\; M_{1N^c}2\lambda_{i\nu}\langle\bar{\Phi}\rangle. \tag{32}$$

Moreover, Eq. (31) has to reproduce the observational result [17]

$$Y_B = (8.697 \pm 0.054) \cdot 10^{-11}. \tag{33}$$

The required $T_{\rm rh}$ in Eq. (31) must be compatible with constraints on the $\widetilde{G}$ abundance, $Y_{3/2}$, at the onset of *nucleosynthesis* (BBN), which is estimated to be

$$Y_{3/2} \simeq 1.9 \cdot 10^{-22}\, T_{\rm rh}/\text{GeV}, \tag{34}$$

where we take into account only thermal production of $\widetilde{G}$, and assume that $\widetilde{G}$ is much heavier than the MSSM gauginos. On the other hand, $Y_{\widetilde{G}}$ is bounded from above in order to avoid spoiling the success of the BBN. For the typical case where $\widetilde{G}$ decays with a tiny hadronic branching ratio, we have

$$Y_{\widetilde{G}} \lesssim \begin{cases} 10^{-14} \\ 10^{-13} \end{cases} \text{for}\;\; m_{3/2} \simeq \begin{cases} 0.69\text{ TeV} \\ 10.6\text{ TeV} \end{cases} \text{implying}\;\; T_{\rm rh} \lesssim 5.3 \cdot \begin{cases} 10^{-2}\text{ EeV}, \\ 10^{-1}\text{ EeV}. \end{cases} \tag{35}$$

The bounds above can be somehow relaxed in the case of a stable $\widetilde{G}$.

Confronting Y_B and $Y_{\widetilde{G}}$—see Eqs. (31) and (34)—with observations we can constrain the parameters of the neutrino sector as function of the inflationary parameters. Indeed, the seesaw formula expresses the light-neutrino masses $m_{i\nu}$ in terms of m_{iD}—involved in Eq. (31)—and $M_i N^c$, generated by the second term in Eq. (20a). We follow the bottom-up approach detailed in Refs. [5, 6], according to which we find the $M_i N^c$'s by using as inputs the m_{iD}'s, a reference mass $m_{\rm r}\nu$ of the ν_i's—$m_{i\nu}$ for *normal ordered* (NO) $m_i\nu$'s, or $m_3\nu$ for *inverted ordered* (IO $m_i\nu$'s—, the two Majorana phases φ_1 and φ_2 of the PMNS matrix, and the best-fit values for the neutrino oscillation parameters, i.e., the differences of the masses-squared (Δm^2_{21} and Δm^2_{31}), the mixing angles (θ_{13}, θ_{23} and θ_{12}) and the Dirac phase δ of PMNS matrix. Namely, we take into account the recently updated best-fit values [21] according to which $\Delta m^2_{21} = 7.5 \cdot 10^{-5}\text{eV}^2$ and $\sin^2\theta_{12} = 0.318$ for both $m_i\nu$ orderings. Also, we take

- For NO $m_i\nu$'s, $\Delta m^2_{31} = 2.55 \cdot 10^{-3}\text{eV}^2$, $\sin^2\theta_{13} = 0.022$, $\sin^2\theta_{23} = 0.574$ and $\delta = 1.08\pi$.
- For IO $m_i\nu$'s, $\Delta m^2_{31} = 2.45 \cdot 10^{-3}\text{eV}^2$, $\sin^2\theta_{13} = 0.0225$, $\sin^2\theta_{23} = 0.578$ and $\delta = 1.58\pi$.

Furthermore, the sum of $m_i\nu$'s is bounded from above at 95% c.l. by the data [17, 21] as follows

$$\sum_i m_i\nu \leq 0.23 \text{ eV for NO } m_i\nu\text{'s} \quad \text{or} \quad \sum_i m_i\nu \leq 0.15 \text{ eV for IO } m_i\nu\text{'s}. \tag{36}$$

The outcome of our computation is presented in Fig. 2, where we depict the allowed values of $m_{3\mathrm{D}}$ versus N for $N_0 = 1$, $\mu = 4$ TeV—which corresponds to $0.9 \lesssim \lambda_\mu/10^{-6} \lesssim 5.2$ as N ranges from 36 to 1—and the remaining parameters shown in the Table of Fig. 2. We use solid, dashed and dot-dashed line when the remaining inputs—i.e. $m_i\nu$, m_3D, φ_1, and φ_2—correspond to the cases A, B and C of the Table of Fig. 2 respectively. We consider NO (cases A and B) and IO (case C) $m_i\nu$'s. In all cases, the current limit in Eq. (36) is safely met. The gauge symmetry considered here does not predict any particular Yukawa unification pattern and so, the m_{iD}'s are free parameters. Care is also taken so that the perturbativity of λ_{iN^c} holds, i.e., $\lambda^2_{iN^c}/4\pi \leq 1$. The inflaton $\widehat{\delta\phi}$ decays mostly into N_1^c's. In all cases $\widehat{\Gamma}_{\delta\phi\to N_i^c} < \widehat{\Gamma}_{\delta\phi\to H}$ and so the ratios $\widehat{\Gamma}_{\delta\phi\to N_i^c}/\widehat{\Gamma}_{\delta\phi}$ in Eq. (31) introduce a considerable reduction in the derivation of Y_B. For the considered cases in Fig. 2, successful nTL requires M_iN^c and m_3D in the ranges $(1 - 10^6)$ EeV—see Table of Fig. 2—and $(265 - 308)$ GeV respectively. Meantime, $Y_{\widetilde{G}}$ and T_{rh} in Eqs. (29) and (34) are confined in the ranges

$$1.6 \lesssim T_{\mathrm{rh}}/10 \text{ PeV} \lesssim 23.3 \quad \text{and} \quad 3.2 \lesssim Y_{\widetilde{G}}/10^{-15} \lesssim 44.3, \tag{37}$$

with both quantities decreasing as N increases. Therefore, the $\widetilde{G}$ constraint can be satisfied even for $m_{3/2} \sim 1$ TeV as deduced comparing Eqs. (35) and (37). As advertised in Sect. 3.1, these values are in nice agreement with the ones needed for the solution of the μ problem within CMSSM in all cases of Table 1.

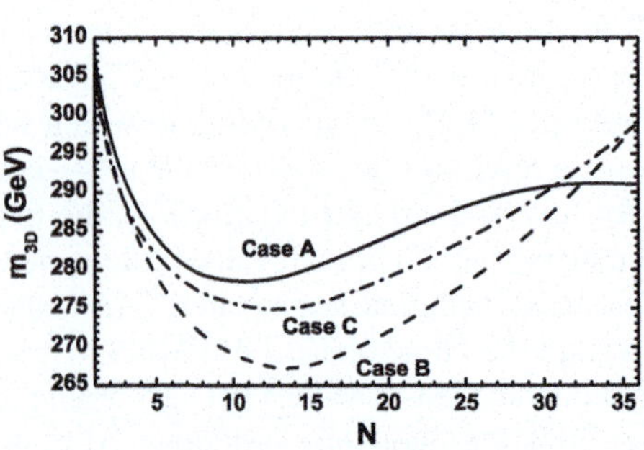

CASES : Hierarchy :	A NO	B NO	C IO
$m_{r\nu}$ / eV	0.02	0.01	0.007
$\Sigma_i m_{i\nu}$ / eV	0.096	0.075	0.11
m_{1D} / GeV	0.4	0.1	1.1
m_{2D} / GeV	1	0.58	0.6
φ_1	$\pi/1.5$	$\pi/4$	$-3\pi/4$
φ_2	0	π	$5\pi/4$
$M_{1N^c}/10^{10}$ GeV	1.3	0.16	1.94
$M_{2N^c}/10^{10}$ GeV	2.5	1.97	2.4
$M_{3N^c}/10^{15}$ GeV	1.8 - 2	1.1 - 1.45	3.9 - 4.8

Fig. 2 Contours, yielding the central Y_B in Eq. (33) consistently with the inflationary requirements, in the $N - m_{3\mathrm{D}}$ plane. We take as inputs $N_0 = 1$, $\mu = 4$ TeV and the values of $m_{i\nu}$, $m_{1\mathrm{D}}$, $m_{2\mathrm{D}}$, φ_1 and φ_2 shown in the Table which correspond to cases A (solid line), B (dashed line) and C (dot-dashed line)

4 Summary

We focused on the simplest among the implementations introduced in Ref. [13] of T-model inflation in SUGRA, using as inflaton a Higgs field which breaks a gauge $U(1)_{B-L}$ symmetry. We employed the super- and Kähler potentials W and K shown in Eqs. (9a) and (9b) which lead to the inflationary potential in Eq. (1) and the $\phi - \widehat{\phi}$ relation in Eq. (5). The model predicts $n_s \sim 0.963$ and r increasing with the prefactor $N \lesssim 36$ in Eq. (9b) without unnatural tuning. We also specified a post-inflationary completion, based on the extra terms in Eqs. (20a) and (20b), which offers a nice solution to the μ problem of MSSM and allows for baryogenesis via nTL with M_{iN^c} in the range $(1 - 10^6)$ EeV.

Acknowledgements I would like to thank N. Mavromatos, Q. Shafi and F. Farakos for interesting discussions and questions. This research work was supported by the Hellenic Foundation for Research and Innovation (H.F.R.I.) under the "First Call for H.F.R.I. Research Projects to support Faculty members and Researchers and the procurement of high-cost research equipment grant" (Project Number: 2251).

References

1. J. Martin, C. Ringeval, V. Vennin, Phys. Dark Univ. **5**, 75 (2014). ([arXiv:1303.3787])
2. D.S. Salopek, J.R. Bond, J.M. Bardeen, Phys. Rev. D **40**, 1753 (1989); J.L. Cervantes-Cota, H. Dehnen, Phys. Rev. D **51**, 395 (1995). [astro-ph/9412032]
3. N. Kaloper, L. Sorbo, J. Yokoyama, Phys. Rev. D **78**, 043527 (2008). ([arXiv:0803.3809])
4. J.L. Cervantes-Cota, H. Dehnen, Nucl. Phys. **B442**, 391 (1995). [astro-ph/9505069]; F.L. Bezrukov, M. Shaposhnikov, Phys. Lett. B **659**, 703 (2008). arXiv:0710.3755
5. C. Pallis, Universe **4**(1), 13 (2018). arXiv:1710.05759; C. Pallis, PoS **CORFU 2017**, 086 (2018). arXiv:1804.07038
6. C. Pallis, Q. Shafi, Eur. Phys. J. C **78**(6), 6, 523 (2018). arXiv:1803.00349; C. Pallis, *Contribution to: CORFU2022*. arXiv:2305.00523
7. Y. Akrami, et al., Planck Collaboration, Astron. Astrophys. **641**, A10 (2020). arXiv:1807.06211
8. M. Tristram et al., Phys. Rev. Lett. **127**, 151301 (2021). ([arXiv:2112.07961])
9. R. Kallosh, A. Linde, D. Roest, J. High Energy Phys. **11**, 198 (2013). ([arXiv:1311.0472])
10. C. Pallis, Eur. Phys. J. C **82**(5), 444 (2022). arXiv:2204.01047
11. J. Ellis, D. Nanopoulos, K. Olive, J. Cosmol. Astropart. Phys. **10**, 009 (2013). ([arXiv:1307.3537])
12. R. Kallosh, A. Linde, J. Cosmol. Astropart. Phys. **07**, 002 (2013). ([arXiv:1306.5220])
13. C. Pallis, J. Cosmol. Astropart. Phys. **05**, 043 (2021). ([arXiv:2103.05534]; C. Pallis, J. Phys. Conf. Ser. 2105, no. 12, 12 (2021) [arXiv:2109.06618])
14. J.J.M. Carrasco, R. Kallosh, A. Linde, D. Roest, Phys. Rev. D **92**(4), 041301 (2015). arXiv:1504.05557
15. R. Kallosh, A. Linde, T. Rube, Phys. Rev. D **83**, 043507 (2011). ([arXiv:1011.5945])
16. C. Pallis, N. Toumbas, J. Cosmol. Astropart. Phys. **05**(05), 015 (2016). arXiv:1512.05657
17. N. Aghanim, et al., Planck Collaboration, Astron. Astrophys. **641**, A6 (2020). Erratum: Astron. Astrophys. **652**, C4 (2021). arXiv:1807.06209
18. G.R. Dvali, G. Lazarides, Q. Shafi, Phys. Lett. B **424**, 259 (1998). [hep-ph/ 9710314]
19. P. Athron, et al., GAMBIT Collaboration, Eur. Phys. J. C **77**(12), 824 (2017). arXiv:1705.07935
20. https://pdg.lbl.gov/2020/reviews
21. P.F. de Salas et al., J. High Energy Phys. **02**, 071 (2021). ([arXiv:2006.11237])

Searches for Supersymmetry Using the Full CMS Run II Data Set

Costas Vellidis

Abstract Supersymmetry (SUSY) remains one of the most appealing theories beyond the standard model (BSM) of elementary particles, due to its completeness, rich phenomenology, and ability to provide answers to fundamental open questions of modern particle physics and cosmology. The sample of proton-proton collisions at 13 TeV center-of-mass energy collected with the Compact Muon Solenoid (CMS) detector during Run II of the CERN Large Hadron Collider (LHC) is a statistically powerful data set to look for new SUSY signatures and extend previous searches into unexplored regions of the parameter space. A number of recent results from SUSY searches with the full CMS Run II data set are presented in this contribution.

Keywords Large hadron collider · Compact muon solenoid · Particle physics · Beyond standard model · Supersymmetry

1 Introduction

The theory of supersymmetry (SUSY) is among the most popular extensions of the standard model (SM) of particle physics, because it can answer some of the fundamental questions which the SM cannot, such as the mass hierarchy problem, the nature of dark matter (DM), and the unification of forces. It postulates a "copy" of the SM particles consisting of new particles, the "superpartners", which differ from their SM partners by $\hbar/2$ unit of spin. This theory has the same gauge and flavor structure as the SM. Therefore, it assumes the same fundamental interactions and symmetries of Nature, with the notable exception that it requires 3 neutral + 2 charged Higgs bosons. And this is because two Higgs doublets are required, one for the "up" quarks U and one for the "down" quarks D. SUSY must be broken at some high energy scale,

On behalf of the CMS Collaboration.

C. Vellidis (✉)
National and Kapodisitrian University of Athens, Zografou University Campus, Athens, Greece
e-mail: Konstantinos.Vellidis@cern.ch
URL: http://scholar.uoa.gr/cvellid

J. Strologas and L. Perivolaropoulos (eds.), *HEP2023*, Springer Proceedings in Physics 333 https://doi.org/10.1007/978-3-031-76856-9_8

implying that the superpatners are heavier than their SM partners, and thus they have not yet been observed. The complete theory has a very rich phenomenology, but also a very large space of free parameters. Consequently, searches for signs of SUSY test simplified models, in which all but very few parameters are assumed to manifest at some inaccessible energy scale. SUSY signals have been searched in all high-energy colliders for more than 30 years: LEP, Tevatron, and LHC. No signs of it have been identified yet, but the situation is far from excluding all scenarios. There are still inaccessible regions of the parameter space, as well as extraneous signatures, such as superpatners with long lifetimes, which are less explored. This article presents recent results of SUSY searches using data from proton-proton collisions at a center-of-mass energy of 13 TeV, collected with the CMS detector [1] during Run II of operations of the CERN LHC in the years 2016–2018.

2 Searches with Hadronic Final States

SUSY search in photon+jets final states. This search [2] is sensitive to gauge-mediated supersymmetry breaking (GMSB) models with a light gravitino as the lightest supersymmetric particle (LSP). Events are selected by requiring at least one photon, at least two jets, and missing transverse momentum (MET). W/Z and Higgs bosons decaying to merged jet pairs are identified as "fat" jets (AK8). The dominant SM backgrounds are estimated from data and involve Wγ+jets and $t\bar{t}\gamma$+jets events with a lost electron or muon, W+jets and $t\bar{t}$+jets events with an electron misidentified as a photon, Z($\nu\bar{\nu}$)γ+jets events, and γ+jets events with spurious MET. No significant signal is observed and limits are set excluding gluino masses up to 2.35 TeV, top squark masses up to 1.43 TeV, and electroweakino next-to-lightest supersymmetric particle (NLSP) masses up to 1.30 TeV. The left plot of Fig. 1 shows the exclusion limit on the cross section as a function of the NLSP mass.

Stealth SUSY search in diphoton+jets final states. A recent search [3] explores a minimal extension of the simple supersymmetric standard model (MSSM) with a scalar singlet S and its superpartner, the singlino $\tilde{S}$. A GMSB scenario is assumed with a light gravitino LSP, so when the mass difference between the scalar and the singlino is small there is no MET requirement in the final state, hence the term "stealth" SUSY. Events are selected by requiring two photons, isolated within a cone of radius 0.3 in the (η, ϕ) space, and at least four jets. SM backgrounds from diphoton+jets events are estimated from data and then adjusted using simulation. No significant signal is observed and limits are set excluding gluino and squark masses up to 2.15 and 1.85 TeV, respectively, extending by about 70% the excluded region of the $(m_{\tilde{q}}, m_{\tilde{\chi}_1^0})$ space shown in the right plot of Fig. 1.

RPV + stealth SUSY search in $t\bar{t}$+jets final states. It is the first search of this signal in top squark pair production at the LHC [4]. It looks for R-parity violating (RPV) SUSY with trilinear Yukawa coupling λ''_{ijk}, $i, j, k = 1, 2$ as well as for stealth SUSY

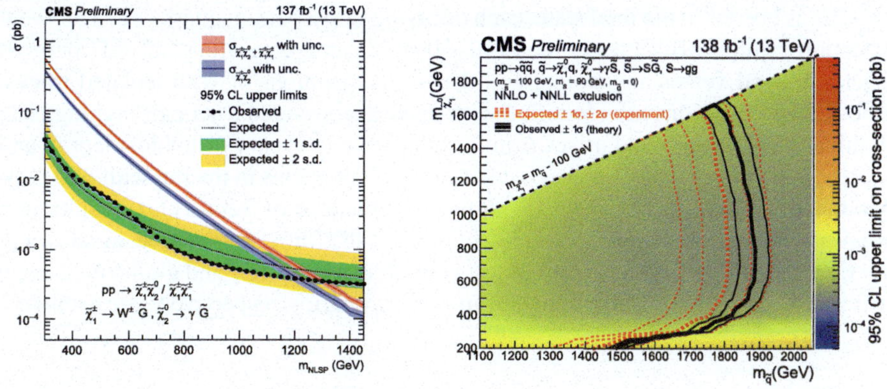

Fig. 1 Exclusion limits from the searches in single photon+jets final states [2] (left) and diphoton + jets final states [3] (right). See the text for details

with an R-parity even singlet scalar S and its singlino superpatner plus a light gravitino LSP. Assuming small mass difference between the scalar and the singlino, there is no MET requirement. The event selection requires one electron or muon plus at least seven jets with at least one of them tagged as originating from a b quark (b-jet). The dominant SM background arises from $t\bar{t}$+jets events and the signal-background discrimination is done using a deep neural network (DNN). Limits are set excluding top squark masses up to 700 (900) GeV in the RPV (stealth) SUSY scenario. The left plot of Fig. 2 shows the exclusion limit on the cross section for top squark pair production as a function of the top squark mass in the stealth SUSY scenario.

Higgsino decays to Higgs boson pairs. This search [5] looks for a signal in the higgsino-dominated GMSB scenario with either a neutralino or a goldstino LSP and

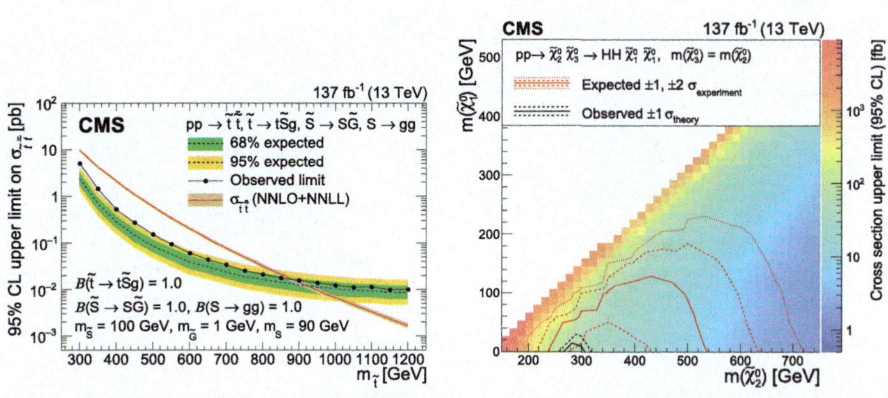

Fig. 2 Exclusion limits from the searches in $t\bar{t}$+jets final states [4] (left) and in Higgs boson pair final states [5] (right). See the text for details

two Higgs bosons in the final state, each decaying into a bottom quark pair. The event selection requires MET greater than 150 (300) GeV in events with resolved (merged) b-jet pairs and vetoes any events with electrons or muons or with isolated tracks. Boosted Higgs bosons are identified with mass-decorrelated b-jet pair tagging. The dominant SM background arises from semileptonic $t\bar{t}$ events with a lost lepton and it is estimated from data. An observed excess of four events in the bin with 300 GeV MET and 3 b-jets, having a local (global) significance of 3.3 (2.1) σ, results in a weaker than expected observed exclusion limit in the $(m_{\tilde{\chi}^2_0},\ m_{\tilde{\chi}^0_1})$ space, as shown in the right plot of Fig. 2. Masses of lightest neutralino up to 1 TeV and gluino masses up to 2.3 TeV for the goldstino and neutralino LSP scenarios, respectively, are excluded.

3 Searches with Heavy Object Taggers—MVA

Top squarks in fully hadronic final states. This search [6] explores top squark pair as well as gluino pair production in both compressed ($\Delta m = m_{\tilde{t}} - m_{LSP} < m_W$) and uncompressed ($\Delta m > m_W$) mass spectra. It is also sensitive to loop-induced flavor-changing neutral currents (FCNC) through the decay chain $\tilde{g} \rightarrow t\tilde{t}(c\tilde{\chi}^0_1)$. Top quarks and W bosons decaying into either resolved or merged jets are identified using deep learning algorithms. The dominant SM background, arising from $Z(\nu\bar{\nu})$+jets events in the compressed scenario and from semileptonic $t\bar{t}$ events with a lost lepton in the uncompressed scenario, are estimated from both data and simulation. No significant signal is observed and limits are set excluding top squark masses up to 1310 GeV, gluino masses up to 2260 GeV, and LSP masses up to 1400 GeV in the gluino production interpretation, pushing previous limits further by 100–300 GeV. The left plot of Fig. 3 shows the exclusion limit in the $(m_{\tilde{g}},\ m_{\tilde{\chi}^0_1})$ space.

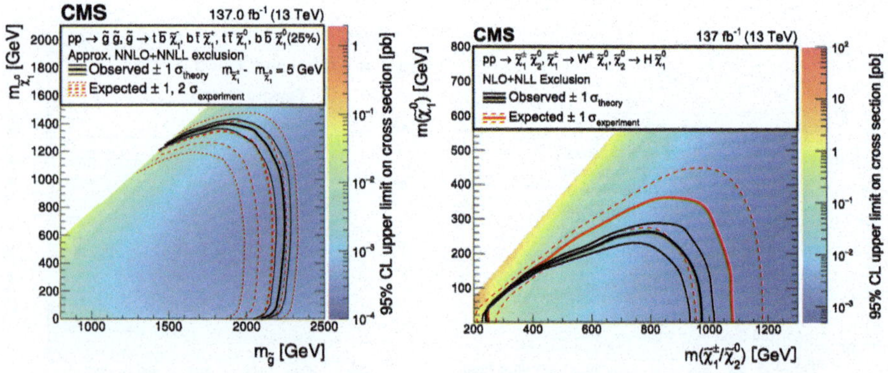

Fig. 3 Exclusion limits from the searches for top squarks [6] (left) and for electroweakinos [7] (right) in fully hadronic final states. See the text for details

Electroweakinos in fully hadronic final states. This search [7] employs deep neural networks (DNNs) to classify heavy objects as bosons (W, Z, H), top quarks, or anything else, and subcategorize decays by flavor. Decays of neutral bosons (Z, H) into bottom quark pairs are reconstructed by combining a b-jet pair tagger, for the resolved case, with a DNN, for the merged case where the bosons are identified as "fat" jets (AK8). Decays of W and Z bosons into lighter quark pairs are reconstructed by combining a vector boson tagger with an adversarial DNN, decorrelated from the jet mass, to discriminate the bosons from QCD jets. A b-jet veto is applied to divide the events into a VV search region (V = W or Z) involving light-flavor jets and a VH search region (SR) involving a reconstructed $H(b\bar{b})$ boson. The dominant SM backgrounds, coming from W($\ell\nu$)/Z($\nu\bar{\nu}$)+jets events in the b-veto SR and from $t\bar{t}$ events in the b-tag SR, are estimated by combined simulation and data-driven methods. For a low-mass LSP, the search excludes NLSP masses up to 670/760/970 GeV in the WW/WZ/WH final states, respectively. The right plot of Fig. 3 shows the exclusion limit in the (m_{NLSP}, m_{LSP}) space for the WH final state.

Gluinos in lepton+jets final states using heavy-object tagging. This search [8] looks for events with four top quarks and significant MET or four light quarks and significant MET, where one W boson coming from a top quark or a chargino decays into an electron or muon and a neutrino. The azimuthal distance between the lepton and the reconstructed W in the plane transverse to the beam is used to suppress the dominant SM backgrounds arising from $t\bar{t}$, W+jets, and QCD multijet events, which are estimated by combined simulation and data-driven methods. The search explores both compressed ($m_{\tilde{g}} \ll m_{\tilde{\chi}_1^0}$) and uncompressed mass spectra, looking for resolved ($p_T < 400,\ 200$ GeV) or boosted ($p_T > 400,\ 200$ GeV) top quarks/W bosons, respectively, using a deep learning tagger. Here p_T is the momentum component transverse to the beam. Limits are set excluding gluino and LSP masses up to 2130 and 1270 GeV in the 4t final state, extending previous limits by 320 and 170 GeV, respectively, and gluino and LSP masses up to 2280 and 1220 GeV in the 4q final state, extending previous limits by 380 and 270 GeV, respectively. The left plot of Fig. 4 shows the exclusion limit in the $(m_{\tilde{g}}, m_{\tilde{\chi}_1^0})$ space.

Top squarks in the corridor. This analysis [9] includes a combination of results from previous searches in final states with 0, 1, or 2 leptons, and a new search using opposite-sign dilepton events in the kinematic regime where the mass splitting between the top squark and the LSP is close to the top quark mass, the so-called "corridor". Such events are kinematically very difficult to distinguish from SM dileptonic $t\bar{t}$ events. A DNN is employed to discriminate this signal from the SM $t\bar{t}$ background. In addition, an alternative model of light fermionic DM, pair-produced via a (pseudo)scalar mediator, is explored. The combined limits from the previous searches exclude top squark masses up to 1325 GeV for a massless LSP and LSP masses up to 700 GeV for a top squark mass of 1150 GeV. In the corridor, top squark masses between 145 and 295 GeV and LSP masses up to 100 GeV are excluded for the first time with CMS data, as shown on the right plot of Fig. 4. The DM model is excluded for mediator masses up to 420 GeV.

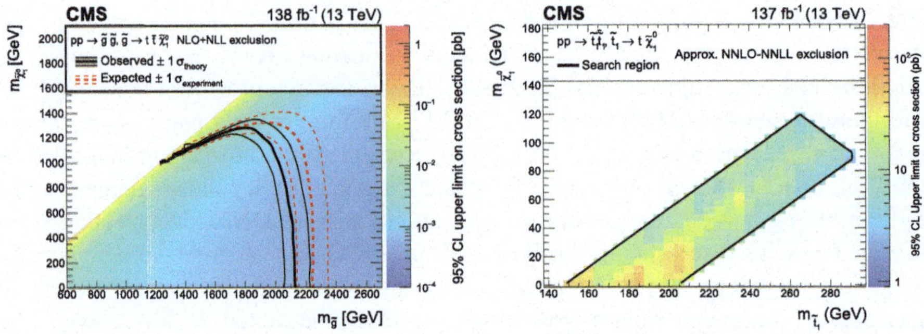

Fig. 4 Exclusion limits from the searches for gluinos in lepton+jets final states [8] (left) and for top squarks in the "corridor" [9] (right). See the text for details

4 Searches with Electrons and Muons

Multi-leptons. This search [10] is sensitive to mass-degenerate wino-like or higgsino-like chargino-neutralino production, and to neutralino-neutralino production in a GMSB scenario with a gravitino LSP. It is optimized in event categories with 2, 3, and 4 leptons based on sign and flavor. The dominant prompt lepton backgrounds arise from photon conversions, WZ events, and ZZ events in the 2-, 3-, and 4-lepton categories, respectively. The dominant non-prompt lepton backgrounds arise from $t\bar{t}$ and Drell-Yan (DY) production. Limits are set excluding chargino masses up to 1450 GeV in 3-lepton events using a parametric NN, up to 970 GeV in 3τ events for a model with right-handed sleptons, and up to 650 (250) GeV in events with a WZ (WH) final state. Neutralino masses are excluded up to 600 GeV in events with a Z or an H boson in the final state. These are the most stringent chargino-neutralino production limits for NLSP−LSP mass splitting close to the Z boson mass. The left plot of Fig. 5 shows the exclusion limit in the (m_{NLSP}, m_{LSP}) space for the chargino-neutralino scenario.

Soft opposite-sign leptons. This search [11] explores electroweakino and top squark pair production in the compressed-mass scenario. It includes the 3-lepton final state which was not considered in the previous round of this analysis that used only the 2016 CMS data. The event selection requires two or three electrons or muons with transverse momentum below 30 GeV in the final state, including one opposite-sign pair, plus an initial-state radiation (ISR) jet to induce significant MET above 125 GeV. The search is optimized in MET bins and looks for the electroweakino production signal in the dilepton mass spectrum, and for the top squark pair production signal in the leading lepton's transverse momentum spectrum. The dominant SM backgrounds from DY, $t\bar{t}$, WW, WZ production, plus non-prompt leptons or misidentified leptons in W+jets (2-lepton final state) and $t\bar{t}$ (3-lepton final state) events are constrained from a same-sign dilepton sample. Limits are set excluding NLSP masses up to 275 GeV for an NLSP−LSP mass splitting of 10 GeV in the wino-bino scenario and

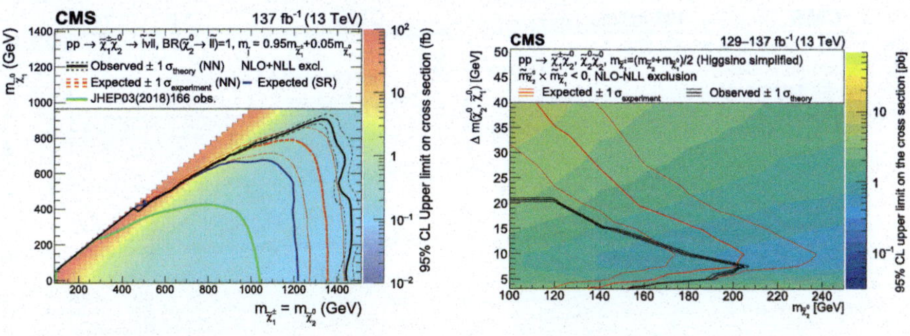

Fig. 5 Exclusion limits from the searches in multi-lepton final states [10] (left) and in soft opposite-sign (SOS) lepton pair final states [11] (right). See the text for details

up to 205 (150) GeV for a splitting of 7.5 (3) GeV in the higgsino scenario. Also, top squark masses up to 540 (480) GeV for a mass splitting of 30 GeV are excluded in the 4-body (chargino-mediated) top squark decay scenario. The right plot of Fig. 5 shows the exclusion limit in the $(m_{NLSP},\ \Delta m)$ space for the higgsino scenario.

Opposite-sign leptons "on the edge". This search [12] looks for various SUSY modes of opposite-sign dilepton production, including cascade $\tilde{\chi}_2^0$ decays analyzed in the dilepton mass spectrum, which shows a peak at the Z boson mass when the decay contains an on-shell Z boson, an "edge" when it contains the chain $\tilde{\chi}_2^0 \to \ell^+\tilde{\ell}(\ell^-\tilde{\chi}_1^0)$, and a continuum when it contains slepton pair decays. A fit of the dilepton mass is done in these scenarios, while MET and jet multiplicity are studied for the other signals. Dominant SM backgrounds arise from $t\bar{t}$ production, DY+jets production with instrumental MET, and VZ production (V = W or Z) with neutrinos in the final state, and they are estimated from data and simulation. Limits are set excluding gluino masses up to 1870 GeV for squark (sbottom) masses up to 1800 (1600) GeV and NLSP masses up to 800 GeV for slepton masses up to 700 GeV, extending previous CMS limits by a few hundred GeV. The left plot of Fig. 6 shows the exclusion limit in the $(m_{NLSP},\ m_{LSP})$ space.

5 Searches with τ Leptons

Direct τ slepton pair production. This search [13] is sensitive to early-Universe neutralino-stau coannihilation models explaining the observed DM relic density. It includes a GMSB-inspired search for long-lived stau with a light gravitino LSP that was not considered in previous stau pair production searches from the LEP and the LHC. Events with opposite-sign hadronically decaying tau pairs are selected with a deep learning tagger to discriminate them from QCD jets. The dominant SM backgrounds originating from DY and misidentified di-tau events are estimated by

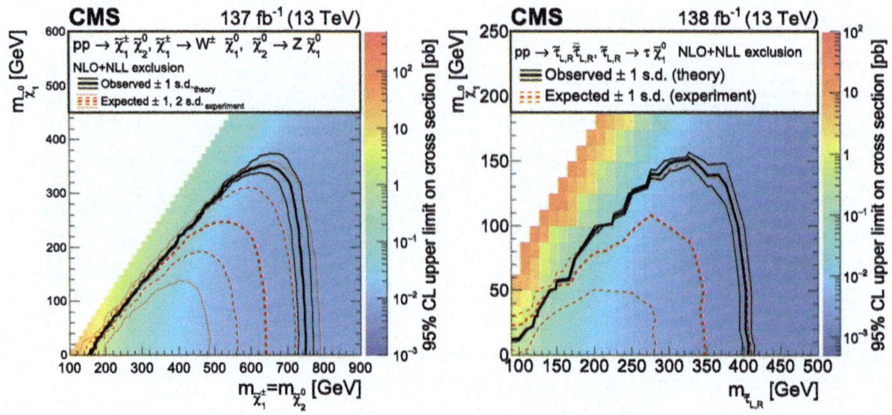

Fig. 6 Exclusion limits from the searches in opposite-sign lepton pair final states "on the edge" [12] (left) and in τ lepton pair final states [13] (right). See the text for details

data-driven methods. Limits are set excluding stau masses between 90 and 400 GeV for a nearly massless LSP, as well as stau masses between 150 and 220 GeV for a stau proper decay length of 0.1 mm. The right plot of Fig. 6 shows the exclusion limit in the $(m_{\tilde{\tau}},\ m_{\tilde{\chi}_1^0})$ space.

6 Combinations

Combination of electroweak SUSY searches. A recent analysis [14] combined exclusion limits from all searches for electroweakino and slepton production with the full Run II sample. These include gaugino-like chargino-neutralino production, higgsino-like neutralino pair production in a GMSB scenario, a higgsino-bino interpretation, and slepton pair production. The combination also included a new interpretation of the soft opposite-sign (SOS) lepton analysis with a slepton pair production model, which is the first search for slepton pair production in the compressed mass scenario. This re-interpretation excludes slepton masses up to 215 GeV for a slepton-LSP mass splitting of 5 GeV. The combination excludes slepton masses up to 700 GeV and NLSP masses up to 800 GeV for LSP masses up to 50 GeV, whereas for the chargino-neutralino production and for the neutralino pair production interpretations it extends previous CMS limits by several hundreds of GeV. Figure 7 shows the exclusion limits for the slepton pair production derived from the combination (left plot) and from the re-interpretation of the SOS lepton analysis (right plot).

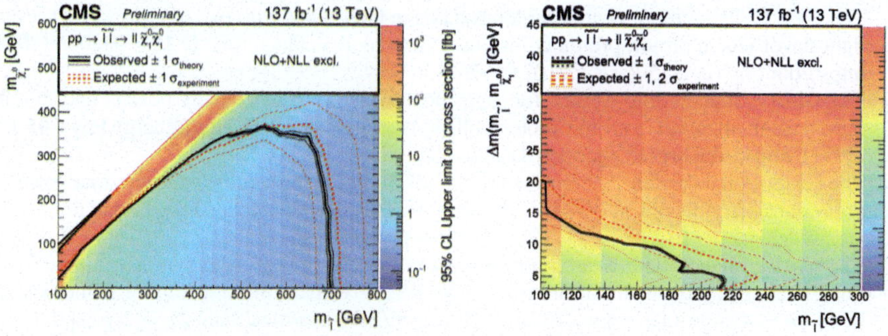

Fig. 7 Exclusion limits on the slepton pair production from the combination of electroweak SUSY searches [14] (left) and from the re-interpretation of the SOS lepton analysis [14] (right), which is included in the combination. See the text for details

7 Conclusions: The Way Ahead

CMS has achieved great progress in SUSY searches using the Run II sample. Exclusion regions of the parameter space have been extended for a large set of simplified models. Many new techniques have been introduced to improve the sensitivity of the searches beyond the crude statistical improvement, such as triggers especially designed for SUSY searches and event classifiers using machine learning algorithms, thus paving the way for analyzing the Run III sample with optimal efficiency. Future searches for SUSY are expected to look at harder topologies, such as top squark decays "in the corridor"; narrower phase space, such as the compressed scenarios examined in the SOS analysis; and LLP signatures, as in the direct τ slepton production search presented here. The double size of the sample expected from LHC Run III operations will improve significantly the sensitivity. Finally, the High-Luminosity LHC (HL-LHC) will open a new era of searches for BSM physics, with all levers advanced to a substantially higher level, including much larger samples, advanced detector and trigger technologies, and more analysis power.

References

1. The CMS Collaboration: The CMS Experiment at the CERN LHC. JINST 3 S08004 (2008). https://cds.cern.ch/record/1129810/files/jinst8_08_s08004.pdf
2. The CMS Collaboration: Search for new physics in multijet events with at least one photon and large missing transverse momentum in proton-proton collisions at 13 TeV. CMS PAS SUS-21-009. https://cds.cern.ch/record/2852844/files/SUS-21-009-pas.pdf
3. The CMS Collaboration: Search for stealth SUSY in final states with two photons, jets, and low missing transverse momentum proton-proton collisions at 13 TeV. CMS PAS SUS-19-001. https://cds.cern.ch/record/2852182/files/SUS-19-001-pas.pdf

4. The CMS Collaboration: Search for top squarks in final states with two top quarks and several light-flavor jets in proton-proton collisions at $\sqrt{s}$=13 TeV. Phys. Rev. D **104**, 032006 (2021). https://doi.org/10.48550/arXiv.2102.06976
5. The CMS Collaboration: Search for higgsinos decaying to two Higgs bosons and missing transverse momentum in proton-proton collisions at $\sqrt{s}$=13 TeV. J. High Energy Phys. **05**, 014 (2022). https://doi.org/10.48550/arXiv.2201.04206
6. The CMS Collaboration: Search for top squark production in fully hadronic final states in proton-proton collisions at $\sqrt{s}$=13 TeV. Phys. Rev. D **104**, 052001 (2021). https://doi.org/10.48550/arXiv.2103.01290
7. The CMS Collaboration: Search for electroweak production of supersymmetric particles in final states containing hadronic decays of WW, WZ, or WH and missing transverse momentum. CMS PAS SUS-21-002. https://cds.cern.ch/record/2779116/files/SUS-21-002-pas.pdf
8. The CMS Collaboration: Search for supersymmetry in final states with a single electron or muon using angular correlations and heavy object tagging in proton-proton collisions at $\sqrt{s}$=13 TeV. CMS PAS SUS-21-007. https://cds.cern.ch/record/2805223/files/SUS-21-007-pas.pdf
9. The CMS Collaboration: Combined searches for the production of supersymmetric top quark partners in proton-proton collisions at $\sqrt{s}$=13 TeV. Eur. Phys. J. C **81**, 970 (2021). https://doi.org/10.48550/arXiv.2107.10892
10. The CMS Collaboration: Search for electroweak production of charginos and neutralinos in proton-proton collisions at $\sqrt{s}$=13 TeV. J. High Energy Phys. **04**, 147 (2022). https://doi.org/10.48550/arXiv.2106.14246
11. The CMS Collaboration: Search for supersymmetry in final states with two or three soft leptons and missing transverse momentum in proton-proton collisions at $\sqrt{s}$=13 TeV. J. High Energy Phys. **04**, 091 (2022). https://doi.org/10.48550/arXiv.2111.06296
12. The CMS Collaboration: Search for supersymmetry in final states with two oppositely charged same-flavor leptons and missing transverse momentumin proton-proton collisions at $\sqrt{s}$=13 TeV. J. High Energy Phys. **04**, 123 (2021). https://doi.org/10.48550/arXiv.2012.08600
13. The CMS Collaboration: Search for direct pair production of supersymmetric partners to the τ lepton in the all-hadronic final state at $\sqrt{s}$=13 TeV. CMS PAS SUS-21-001. https://cds.cern.ch/record/2777046/files/SUS-21-001-pas.pdf
14. The CMS Collaboration: Combined search for electroweak production of winos, binos, higgsinos, and sleptons in proton-proton collisions at $\sqrt{s}$=13 TeV. CMS PAS SUS-21-008. https://cds.cern.ch/record/2853345/files/SUS-21-008-pas.pdf

Dimension-8 EFT Interpretation for the EWK Production of ZZjj in the Four-Lepton Channel

A. Marantis, A. Leisos, K. Kordas, I. Maznas, and A. Tsirigotis

Abstract The present study focuses on Vector Boson Scattering (VBS) processes, which offer valuable insights into the Quartic Gauge Couplings (QGCs). While the Standard Model allows self-interactions of charged vector gauge bosons, interactions involving only neutral bosons are prohibited. The analysis utilizes Monte Carlo (MC) samples containing VBS events featuring two Z-bosons in association with two jets in the final state. The objective is to estimate the constraints on dimension-8 Effective Field Theory coefficients for the $\mathcal{O}_{T,j}$ operators in the ZZjj channel at proton-proton collisions at center-of-mass energy of 13 TeV, using data of an integrated luminosity of 139 fb^{-1}. To mimic the geometry of the ATLAS detector, fully reconstructed MC samples with final-state configurations of 4 leptons and 2 jets are employed. This preliminary study paves the way for further analysis and provides valuable information on the behavior of QGCs in VBS processes.

Keywords Vector boson scattering · Effective field theory · Quartic gauge boson couplings

1 Introduction

The structure of Quartic Gauge Couplings (QGCs) is determined by the gauge symmetry of the Standard Model (SM), specifically the $SU(2)_L \otimes U(1)_Y$ gauge symmetry. The study of QGCs is valuable for testing the predictions of the SM as well as exploring possible indications of physics Beyond the Standard Model (BSM). QGC processes are investigated in the context of Vector Boson Scattering (VBS), which involves the production of pairs of electroweak vector bosons [1, 2]. This multiboson production, particularly VBS, provides a powerful tool to probe the

A. Marantis (✉) · A. Leisos · A. Tsirigotis
Physics Laboratory, School of Science and Technology, Hellenic Open University, Patras, Greece
e-mail: alexandros.marantis@cern.ch

K. Kordas · I. Maznas
Department of Physics, Aristotle University of Thessaloniki, Thessaloniki, Greece

J. Strologas and L. Perivolaropoulos (eds.), *HEP2023*, Springer Proceedings in Physics 333 https://doi.org/10.1007/978-3-031-76856-9_9

Electroweak Symmetry Breaking (EWSB) sector of the SM [3]. In the SM, cancellations of contributions from particular Feynman diagrams prevent the scattering amplitudes involving charged gauge bosons from growing excessively with increasing interaction energy. As a result, there are no unphysical enhancements of cross sections and no violation of unitarity bounds. However, the couplings between neutral gauge bosons, such as ZZZZ, ZZZA, ZZAA, ZAAA, and AAAA, (where Z is the Z boson and A the photon) are forbidden in the SM, and deviations from the SM predictions can lead to the unphysical growth of scattering amplitudes with interaction energy [4]. Therefore, "anomalous" quartic gauge couplings (aQGCs) can lead to deviations from SM. Note that aQGCs can exist in the scattering of charged vector bosons as well. In the case where the $SU(2)_L \otimes U(1)_Y$ symmetry is linearly realized, which is the focus of this article, the lowest order QGCs are described by dimension-8 operators known as genuine QGC operators [5]. These operators are part of a low-energy Effective Field Theory (EFT) framework [6, 7]. The EFT can be defined as a parameterization given by the equation:

$$\mathcal{L}_{\mathcal{EFT}} = \mathcal{L}_{\mathcal{SM}} + \sum_{d>4} \sum_{i} \frac{f_i}{\Lambda^{d-4}} \mathcal{O}_i \qquad (1)$$

Here, $\mathcal{L}_{SM}$ represents the Lagrangian of the SM, f_i are the Wilson coefficients, d is the dimension of the operators $\mathcal{O}_i$, which involve various fields such as Higgs doublets, gauge bosons, fermionic fields, and covariant derivatives. It is worth noting that only the operators with the lowest dimensions have significant contributions below the scale of new physics, denoted as Λ. In the limit where Λ approaches infinity, the Lagrangian is solely composed of SM terms. As previously mentioned, this study focuses on genuine QGC operators, i.e. operators that specifically contribute to QGCs without being associated with Triple Gauge Couplings (TGCs). In particular, the investigation centers around dimension-8 QGC operators, which are listed in Table 1. The article is organized as follows. Section 2 introduces the VBS process and describes the experimental phase space, also known as the "fiducial region", which is utilized to investigate the signatures of QGCs. Section 3 provides a study for setting

Table 1 Quartic vertices induced by each dim-8 operator

	WWWW	WWZZ	ZZZZ	WWAZ	WWAA	ZZZA	ZZAA	ZAAA	AAAA
$\mathcal{O}_{S,0}$, $\mathcal{O}_{S,1}$	x	x	x						
$\mathcal{O}_{M,0}$, $\mathcal{O}_{M,1}$, $\mathcal{O}_{M,6}$, $\mathcal{O}_{M,7}$	x	x	x	x	x	x	x		
$\mathcal{O}_{M,2}$, $\mathcal{O}_{M,3}$, $\mathcal{O}_{M,4}$, $\mathcal{O}_{M,5}$		x	x	x	x	x	x		
$\mathcal{O}_{T,0}$, $\mathcal{O}_{T,1}$, $\mathcal{O}_{T,2}$	x	x	x	x	x	x	x	x	x
$\mathcal{O}_{T,5}$, $\mathcal{O}_{T,6}$, $\mathcal{O}_{T,7}$		x	x	x	x	x	x	x	x
$\mathcal{O}_{T,8}$, $\mathcal{O}_{T,9}$			x			x	x	x	x

limits on the Wilson coefficients f_i by using MC samples that have been generated using a simulation of the ATLAS detector geometry. The MC events incorporate the detector response, including the effects of tracking, calorimetry, and particle identification. This allows us to model the experimental conditions of the ATLAS detector. Finally, the findings of this study are discussed in Sect. 4.

2 Definition of the Fiducial Phase Space

Extensive studies on the effects of the QGCs in VBS processes have been conducted in recent years at the LHC experiments [8–16]. In this article, we focus specifically on the $ZZjj$ channel produced at proton-proton collisions at a center-of-mass energy ($\sqrt{s}$) of 13 TeV corresponding to an integrated luminosity of 139 fb^{-1}. The $pp \rightarrow ZZjj$ events are generated using the MadGraph5_aMC@NLO generator [17], employing the Eboli-Gonzales-Carcia model [18, 19]. The sensitivity study (Sect. 3) is performed using "truth-level" information, without simulating the hadronization process at leading-order (LO) accuracy. For the estimation of the constraints on dimension-8 EFT coefficients the MC samples contain events with "detector-level" information with LO accuracy. The final state being studied consists of four charged leptons originating from the Z-boson pair and two jets arising from the initial quark scattering. Figure 1 illustrates representative Feynman diagrams for the Electroweak (EWK) $ZZjj$ production. To emulate the realistic kinematic and geometrical requirements of the ATLAS $ZZjj$ analysis, we define

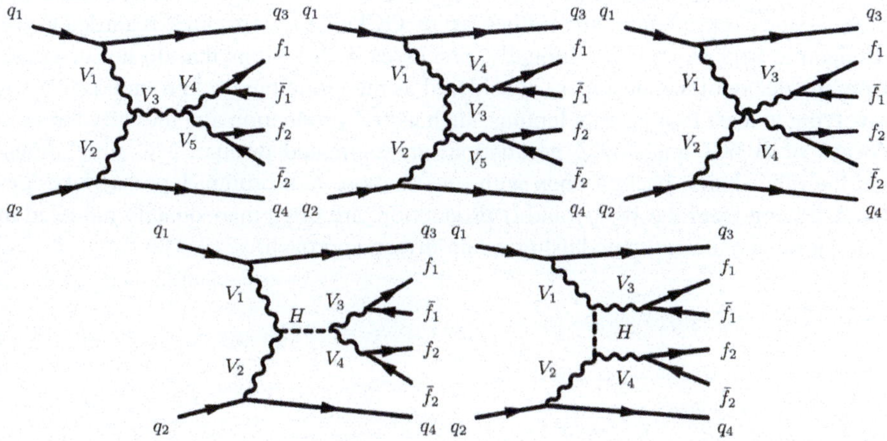

Fig. 1 Representative Feynman diagrams of the EWK $ZZjj$ production. The scattering topology includes either a TGC vetex with the production of a W/Z boson in the s-channel (top left) or in the t-channel (top middle), a QGC vertex (top right) or the exchange of a Higgs boson either in the s-channel (bottom left) or in the t-channel (bottom right). The labels V_i refer to the vector bosons (V=W, Z) and f_i to fermions

a "fiducial" phase space that includes optimized kinematic cuts designed to select EWK $ZZjj$ events, while minimizing the contribution from events produced by Quantum Chromodynamics (QCD) vertices.

The signal phase space is determined based on the criteria outlined below. The transverse momenta requirements for the leading and sub-leading lepton are set at $p_T > 20$ GeV/c. The absolute pseudorapidity ($|\eta|$) is limited to $|\eta| < 2.47$ for electrons and $|\eta| < 2.7$ for muons. The pseudorapidity is defined as $\eta = -\ln\tan(\theta/2)$, where θ is the polar angle with respect to the proton beam axis.[1] Moreover, a separation criterion between any two different-flavor leptons is imposed, requiring $\Delta R(l, l') > 0.05$, where $\Delta R(l, l') = \sqrt{(\Delta\eta)^2 + (\Delta\phi)^2}$. Quadruplets of leptons are constructed by pairing same-flavor opposite-sign (SFOS) leptons to form Z boson candidates, i.e. SFOS pairs which have a mass within the range of 66 GeV/c^2 to 116 GeV/c^2. In cases where multiple possible quadruplets exist, the combination of leptons that minimizes the sum of the absolute differences between the Z boson candidate mass and the Z boson pole mass, $|m_{ll} - m_Z| + |m_{l'l'} - m_Z|$, is selected. The Z boson pole mass is taken as $m_Z = 91.19$ GeV/c^2. Additionally, the invariant mass of all possible pairs of SFOS leptons must exceed $m_{ll} > 5$ GeV/c^2 and the invariant mass of the four-lepton system must satisfy $m_{4l} > 130$ GeV/c^2.

Regarding the jets, a transverse momentum threshold of $p_T > 40$ GeV/c is applied for the leading jet, along with an absolute pseudorapidity restriction of $|\eta| < 4.5$. The two jets are required to have opposite pseudorapidities, i.e. to move to opposite polar directions, characterized by $\eta_1 \cdot \eta_2 < 0$, and to have a significant rapidity separation of $\Delta y_{jj} > 2$. Finally, the invariant mass of the jet system must satisfy $m_{jj} > 300$ GeV/c^2.

The above selection criteria collectively define the fiducial phase space utilized in the analysis in order to minimize the background contribution. The background in the present study is predominantly influenced by QCD $ZZjj$ processes, regarded as the irreducible background. Additionally, processes with leptons that do not originate from the decay of a Z boson are considered as the reducible background: processes involving at least four prompt leptons, such as $t\bar{t}Z$ production and the fully leptonic decays of WWZ and WZZ production, are estimated through MC simulations. On the other hand, backgrounds with one or more non-prompt leptons (fake leptons), arising from the $WZjj$ and $t\bar{t}$ production, are often inadequately modeled in simulation and are estimated using a data-driven approach.

[1] ATLAS uses a right-handed coordinate system, positioning its origin at the nominal interaction point (IP) situated in the center of the detector, with the z-axis aligned along the beam pipe. The x-axis extends from the IP towards the center of the LHC ring, while the y-axis points upwards. Cylindrical coordinates (r, ϕ) in the transverse plane are used, where ϕ represents the azimuthal angle around the z-axis.

3 Expected Dimension-8 EFT Limits of the f_T Wilson Coefficients

The Lagrangian for the EFT of Eq. (1), considering only dimension-8 operators that modify the interactions among electroweak gauge bosons can be written as:

$$\mathcal{L}_{\mathcal{EFT}} = \mathcal{L}_{\mathcal{SM}} + \sum_{j=0,1} \frac{f_{S,j}}{\Lambda^4}\mathcal{O}_{S,j} + \sum_{j=0,\ldots,9} \frac{f_{T,j}}{\Lambda^4}\mathcal{O}_{T,j} + \sum_{j=0,\ldots,7} \frac{f_{M,j}}{\Lambda^4}\mathcal{O}_{M,j} \quad (2)$$

In this equation, the terms $\frac{f_{S,j}}{\Lambda^4}\mathcal{O}_{S,j}$, $\frac{f_{T,j}}{\Lambda^4}\mathcal{O}_{T,j}$, and $\frac{f_{M,j}}{\Lambda^4}\mathcal{O}_{M,j}$ correspond to the contributions of dimension-8 operators, denoted as $\mathcal{O}_{S,j}$, $\mathcal{O}_{T,j}$ and $\mathcal{O}_{M,j}$, respectively. These operators are classified into three groups based on the number of gauge-boson strength fields they contain. A comprehensive description of the three operator groups is provided in Ref. [20]. These genuine QGC operators have the ability to generate quartic couplings among neutral gauge bosons.

A previous study [21] conducted to assess the impact of different dimension-8 EFT Wilson coefficients on the production cross section within the VBS $ZZjj$ process, revealed that the f_T Wilson coefficients exhibit the highest sensitivity; among the 17 coefficients of the genuine QGC operators considered (see Table 1), the f_T coefficients have the greatest influence on the cross section of the $ZZjj$ production. Table 2 presents the ranking of all coefficients based on their contribution to the total cross section of the EWK $ZZjj$ process. Consequently, in the present study, we focus on estimating the expected ("Asimov") limits for the coefficients associated with the $\mathcal{O}_{T,j}$ operators, since these operators have the most significant impact on the cross section of the VBS $ZZjj$ process.

The total amplitude of the process is given by the sum of the SM component and the contributions from the EFT operators:

$$|A_{SM} + \sum_i c_i A_i|^2 = |A_{SM}|^2 + \sum_i c_i 2Re(A^*_{SM} A_i) + \sum_i c_i^2 |A_i|^2 + \sum_{i,j,i\neq j} c_i c_j 2Re(A^*_i A_j) \quad (3)$$

Here, A_{SM} and A_i represent the amplitudes for the SM and EFT operators respectively. The term $c_i 2Re(A^*_{SM} A_i)$ represents the interference between the SM and the EFT operator i (linear term), $c_i^2|A_i|^2$ represents the pure EFT contribution (quadratic term), and $c_i c_j 2Re(A^*_i A_j)$ represents the interference between the EFT coefficients i and j (cross terms). In the present study only one non-zero Wilson coefficient is considered at a time, meaning that the interference between the EFT operators (cross terms) is zero.

We use the invariant mass of the 4-lepton system m_{4l} as an observable for the extraction of the coefficient limits, since it is the most sensitive among the kinematical variables that are used in the analysis.[2] A detailed study regarding the sensitivity of

[2] Other sensitive variables that are functions of kinematical variables i.e. optimal observables [22] are not considered in this study.

Table 2 Ranking of the QGC coefficients according to their contribution to the total cross section of the EWK $ZZjj$ process [21]

Rank	QGC coefficient λ (TeV^{-4})	σ_{tot} (fb) for $\lambda = 1$ TeV^{-4}
1.	f_{T0}/Λ^4	92.995
2.	f_{T1}/Λ^4	58.135
3.	f_{T8}/Λ^4	27.153
4.	f_{T5}/Λ^4	17.430
5.	f_{T2}/Λ^4	15.112
6.	f_{T6}/Λ^4	7.309
7.	f_{T9}/Λ^4	5.931
8.	f_{M0}/Λ^4	3.262
9.	f_{M2}/Λ^4	1.944
10.	f_{T7}/Λ^4	1.882
11.	f_{M1}/Λ^4	0.377
12.	f_{M4}/Λ^4	0.251
13.	f_{M3}/Λ^4	0.174
14.	f_{M5}/Λ^4	0.130
15.	f_{M7}/Λ^4	0.036
16.	f_{S1}/Λ^4	0.035
17.	f_{S0}/Λ^4	0.0008

the kinematical variables to the aQGC effects can be seen in Ref. [21]. The event selection follows the criteria of Sect. 2 for centrality $\zeta < 0.4$ where

$$\zeta = \frac{|y_{quadruplet} - 0.5 \cdot (y_{leading\ jet} + y_{sub-leading\ jet})|}{|y_{leading\ jet} - y_{sub-leading\ jet}|} \tag{4}$$

and where $y_{quadruplet}$, $y_{leading\ jet}$ and $y_{sub-leading\ jet}$ are the rapidities of the lepton quadruplet, the leading (most energetic) jet and the sub-leading jet respectively. The distribution of the expected yield as a function of m_{4l} can be seen in Fig. 2.

The limits on the aQGCs are determined through a binned likelihood fit, taking into account the expected number of events as given by the MC estimates. The comparison between the expected observation from SM alone ("Asimov dataset") and the predicted number of events, including the aQGC contribution, is based on Poisson statistics. The formulation of the profile likelihood function is given by:

$$L = \prod_{i=1}^{N_{bins}} L_{poiss}(n_i|\mu_i(\lambda, \theta)) \times L_{gauss}(\theta_i) \tag{5}$$

where, n_i represents the "observed" number of events in the ith bin, $\mu_i(\lambda, \theta)$ is the predicted number of events in that bin, $\lambda = f_T/\Lambda^4$ is the value of the QGC

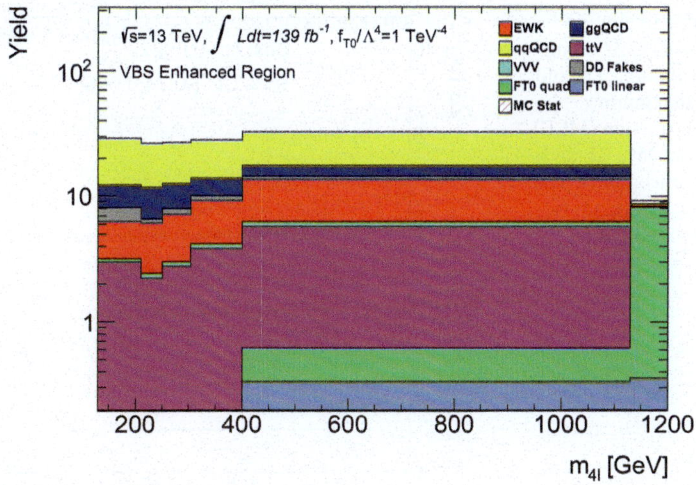

Fig. 2 Yield distribution for the MC SM and the f_{T0} aQGC (sum of quadratic and linear term) as a function of the invariant mass of the 4-lepton system

coefficient being examined, and θ represents a list of nuisance parameters that account for systematic uncertainties.[3] The likelihood function is constructed by multiplying the Poisson likelihood L_{poiss} for the "observed" and the predicted events, with the Gaussian likelihood L_{gauss} to constrain the nuisance parameters affecting the event yield.

The expected limits on the aQGC values, are determined using confidence intervals at a 95% confidence level (CL), by using the SM prediction as a hypothetical observation (Asimov dataset). To obtain these limits, first, the likelihood function is maximized simultaneously for the parameter of interest, λ, and the nuisance parameters, θ, resulting in the maximum likelihood $L(\hat{\lambda}, \hat{\theta})$. Next, the likelihood function is scanned for different values of the parameter of interest λ, while the nuisance parameters which maximize the likelihood L are selected at each step, resulting in the profile likelihood, $L(\lambda, \hat{\hat{\theta}})$. The likelihood ratio, defined as the ratio of the profile likelihood to the maximum likelihood, is then computed:

$$q(\lambda) = \frac{L(\lambda, \hat{\hat{\theta}})}{L(\hat{\lambda}, \hat{\theta})} \tag{6}$$

[3] Apart from the theoretical uncertainties which primarily accounts for the renormalization/factorization scale dependence and the parton distribution functions, a variety of experimental uncertainties arise from differences between data and simulation in the trigger efficiency and deficiencies in detector simulation and lepton/jet reconstruction, as well as in the luminosity determination, and the average number of proton-proton interactions per beam bunch crossing.

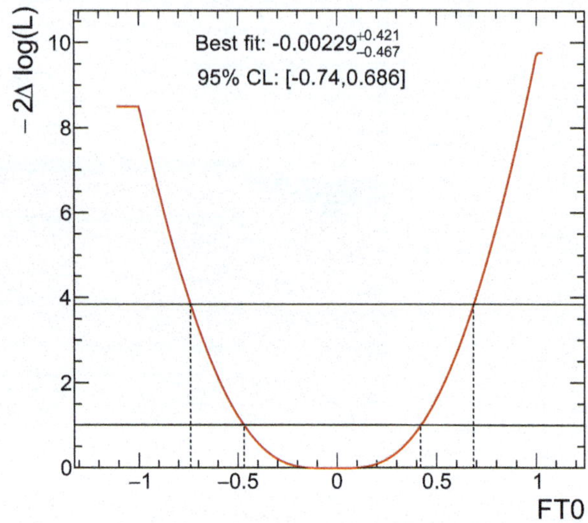

Fig. 3 Profile likelihood scan for $\lambda = f_{T0}/\Lambda^4$ using the m_{4l} distribution. The plot depicts the maximum likelihood estimation (MLE) along with the 95% CL and the 1-sigma uncertainty. The distribution corresponds to the case of no unitarity constraints

Table 3 Expected 95% CL limits for the dimension-8 Wilson coefficients at detector-level, corresponding to an integrated luminosity of 139 fb^{-1}

Wilson coefficient	Expected (Asimov) 95% CL Limit
	m_{4l}
f_{T0}	[−0.74, 0.69]
f_{T1}	[−0.85, 0.84]
f_{T2}	[−1.78, 1.69]
f_{T5}	[−1.78, 1.69]
f_{T6}	[−2.67, 2.65]
f_{T7}	[−5.85, 5.56]
f_{T8}	[−1.45, 1.45]
f_{T9}	[−3.13, 3.14]

By examining the profile likelihood ratio as a function of λ, we determine the boundaries of the confidence interval. Figure 3 portraits the likelihood scan for the case of $\lambda = f_{T0}/\Lambda^4$, where the X-axis represent the scanning Wilson coefficient and Y-axis is the test statistic $-2log(q) = -2\Delta log(L)$. Table 3 presents the expected limits on the dimension-8 Wilson coefficients.

The EFTs which are used to make predictions of observable rates can often result in unitarity violations at high energies. However, predictions tested at colliders must be fully unitary in order to be consistent with the underlying quantum field theory. To overcome this issue, unitarization methods are used to convert non-unitary predictions into reliable, unitary ones that can be compared to experimental data. In the present analysis we use the cutoff method, also known as the clipping scan, in order to study the impact of the system energy on the limit setting. The clipping scan is not considered a unitarization method in itself, but a way of achieving unitarity. It is

Table 4 Expected 95% CL limits on the selected dimension-8 Wilson coefficients, for various m_{4l} cutoff values. The limits are obtained using the "observed" event yield as a function of the invariant-mass of the 4-lepton system (m_{4l}) at detector-level

Wilson coefficient	Expected (Asimov) 95% CL limit				
	2 TeV	3 TeV	5 TeV	10 TeV	∞
f_{T0}	[–1.46, 1.32]	[–0.93, 0.85]	[–0.76, 0.70]	[–0.74, 0.69]	[–0.74, 0.69]
f_{T1}	[–1.77, 1.74]	[–1.12, 1.10]	[–0.88, 0.87]	[–0.85, 0.84]	[–0.85, 0.84]
f_{T2}	[–3.64, 3.35]	[–2.32, 2.17]	[–1.85, 1.74]	[–1.78, 1.69]	[–1.78, 1.69]
f_{T5}	[–3.64, 3.37]	[–2.31, 2.17]	[–1.84, 1.75]	[–1.78, 1.69]	[–1.78, 1.69]
f_{T6}	[–4.70, 4.65]	[–3.22, 3.19]	[–2.73, 2.70]	[–2.67, 2.65]	[–2.67, 2.65]
f_{T7}	[–11.46, 10.67]	[–7.44, 7.00]	[–6.02, 5.71]	[–5.85, 5.56]	[–5.85, 5.56]
f_{T8}	[–2.90, 2.90]	[–1.86, 1.86]	[–1.49, 1.49]	[–1.45, 1.45]	[–1.45, 1.45]
f_{T9}	[–6.17, 6.17]	[–3.97, 3.97]	[–3.23, 3.23]	[–3.13, 3.14]	[–3.13, 3.14]

a way to obtain unitary amplitudes by discarding any predictions made for energy values above the lowest value of $\sqrt{s}$ at which any partial wave crosses the unitarity bound [23]. Consequently, we introduce different upper thresholds for the invariant mass of the four-lepton system m_{4l} and we discard the events for which the m_{4l} exceeds these thresholds. The clipping scan is performed at the m_{4l} values of 2 TeV, 3 TeV, 5 TeV and 10 TeV for the m_{4l} observable. As a result, the expected limits on the dimension-8 Wilson coefficients are presented in Table 4 and Fig. 4.

4 Conclusions

The focus of this article is on anomalous QGCs, where the interactions involving gauge bosons deviate from the predictions of the SM. These deviations are captured by dimension-8 operators within an EFT framework, the genuine QGC operators. These operators are of particular interest as they lead to QGCs without TGC counterparts. In order to investigate the QGCs in VBS processes, the study focuses on the $pp \rightarrow ZZjj$ process at a center-of-mass energy of 13 TeV using data of an integrated luminosity of 139 fb^{-1}. The experimental phase space related to the ATLAS detector is defined with optimized kinematic cuts to select $ZZjj \rightarrow 4ljj$ events produced via EWK vertices, while minimizing the production of such events via QCD vertices.

The determination of the expected limits for the dimension-8 Wilson coefficients associated with the QGC operators is achieved through a binned likelihood fit of the "observed" event yield as a function of the invariant mass of the four-lepton system (m_{4l}). The analysis accounts for the number of expected events, employing Poisson statistics and a Gaussian likelihood for the nuisance parameters. At a 95% CL, the expected limits on the dimension-8 Wilson coefficients are derived using the SM prediction as a hypothetical observation (Asimov dataset). The expected

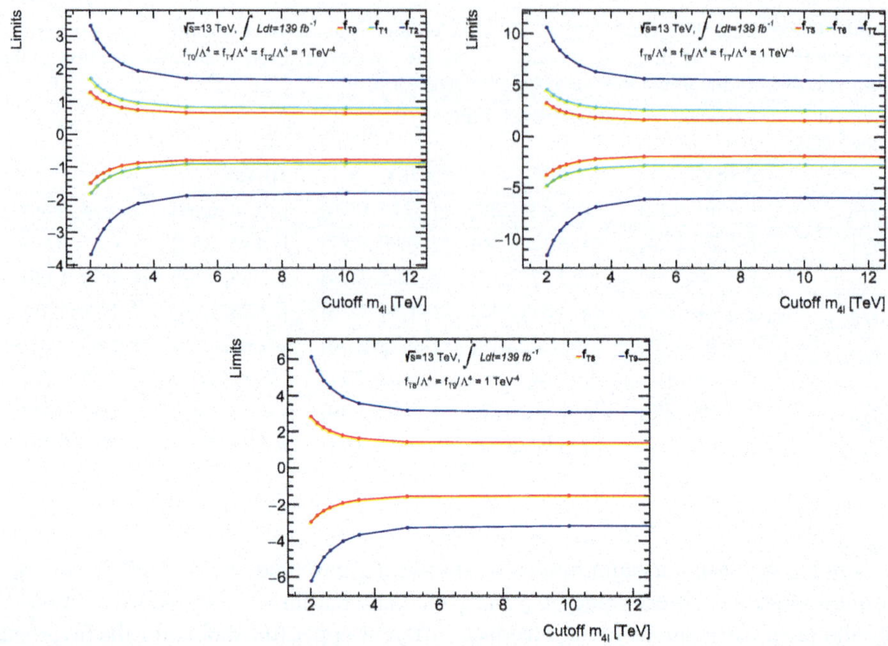

Fig. 4 Expected 95% CL limits on the selected dimension-8 Wilson coefficients, as a function of the clipping energy on m_{4l}. The limits are obtained using the "observed" detector-level event yield as a function of the invariant-mass of the 4-lepton system (m_{4l})

95% CL limits without unitarity constraints are presented in Table 3. Furthermore, the study addresses unitarization challenges arising from the theoretical nature of EFTs. The clipping scan method is employed to ensure unitary amplitudes when making predictions at high energies. The expected limits on the dimension-8 Wilson coefficients are presented in Table 4 for different cutoff values on the invariant mass of the four-lepton system.

References

1. O.J.P. Eboli, M.C. Gonzalez-Garcia, S.M. Lietti, S.F. Novaes, Phys. Rev. D **63**, 075008 (2001). https://doi.org/10.1103/PhysRevD.63.075008. ([arXiv:hep-ph/0009262 [hep-ph]].)
2. O.J.P. Eboli, M.C. Gonzalez-Garcia, S.M. Lietti, Phys. Rev. D **69**, 095005 (2004). https://doi.org/10.1103/PhysRevD.69.095005. ([arXiv:hep-ph/0310141 [hep-ph]].)
3. I. Brivio, T. Corbett, O.J.P. Éboli, M.B. Gavela, J. Gonzalez-Fraile, M.C. Gonzalez-Garcia, L. Merlo, S. Rigolin, JHEP **03**, 024 (2014). https://doi.org/10.1007/JHEP03(2014)024. [arXiv:1311.1823 [hep-ph]]
4. J.M. Cornwall, D.N. Levin, G. Tiktopoulos, Phys. Rev. D **10**, 1145 (1974). [erratum: Phys. Rev. D **11**, 972 (1975)]. https://doi.org/10.1103/PhysRevD.10.1145

5. O.J.P. Eboli, M.C. Gonzalez-Garcia, J.K. Mizukoshi, Phys. Rev. D **74**, 073005 (2006). https://doi.org/10.1103/PhysRevD.74.073005. ([arXiv:hep-ph/0606118 [hep-ph]].)
6. H. Georgi, Ann. Rev. Nucl. Part. Sci. **43**, 209–252 (1993). https://doi.org/10.1146/annurev.ns.43.120193.001233
7. C. Degrande, N. Greiner, W. Kilian, O. Mattelaer, H. Mebane, T. Stelzer, S. Willenbrock, C. Zhang, Ann. Phys. **335**, 21–32 (2013). https://doi.org/10.1016/j.aop.2013.04.016. [arXiv:1205.4231 [hep-ph]]
8. G. Aad et al., [ATLAS], Phys. Rev. Lett. **113**(14), 141803 (2014). https://doi.org/10.1103/PhysRevLett.113.141803. [arXiv:1405.6241 [hep-ex]]
9. V. Khachatryan et al., [CMS], Phys. Rev. Lett. **114**(5), 051801 (2015). https://doi.org/10.1103/PhysRevLett.114.051801. [arXiv:1410.6315 [hep-ex]]
10. A.M. Sirunyan et al., CMS. JHEP **06**, 076 (2020). https://doi.org/10.1007/JHEP06(2020)076. ([arXiv:2002.09902 [hep-ex]].)
11. M. Aaboud et al., ATLAS. JHEP **07**, 107 (2017). https://doi.org/10.1007/JHEP07(2017)107. ([arXiv:1705.01966 [hep-ex]].)
12. A.M. Sirunyan et al., [CMS], Phys. Lett. B **774**, 682–705 (2017). https://doi.org/10.1016/j.physletb.2017.10.020. [arXiv:1708.02812 [hep-ex]]
13. A.M. Sirunyan et al., CMS. Phys. Lett. B **812**, 135992 (2021). https://doi.org/10.1016/j.physletb.2020.135992. ([arXiv:2008.07013 [hep-ex]].)
14. G. Aad et al., [ATLAS], Phys. Rev. D **100**(3), 032007 (2019). https://doi.org/10.1103/PhysRevD.100.032007. [arXiv:1905.07714 [hep-ex]]
15. A.M. Sirunyan et al., [CMS], Phys. Lett. B **795**, 281–307 (2019). https://doi.org/10.1016/j.physletb.2019.05.042. [arXiv:1901.04060 [hep-ex]]
16. A.M. Sirunyan et al., [CMS], Phys. Rev. Lett. **120**(8), 081801 (2018). https://doi.org/10.1103/PhysRevLett.120.081801. [arXiv:1709.05822 [hep-ex]]
17. J. Alwall, M. Herquet, F. Maltoni, O. Mattelaer, T. Stelzer, JHEP **06**, 128 (2011). https://doi.org/10.1007/JHEP06(2011)128. [arXiv:1106.0522 [hep-ph]]
18. A.S. Belyaev, O.J.P. Eboli, M.C. Gonzalez-Garcia, J.K. Mizukoshi, S.F. Novaes, I. Zacharov, Phys. Rev. D **59**, 015022 (1999). https://doi.org/10.1103/PhysRevD.59.015022. ([arXiv:hep-ph/9805229 [hep-ph]].)
19. O.J.P. Éboli, M.C. Gonzalez-Garcia, Phys. Rev. D **93**(9), 093013 (2016). https://doi.org/10.1103/PhysRevD.93.093013. [arXiv:1604.03555 [hep-ph]]
20. M. Baak, A. Blondel, A. Bodek, R. Caputo, T. Corbett, C. Degrande, O. Eboli, J. Erler, B. Feigl, A. Freitas et al., [arXiv:1310.6708 [hep-ph]]
21. A. Marantis, I. Maznas, K. Kordas, A. Leisos, A. Tsirigotis, J. Phys. Conf. Ser. **2105**, 012014 (2021). https://doi.org/10.1088/1742-6596/2105/1/012014
22. A. Marantis, A. Leisos, K. Kordas, A. Tsirigotis, I. Maznas, Optimal observables for the measurement of anomalous quartic gauge boson couplings. Phys. Scr. **95**, 084013 (2020). https://doi.org/10.1088/1402-4896/ab9790
23. C. Garcia-Garcia, M. Herrero, R.A. Morales, Phys. Rev. D **100**(9), 096003 (2019). https://doi.org/10.1103/PhysRevD.100.096003. [arXiv:1907.06668 [hep-ph]]

Anomalies in Effective Theories, Anomalous Couplings and the $g-2$ of the Muon

Pascal Anastasopoulos, Kunio Kaneta, Elias Kiritsis, and Yann Mambrini

Abstract This proceeding is based on the work [1], where we discuss quantum anomalies and their effects on model building. We study an extension of the Standard Model by an (anomalous) Z' boson and analyze phenomenological implications. We focus on the anomalous magnetic moment $(g-2)$ of the muon.

Keywords Anomalies · Effective field theories · Z' · Heavy fermions

1 Introduction

While the Standard Model (SM) is currently the most accurate theory available, it is not the final theory. Incorporating right-handed neutrinos (since neutrinos are massive), dark matter ($\sim 85\%$ more than the observable matter in the Universe), and a quantum manifestation of gravity, among other factors, will be necessary to understand Nature better. As it stands, the SM is considered an effective theory.

During this proceeding, we explore an extension of the Standard Model by adding an extra gauge boson Z'. However, we will focus on an unusual variation of this field, specifically, an anomalous Z' [1, 2].

Anomalies are violations of symmetries due to quantum effects. They can be categorized into two types: violations of global symmetries and violations of gauge

P. Anastasopoulos (✉)
Institute of High Energy Physics, Austrian Academy of Sciences, Vienna, Austria
e-mail: Paschalis.Anastasopoulos@oeaw.ac.at

K. Kaneta
Faculty of Education, Niigata University, Niigata, Japan

E. Kiritsis
Université Paris Cité, CNRS, Astroparticule et Cosmologie, Paris, France

Department of Physics, University of Crete, Heraklion, Greece

Y. Mambrini
Université Paris-Saclay, CNRS/IN2P3, IJCLab, Orsay, France

J. Strologas and L. Perivolaropoulos (eds.), *HEP2023*, Springer Proceedings in Physics 333 https://doi.org/10.1007/978-3-031-76856-9_10

symmetries. Global symmetry violations are allowed as they don't lead to any consistency problems. An example is the decay of π^0 into $\gamma\gamma$, which is explained by a global anomaly. On the other hand, violations of gauge symmetries should be avoided in fundamental theories as they can cause problems of unitarity and renormalizability. In this proceeding, we will discuss anomalies in effective theories and argue that even if the "parent" fundamental theory is anomaly-free, an effective version below a certain scale can appear anomalous [1–7]. From this viewpoint, we can examine anomalous effective theories, which contain new anomaly-generated effective interactions—also known as anomalous coupling—with interesting phenomenology. We will additionally contend that anomalous coupling could be substantial and serve as the leading interaction on the loop level.

This report is organized as follows: Sect. 2 thoroughly examines a toy model that may seem anomalous at a certain scale, but in fact, it is completely anomaly-free thanks to the presence of effective terms, such as axionic and GCS, that cancel out any anomaly present. These terms are unique to models containing anomalies and absent in anomaly-free models. Section 3, then delves into these new terms and effectively combines them to form an "anomalous coupling" directly proportional to the anomaly of the low energy theory. These terms can be significant when the anomaly is large. Moving on to Sect. 4, we discuss an anomalous extension of the SM based on the properties of the previously mentioned toy model. Finally, in Sect. 5, our findings are applied to assess the contribution to the $g-2$ of the muon.

2 Anomalies and Effective Theories: A Toy Model

To support our arguments, we present a toy model consisting of three chiral fermions, namely ψ_L, ψ_R, and χ_L, along with a scalar, S, charged under a gauge field, A^μ. These fields transform under the gauge transformations as follows:

$$A_\mu \to A_\mu - \frac{1}{q}\partial_\mu\epsilon\,, \quad \psi_L \to e^{iq\varepsilon}\psi_L\,, \quad \chi_L \to e^{-iq\varepsilon}\chi_L\,, \quad S \to e^{iq\varepsilon}S\,. \tag{2.1}$$

Notice that ψ_L, χ_L have opposite charges, and ψ_R is neutral under A^μ. The action is[1]

$$\begin{aligned}\mathcal{L} = &-\frac{1}{4}F^{\mu\nu}F_{\mu\nu} - \bar{\psi}_L(i\partial\!\!\!/ + g_A q A\!\!\!/)\psi_L + \bar{\psi}_R i\partial\!\!\!/\psi_R + \bar{\chi}_L(i\partial\!\!\!/ - g_A q A\!\!\!/)\chi_L\\ &+(\partial_\mu + i g_A q A_\mu)S(\partial^\mu - i g_A q A^\mu)S^* + Y\bar{\psi}_L\psi_R S + c.c. + V(|S|^2)\end{aligned} \tag{2.2}$$

[1] It is worth noting that a Majorana term, $M_{maj}\psi_R\psi_R$ (where ψ_R is real), could be included in 2.2 with $M_{maj} > Yv$. This term would trigger a see-saw mechanism for the fields χ_L and χ_S, leading to heavy masses for both fermions once the mass term is diagonalized. Adding this term does not change the fact that the model is superficially anomalous after integrating out these heavy fermions. Nonetheless, the toy model presented in this section clarifies our statement that anomalous effective models can be consistent, and their phenomenological implications can be significant.

where g_A is the gauge coupling constant for the field A^μ. By rewriting $S = re^{ia/v}$, the last transformation is translated to the

$$a \to a + qv\varepsilon \tag{2.3}$$

and the phase of the scalar transforms as an *axion* under (2.1). The above model is anomaly-free since only two fermions are charged, and their contributions to the anomaly cancel out:

$$\mathcal{A} \sim Tr[Q^3] = q^3 + (-q)^3 = 0 \tag{2.4}$$

Next, the potential $V(|S|^2)$ breaks the symmetry, and denoting $\langle S \rangle = v$, we have a theory with a massless fermion χ_R, a Dirac massive fermion ψ, a massive gauge field A^μ, a massive scalar r, and a Goldstone boson a.

$$m_\psi = Yv\,, \quad m_\chi = 0\,, \quad m_a = 0\,, \quad m_A = g_A q v\,, \quad m_r = \sqrt{2\lambda}v \tag{2.5}$$

Lets assume here that $Y \gg g_A q \sim \sqrt{2\lambda}$. We want to study the effective field theory of this model at energies E, between the masses of A^μ and the heavy fermion ψ

$$m_A < E < m_\psi \tag{2.6}$$

At this limit, the heavy fermions are integrated out, and the effective theory has just one massless fermion field χ. In this case, the anomaly is

$$\mathcal{A} \sim Tr[Q^3] = (-q)^3 \neq 0 \tag{2.7}$$

This anomaly is canceled from effective terms after integrating out heavy fermions. More precisely, the triangle diagram with three external gauge fields and a heavy fermion in the loop provides a generalized Chern-Simons term, and the triangle diagram of two gauge bosons and a provides the coupling [3]

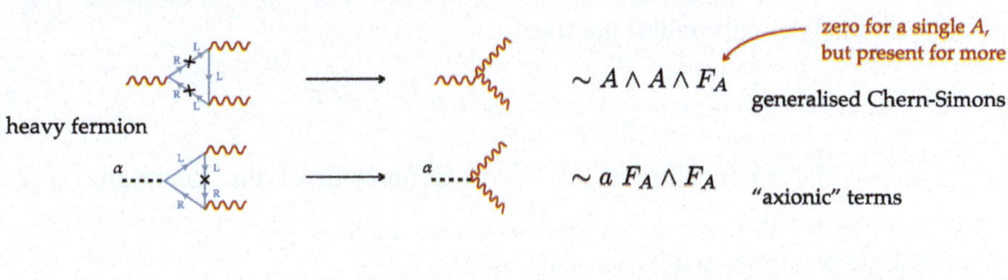

We should also mention that there is an additional coupling between the gauge field A^μ and the field a, coming from the kinetic term of S after the symmetry-breaking

$$|D^\mu S|^2 \quad \to \quad \frac{1}{2}(\partial^\mu a + m_A A^\mu)^2 + \ldots \quad \to$$

Therefore, after the symmetry breaking and having the heavy fermions integrated out, the effective terms are

$$\begin{aligned}\mathcal{L}_{\text{effective}} &= -\frac{1}{4}F_A^2 + \bar{\chi}_L(i\partial\!\!\!/ - g_A q A\!\!\!/)\chi_L \\ &+\frac{1}{2}(\partial a + m_A A^\mu)^2 + \frac{c}{24\pi^2} a F_A \wedge F_A + \frac{E}{24\pi^2} \cancel{A \wedge A \wedge F_A}\end{aligned} \tag{2.8}$$

and the Goldstone mode a behaves like an axionic field. Notice that the last term in (2.8) is vanishing due to the antisymmetry of the wedge symbol. However, it would be present if we had more than one gauge field in our model.

For the purposes of our work on effective field theories, we will refer to the field a as *the axion*, rather than *the Goldstone boson*, and relate the effective action (2.8) to the anomalous actions studied in various works in the past [1–4].

Under the transformations (2.1), (2.3), the action is not gauge invariant. However, there is also a non-invariant contribution from the 1-loop diagram (the anomaly) and the two contributions cancel each other, fixing the unknown parameter c

$$\left.\begin{aligned}\delta\mathcal{L}_{\text{effective}} &= -\frac{m_A c}{24\pi^2}\varepsilon F_A \wedge F_A \\ \delta\mathcal{L}_{\text{1-loop}} &= \frac{q^3}{24\pi^2}\varepsilon F_A \wedge F_A\end{aligned}\right\} \delta\mathcal{L}_{\text{effective}} + \delta\mathcal{L}_{\text{1-loop}} = 0 \;\rightarrow\; c = \frac{q^3}{m_A} \tag{2.9}$$

In a generic theory, with several gauge fields A_i^μ and several axions a^I, the effective action has the form

$$\begin{aligned}\mathcal{L}_{\text{effective}} &= -\frac{1}{4}F_i^2 + \bar{\chi}_f(i\partial\!\!\!/ - g_i q_f^i A\!\!\!/_i)\chi_f \\ &+\frac{1}{2}(\partial a^I + m_i^I A_i^\mu)^2 + \frac{c_{ij}^I}{24\pi^2} a^I F_i \wedge F_j + \frac{E^{ij,k}}{24\pi^2} A_i \wedge A_j \wedge F_k\end{aligned} \tag{2.10}$$

where χ_f all the low energy chiral fermions with charges q_f^i under the gauge fields A_i^μ. The anomaly condition has the form

$$t_{ijk} + E_{ij,k} + E_{ik,j} + m_i^I C_{jk}^I = 0 \tag{2.11}$$

and fixes all the coefficients $E_{ij,k}$, C_{ij}^I as a function of the "anomaly" $t_{ijk} = Tr[q_i q_j q_k] = \sum_f q_i^f q_j^f q_k^f$ [3]. The same anomaly cancellation procedure is described via diagrams in the following figure

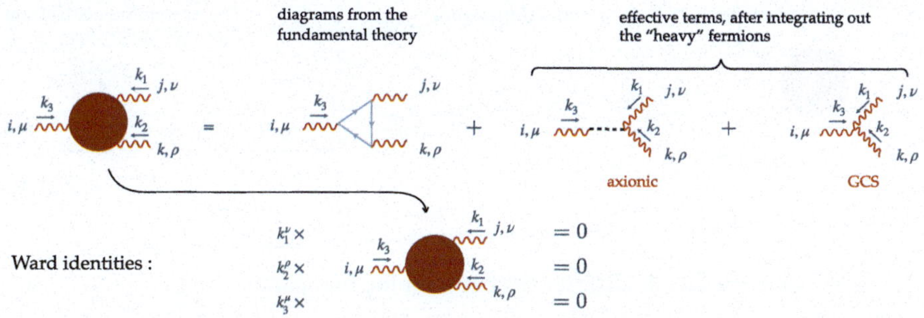

It is important to note that canceling anomalies in the Fundamental theory differs from that of the effective theory. The former cancels out all triangle diagrams, while in the latter, the Ward identities vanish (the currents are conserved), and the diagrams may still have significant phenomenological implications.

3 The Anomalous Coupling

This section will discuss the construction of the "anomalous coupling", which is present in anomalous theories but absent in anomaly-free theories.

Let's consider a massive fermion with non-zero vectorial $g_A q_V \gamma^\mu$ and axial $g_A q_A \gamma_5 \gamma^\mu$ couplings with the gauge field A^μ. The diagram of this fermion in a loop with three external gauge bosons A^μ splits into mass-dependent and mass-independent parts (3.14).

If we have multiple fermions in the model, we must add all individual contributions. The mass-dependent part is proportional to each fermion's individual charges/couplings, and lighter fermions will contribute more than heavier fermions. On the other hand, the mass-independent part is common for all fermions, and the sum over all contributions factorizes to a term proportional to the anomaly t_{ijk} (3.14).

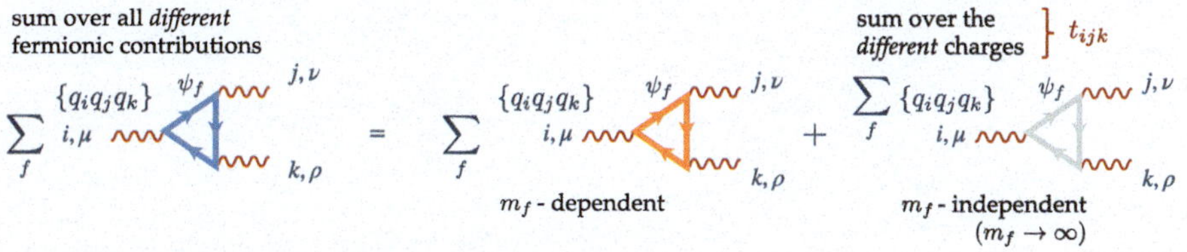

The orange triangle loop depicts the mass-dependent contribution of the triangle diagram, and the gray loop is the mass-independent one. Since the axionic and GCS terms do not depend on the masses of the low energy fermions, and they depend on the anomaly t_{ijk}, the total three-point diagram can be written in the following form where the anomalous coupling contains the mass-independent part of the triangle diagrams, the axionic and the GCS terms. Schematically

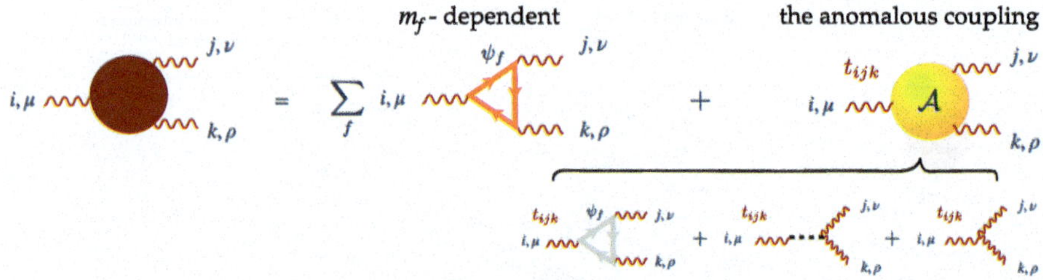

There are some noticeable differences between the two parts.

- m_f-dependent part
  - This diagram is always present. Independently on whether the anomaly is anomalous or not.
  - It is suppressed by the mass of the fermion in the loop, therefore, the lighter the fermion the larger the contribution.
  - Each fermionic triangle depends on the individual charges of the fermion in the loop.
- The anomalous coupling
  - is absent in anomaly-free models.
  - It does not depend on any mass. But, there is a cutoff [1, 6].
  - It depends on the full anomaly. If the anomaly is large (for example, many heavy fermions contribute), the anomalous coupling can be dominant.

The new anomalous couplings have a similar structure to the standard fermionic triangle diagrams. As a result, they introduce new diagrams that are not present in theories that are free of anomalies. For instance, if, at a particular level of perturbation theory, we need to incorporate diagrams with a massive fermion in the loop as a sub-diagram, we should also include diagrams that substitute the triangle with the anomalous coupling.

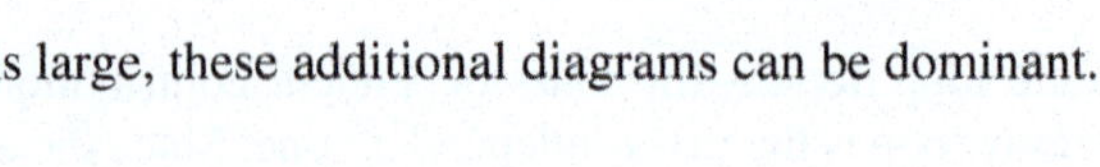

If the anomaly is large, these additional diagrams can be dominant.

4 An Anomalous Extension of the Standard Model

We utilize the concepts discussed earlier, where the Standard Model is a representation of a larger and more fundamental theory. This fundamental theory comprises matter fields, which include the Standard Model's matter fields (quarks, leptons, and the SM Higgs H), various chiral fermions ψ, and an additional scalar field S. The scalar field only has Yukawa couplings with the fermions ψ. The matter fields transform under a gauge group of $SU(3) \times SU(2) \times U(1)_Y \times U(1)_A$.

The SM and additional fermions have charges under the whole group. The total model is free of anomalies since the anomalies of the SM and additional fermions cancel each other out. However, the two sectors have separate anomalies

$$t_{ijk}^{\text{total}} = t_{ijk}^{\text{SM}} + t_{ijk}^{\text{extra}} = 0 \quad \text{with} \quad t_{ijk}^{\text{SM}} = -t_{ijk}^{\text{extra}} \neq 0 \tag{4.1}$$

After S obtains a vev V, the extra fermions acquire masses of order $m_\psi = YV$, and assuming that they are much heavier than our energies, they are integrated out. The theory formed as a result comprises the following elements: the fields of the Standard Model that are charged under the additional field A^μ.

	$SU(3)$	$SU(2)$	$U(1)_Y$	$U(1)_A$
Q_L^i	**3**	**2**	1/6	q_i^{Q-A}
$u_R^{i,c}$	$\bar{\mathbf{3}}$	**1**	−2/3	q_i^{u-A}
$d_R^{i,c}$	$\bar{\mathbf{3}}$	**1**	1/3	q_i^{d-A}
L_L^i	**1**	**2**	−1/2	q_i^{L-A}
$l_R^{i,c}$	**1**	**1**	1	q_i^{l-A}
H	**1**	**2**	1/2	q^{H-A}

an extra massive gauge field A^μ with mass $m_A = g_A q V$, and an axionic field a (Goldstone mode) [1, 2].

It is worth noting that this model is anomalous because the anomalies that contain the field A^μ do not vanish. To counterbalance this, axionic and GCS terms cancel out the anomalies. Figure 1 visually represents these components. The effective action consists of the SM action and an additional part containing the extra fields.

$$\begin{aligned}
\mathcal{L}_{\text{total}} &= \mathcal{L}_{\text{SM}} + \mathcal{L}_{\text{extras}} \qquad\qquad (4.2)\\
\mathcal{L}_{\text{extras}} &= -\frac{1}{4}F_{\mu\nu}^A F_A^{\mu\nu} + \frac{1}{2}(\partial_\mu a + M A_\mu)^2 + g_A q^{\psi-A} A_\mu \bar{\psi}\gamma^\mu\psi \\
&+2i g_A q^{H-A} H^\dagger A^\mu \Big(\partial_\mu - i g_2 T^\alpha W_\mu^\alpha - i g_Y q^{H-Y} Y_\mu - \frac{i}{2} g_A q^{H-A} A_\mu\Big) H \\
&+\frac{1}{24\pi^2} a \Big(C_{YY}\, F_Y \wedge F_Y + C_{YA}\, F_Y \wedge F_A + C_{AA}\, F_A \wedge F_A + \sum_{i=2,3} D_i\, Tr_i[G \wedge G]\Big) \\
&+\frac{1}{24\pi^2} A \wedge Y \wedge \Big(E_{AY,A}\, F_A + E_{AY,Y}\, F_Y\Big) \\
&+\frac{1}{24\pi^2} \sum_{i=2,3} Z_i\, A \wedge Tr_i\Big[A \wedge (dG_i - \frac{2}{3} G_i \wedge G_i)\Big]
\end{aligned}$$

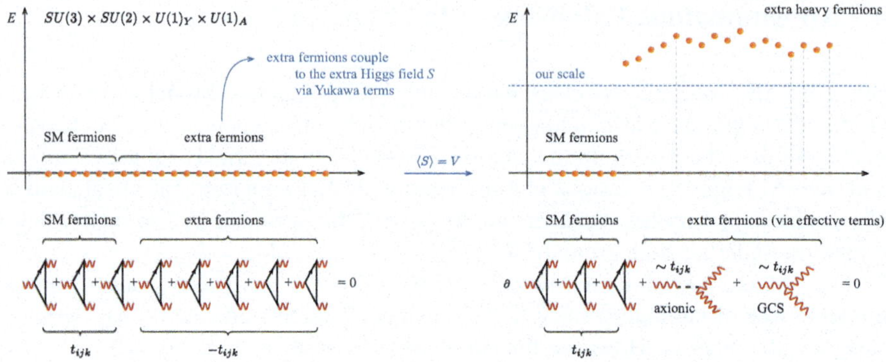

Fig. 1 The effective anomalous model. The E in the y-axis stands for the energy scale of the theory

The action is not gauge invariant under the transformations $A^\mu \to A^\mu + \partial^\mu \varepsilon$ and $Y^\mu \to Y^\mu + \partial^\mu \zeta$. However, these non-vanishing terms cancel the non-gauge invariant terms from the 1-loop (the anomalies). Finally, we have

$$\delta S_{\text{1-loop}} + \delta S_{\text{extra fields}} = 0 \to \begin{cases} MC_{YY} = -2t_{AYY} & MC_{YA} = -2t_{YAA} \\ MC_{AA} = -t_{AAA} & 2MD_i = -3Z_i = -6T_{A,i} \\ E_{AY,A} = t_{YAA} & E_{AY,Y} = t_{AYY} \end{cases} \tag{4.3}$$

and all unknown coefficients $C's$ and $E's$ are expressed via the anomalies.

Next, we have the standard EW symmetry breaking and assuming that the Higgs field is charged under the additional A^μ, the Y_μ, W^3_μ and A_μ mix and they give the photon, Z_μ and Z'_μ with masses that depend on the vevs of both fields S and H [1, 2].

Anomaly cancellation requires a variety of anomalous couplings to eliminate mixed anomalies between A^μ and the known SM gauge groups. New anomalous couplings emerge as we go from the unbroken hypercharge phase to the broken photon phase [1]

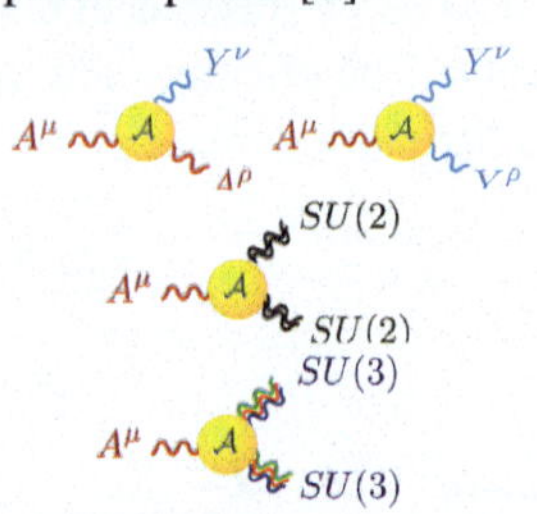

$$\to \gamma\gamma Z,\ \gamma ZZ,\ \gamma\gamma Z',\ \gamma ZZ',\ \gamma Z'Z',\ ZZZ',\ ZZ'Z'$$

$$\to ZW^+W^-,\ Z'W^+W^-,\ \gamma ZW^+W^-,\ \gamma Z'W^+W^-$$

$$\to Z \text{ gluon gluon},\ Z' \text{ gluon gluon},\ \ldots$$

It should be noted that these couplings are linked to anomalies and are not present in anomaly-free extensions of the SM with additional gauge bosons (Z'). Additionally, it is worth mentioning that these couplings can exist off-shell. Of particular importance are the couplings involving only SM particles ($\gamma\gamma Z$, $\gamma\gamma Z'$, ZW^+W^-, γZW^+W^-, Z gluon gluon), which can be tested currently, even before the observation of a Z' [1].

5 Applications: The $g-2$ of the Muon

Our next step is to apply all the above to the $g-2$ of the muon [8–13].

$$\Delta a_\mu = a_\mu^{\text{exp}} - a_\mu^{\text{SM}} = 251(59) \times 10^{-11}\,. \tag{5.1}$$

We aim to assess the impact of an anomalous Z', specifically its anomalous couplings, on the muon's $g-2$. We will concentrate on the extra diagrams that come into play once an additional anomalous Z' is present.

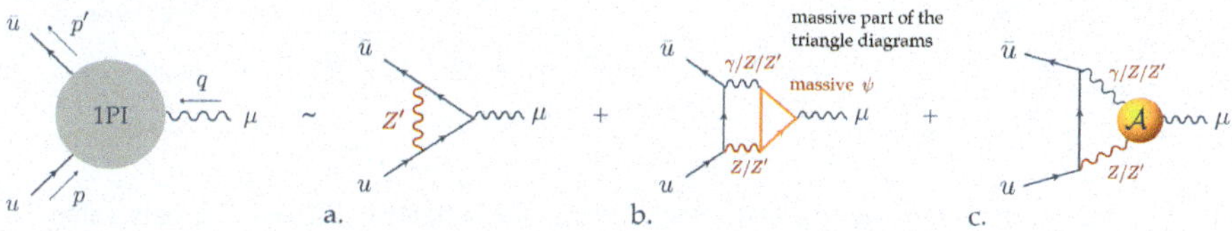

a. The first diagram appears at 1-loop. That diagram consists of two competing terms. One proportional to $(q_V^{\mu-Z'})^2$ minus another term proportional to $(q_A^{\mu-Z'})^2$ [14]. Depending on the values of these charges, the 1-loop contribution can be tiny or zero, smaller than the 2-loop contributions. In this case, the triangle diagram and the anomalous coupling can dominate.
b. At 2-loop, the leading diagram is the one that involves the triangle loop with the massive fermion. This diagram depends on the charges of the individual fermion in the loop, and the mass of the fermion suppresses it—the lightest the fermion, the larger the contribution.
c. In an anomalous theory, we should add another diagram that emerges by substituting the triangle sub-diagram with the anomalous coupling.

   As shown in [1], the diagram with the anomalous coupling is dominant at 2-loops. Especially when the anomaly is large, this contribution can be several orders of magnitude more than the analogous diagram "b".

In the plots below, we provide the contribution of the 1-loop diagram versus the diagram with the anomalous coupling for various anomaly values. We choose a low Z' mass model with $m_{Z'} \in 10 - 100$ MeV. At that range, the vectorial/axial couplings are chosen $g_{Z'} q_V^{\mu-Z'} = 10 g_{Z'} q_A^{\mu-Z'} = 8 \times 10^{-4}$. The dashed line is the δa, and the green zone is the 1σ uncertainty. The plot 2 compares the 1-loop and the "anomalous" diagram contributions (for different anomaly values). The plot 3 gives the sum of these contributions to the $g-2$ of the muon.

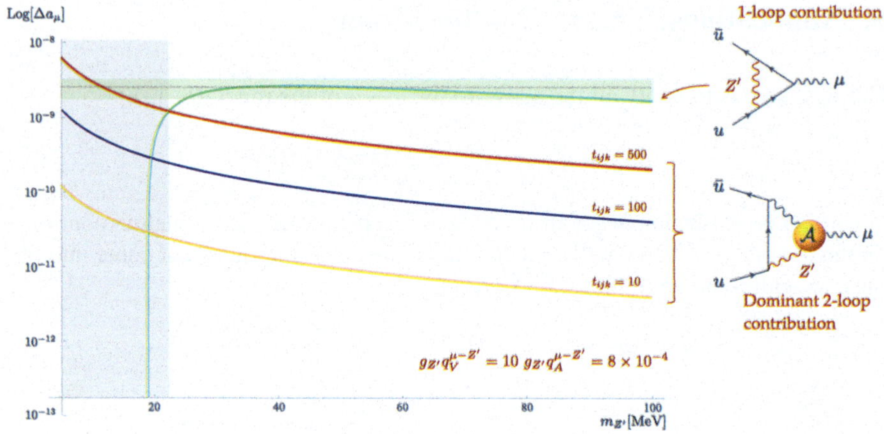

Fig. 2 Comparison of the 1-loop and the anomalous contribution for different anomaly values

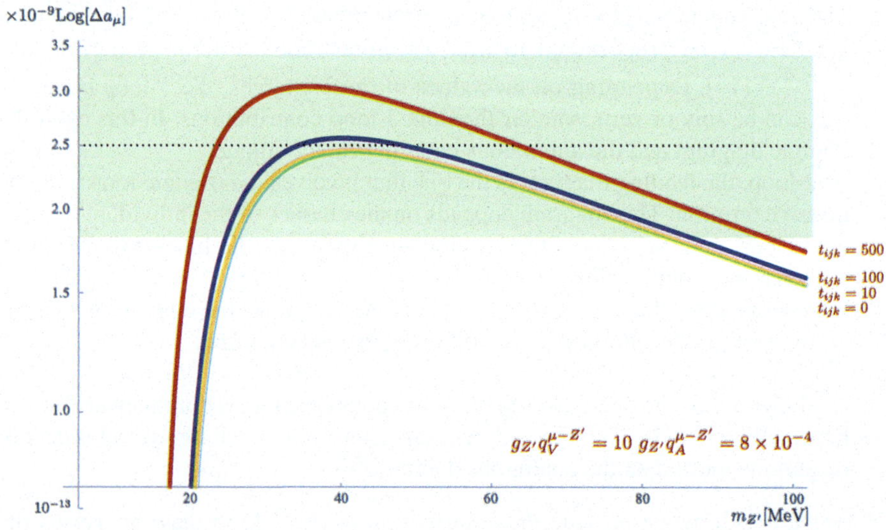

Fig. 3 Corrections to the theoretical value of the $g - 2$ of the muon in anomalous effective SM

6 Conclusions

In this proceeding, we present an effective Standard Model that is anomalous. We argue that these models naturally arise at specific scales when the Fundamental Theory contains additional heavy fermionic states, which primordially contribute to the anomaly cancellation. We can remedy the anomalies by including effective terms from the heavy part of the theory.

It is possible that the existence of another boson, referred to as Z', could be considered anomalous. With this, we can avoid excessive and impractical fine-tuning to eliminate anomalies.

The heavy sector may significantly contribute to the anomalies, resulting in extensive phenomenological consequences. In particular, we examined the impact of the anomalous coupling present in the effective theory due to the integrated out heavy sector on the muon's $g-2$. Our analysis revealed that certain parameter ranges may lead to dominance of the anomalous coupling, even in comparison to lower loop contributions.

This work has opened up opportunities for further research in both theoretical and phenomenological directions. For instance, we can generalize the anomalous cancellation mechanism in effective theories in higher dimensions (6D), study the cutoff scale of the effective actions in comparison to the weak gravity conjecture [15], analyze anomalous Z' models with the new anomalous couplings, especially the ones containing only SM gauge fields, and investigate effective models with the features mentioned above, where total anomaly cancellation is achieved, but not individually in each sector [16]. Additionally, we can study the cosmological implications of the new anomalous couplings [17].

Acknowledgements P.A. was supported by FWF Austrian Science Fund via the SAP P 36423-N.

References

1. P. Anastasopoulos, K. Kaneta, E. Kiritsis, Y. Mambrini, JHEP **02**, 051 (2023). ([arXiv:2209.12947 [hep-ph]])
2. C. Coriano, N. Irges, E. Kiritsis, Nucl. Phys. B **746**, 77–135 (2006). https://doi.org/10.1016/j.nuclphysb.2006.04.009. [ArXiv:hepph/0510332]
3. P. Anastasopoulos, M. Bianchi, E. Dudas, E. Kiritsis, JHEP **11**, 057 (2006). [ArXiv:hepph/0605225]. P. Anastasopoulos, Fortsch. Phys. **55**, 633–638 (2007). [ArXiv:hepph/0701114]
4. P. Anastasopoulos, F. Fucito, A. Lionetto, G. Pradisi, A. Racioppi, Y.S. Stanev, Phys. Rev. D **78**, 085014 (2008). ([arXiv:0804.1156 [hep-th]])
5. A. Dedes, K. Suxho, Phys. Rev. D **85**, 095024 (2012). ([arXiv:1202.4940 [hep-ph]])
6. J. Preskill, Ann. Phys. **210**, 323–379 (1991). https://doi.org/10.1016/0003-4916(91)90046-B
7. J.A. Dror, R. Lasenby, M. Pospelov, Phys. Rev. D **96**(7), 075036 (2017). [arXiv:1707.01503 [hep-ph]]
8. T. Aoyama et al., Phys. Rept. **887**, 1–166 (2020). [arXiv:2006.04822]
9. G.W. Bennett et al., Muon g-2 collaboration. Phys. Rev. D **73**, 072003 (2006). [hep-ex/0602035]
10. B. Abi et al., Muon g-2 collaboration. Phys. Rev. Lett. **126**(14), 141801 (2021). [arXiv:2104.03281]
11. S. Borsanyi et al., arXiv:2002.12347
12. J. Grange et al., Muon g-2 collaboration. arXiv:1501.06858
13. M. Abe et al., PTEP **2019**(5), 053C02 (2019). [arXiv:1901.03047]
14. A. Bodas, R. Coy, S.J.D. King, Eur. Phys. J. C **81**(12), 1065 (2021). https://link.springer.com/article/10.1140/epjc/s10052-021-09850-x. [arXiv:2102.07781 [hep-ph]]
15. N. Craig, I. Garcia Garcia, G.D. Kribs, JHEP **11**, 063 (2020). [arXiv:1912.10054 [hep-ph]]
16. P. Anastasopoulos, I. Antoniadis, K. Benakli, M. Goodsell, F. Rondeau, to appear
17. S. Cléry, P. Anastasopoulos, Y. Mambrini, [arXiv:2307.06011 [hep-th]]

Observable Primordial Gravitational Waves from Cosmic Inflation

Konstantinos Dimopoulos

Abstract I will review briefly how inflation is expected to generate a stochastic background of primordial gravitational waves (GWs). Then, I will discuss how such GWs can be enhanced by a stiff period following inflation, enough to be observable. I will present examples of this in the context of hybrid inflation with α-attractors, or a period of hyperkination in Palatini gravity.

Keywords Primordial gravitational waves · Cosmic inflation · α-attractors · Palatini modified gravity

1 Introduction

The history of the Universe requires special initial conditions which are arranged by cosmic inflation. In a nutshell, cosmic inflation can be defined as a period of accelerated expansion in the Early Universe. Inflation results in the Universe being large, spatially flat and uniform, in accordance to observations [1]. Inflation also generates the primordial density perturbations (PDPs), which are necessary for galaxies and galaxy clusters to form [2]. The PDPs reflect themselves on the cosmic microwave background (CMB) radiation, through the Sachs-Wolfe effect. Agreement with CMB observations is spectacular, when inflation is quasi-de Sitter and space expands exponentially [3].

However, there is another generic prediction of inflation beyond the acoustic peaks observed in the CMB, namely the generation of a stochastic spectrum of primordial gravitational waves (GWs) [4]. This prediction will soon be tested, either indirectly, through CMB polarisation observations, or directly through interferometers (Advanced LIGO, LISA).

WWW home page: https://www.lancaster.ac.uk/physics/about-us/people/konstantinos-dimopoulos.

K. Dimopoulos (✉)
Consortium for Fundamental Physics, Physics Department, Lancaster University, Lancaster LA1 4YB, UK
e-mail: k.dimopoulos1@lancaster.ac.uk

J. Strologas and L. Perivolaropoulos (eds.), *HEP2023*, Springer Proceedings in Physics 333 https://doi.org/10.1007/978-3-031-76856-9_11

How does inflation produce these GWs? Below, I attempt a brief overview. I use natural units with $c = \hbar = 1$ and $8\pi G = m_P^{-2}$, with $m_P = 2.43 \times 10^{18}$ GeV being the reduced Planck mass. The signature of the metric is positive.

2 Particle Production of Gravitational Waves During Cosmic Inflation

Following the recipe of linearised gravity, we consider a comoving perturbation of the metric h_{ij}, such that the line element of spatially flat FRW spacetime is

$$\mathrm{d}s^2 = a^2(\tau)[-\mathrm{d}\tau^2 + (\delta_{ij} + h_{ij})\mathrm{d}x^i\mathrm{d}x^j]\,, \tag{1}$$

where $a(\tau)$ is the scale factor of the Universe as a function of conformal time τ and δ_{ij} is equal to one when $i = j$ and to zero otherwise, with $i, j = 1, 2, 3$. The metric perturbation is symmetric $h_{ij} = h_{ji}$, traceless $h^i_i = 0$ and transverse $\nabla_i h^{ij} = 0$. This means that it is corresponds to two degrees of freedom $(6 - 1 - 3 = 2)$, which are the two polarisations, $\oplus$ and $\otimes$, of the GWs.

Thus, by Fourier transform, we can write

$$h_{ij}(\tau, \mathbf{x}) = \sqrt{16\pi G}\int \frac{\mathrm{d}^3k}{(2\pi)^{3/2}} h_{ij}(\tau, \mathbf{k})\, e^{i\mathbf{k}\cdot\mathbf{x}}\,, \tag{2}$$

with

$$h_{ij}(\tau, \mathbf{k}) = \sum_{s=\oplus,\otimes} h^s_k e^s_{ij}(\mathbf{k})\,, \tag{3}$$

where e^s_{ij} is symmetric $e^s_{ij} = e^s_{ji}$, traceless $e^{s\,i}_{i} = 0$ and transverse $k^i e^s_{ij} = 0$, with k^i being the 3-wavevector of the GWs.

The second-order GW action is

$$S_{\rm GW} = \frac{1}{64\pi G}\int \mathrm{d}^4x\sqrt{-g}\, g^{\mu\nu}\partial_\mu h_{ij}\,\partial_\nu h^{ij}\,, \tag{4}$$

where g is the determinant of the metric $g_{\mu\nu}$ and $\mu, \nu = 0, 1, 2, 3$. From the above action one obtains the equation of motion (EoM), which reads

$$h''_{ij} + 2\frac{a'}{a}h'_{ij} - \nabla^2 h_{ij} = 0 \Rightarrow h^{s\,\prime\prime}_k + 2\frac{a'}{a}h^{s\,\prime}_k + k^2 h^s_k = 0\,, \tag{5}$$

where the prime denotes derivative with respect to conformal time.

The above equations show that the metric perturbation polarisations behave as free massless scalar fields $\psi^s_k(\tau)$ which can be written as $h^s_k(\tau) = \sqrt{16\pi G}\psi^s_k(\tau)$. To study particle production, we introduce the Muchanov-Sasaki variable $v^s_k(\tau) \equiv$

$a(\tau)\psi_k^s(\tau)$. In terms of this variable, the EoM becomes the well-known Muchanov-Sasaki equation

$$v_k^{s\,\prime\prime} + \left(k^2 - \frac{a''}{a}\right) v_k^s = 0\,, \tag{6}$$

where $v_k^2 = a h_k^s m_P / \sqrt{2}$.

To proceed we quantize the metric perturbations (second quantization) be expanding them in terms of creation and annihilation operators as

$$v^s(\tau, \mathbf{x}) = \int \frac{\mathrm{d}^3 k}{(2\pi)^{3/2}} \left[v_k^s \hat{a}_{\mathbf{k}}^s \, e^{i\mathbf{k}\cdot\mathbf{x}} + (v_k^s)^* \hat{a}_{\mathbf{k}}^{s\,\dagger} \, e^{-i\mathbf{k}\cdot\mathbf{x}} \right], \tag{7}$$

where $\hat{a}_{\mathbf{k}}^s$ and $\hat{a}_{\mathbf{k}}^{s\,\dagger}$ are creation and annihilation operators respectively, which satisfy the algebra

$$[\hat{a}_{\mathbf{k}}^s, \hat{a}_{\mathbf{q}}^{r\,\dagger}] = \delta^{sr} \delta^{(3)}(\mathbf{q} - \mathbf{k}) \quad \text{and} \quad [\hat{a}_{\mathbf{k}}^s, \hat{a}_{\mathbf{q}}^r] = [\hat{a}_{\mathbf{k}}^{s\,\dagger}, \hat{a}_{\mathbf{q}}^{r\,\dagger}] = 0\,, \tag{8}$$

where the value of δ^{sr} is unity when $s = r$ or zero otherwise.

Inserting the above in Eq. (6), the solution for the mode functions is

$$v_k^s(\tau) = \frac{1}{\sqrt{2k}} \left(1 - \frac{i}{k\tau}\right) e^{-ik\tau}\,. \tag{9}$$

The above solution, in the subhorizon limit $-k\tau \to +\infty$ becomes $v_k^s \to e^{-ik\tau}/\sqrt{2k}$, which is the well-known Bunch-Davis vacuum [5]. In the superhorizon limit $-k\tau = \frac{k}{aH} \to 0$, the above solution becomes $v_k^s \to \frac{i}{\sqrt{2k}} \frac{aH}{k}$, where H is the Hubble parameter. Thus, in the superhorizon limit we find

$$h_k^s = \frac{\sqrt{2}\, v_k^s}{a m_P} = \frac{iH}{m_P k^{3/2}} \Rightarrow |h_k^s|^2 = \frac{H^2}{m_P^2 k^3} \approx \text{constant}\,, \tag{10}$$

where we considered that $H \approx$ constant in quasi-de Sitter inflation.

Therefore, we see that on superhorizon scales inflation generates a spectrum of primordial GWs. The value of their spectrum is

$$\mathcal{P}_h(k) = \frac{k^3}{2\pi^2} \langle h_{ij}(k)\, h^{ij}(k) \rangle = \frac{k^3}{\pi^2} \sum_{s=\oplus,\otimes} |h_k^s|^2 = \frac{2H^2}{\pi^2 m_P^2} = 64\pi G \left(\frac{H}{2\pi}\right)^2. \tag{11}$$

The scalar perturbations corresponding to the PDPs have spectrum

$$\mathcal{P}_\zeta(k) = \frac{H^2}{8\pi^2 \epsilon m_P^2}\,, \tag{12}$$

where $\epsilon \equiv -\dot{H}/H^2$, with the dot denoting derivative with respect to the cosmic time. From the above, we find the consistency equation $r \equiv \mathcal{P}_h/\mathcal{P}_\zeta = 16\epsilon$, which can be tested in the near future. At the moment, the CMB observations impose only an upper bound on r: $0 \leq r < 0.036$ [6].

3 Density of the Gravitational Waves

In view of Eq. (4), the energy-momentum tensor of the GWs is

$$T_{\mu\nu}^{\rm GW} = -\frac{2}{\sqrt{-g}}\frac{\delta S_{\rm GW}}{\delta g^{\mu\nu}} = \frac{\langle \nabla_\mu h_{ij} \nabla_\nu h^{ij} \rangle}{32\pi G}\,, \tag{13}$$

from which we can read the density $\rho_{\rm GW} = a^{-2}T_{00}$. Using Eq. (5), we can write

$$\langle \rho_{\rm GW} \rangle = \int_0^{+\infty} \frac{k^3}{2\pi^2} \sum_{s=\oplus,\otimes} \frac{|h_k^{s\prime}|^2 + k^2|h_k^s|^2}{a^2}\, {\rm d}(\ln k)\,. \tag{14}$$

Similarly, we find the isotropic pressure $p_{\rm GW} = a^{-2}T_i^{\,i}$, for which

$$\langle p_{\rm GW} \rangle = \int_0^{+\infty} \frac{k^3}{2\pi^2} \sum_{s=\oplus,\otimes} \frac{|h_k^{s\prime}|^2 - \frac{1}{3}k^2|h_k^s|^2}{a^2}\, {\rm d}(\ln k)\,. \tag{15}$$

The above integrals are dominated by the subhorizon limit $k \gg aH$. In this limit we have $h_k^s \propto e^{-ik\tau}$, which implies $h_k^{s\prime} = -ik\, h_k^s \Rightarrow |h_k^{s\prime}|^2 = k^2|h_k^s|^2$. Therefore, for the barotropic parameter we find

$$w_{\rm GW} = \frac{p_{\rm GW}}{\rho_{\rm GW}} = \frac{\frac{2}{3}k^2|h_k^s|^2}{2k^2|h_k^s|^2} = \frac{1}{3}\,. \tag{16}$$

Thus, we find that the density of the gravitational waves redshifts as radiation with the Universe expansion $\rho_{\rm GW} \propto a^{-3(1+w_{\rm GW})} = a^{-4}$.

In view of the above, the density parameter of the GWs per logarithmic momentum interval is

$$\Omega_{\rm GW}(k) \equiv \frac{1}{\rho_c}\frac{{\rm d}\rho_{\rm GW}}{{\rm d}\ln k} = \frac{k^3}{2\pi^2}\frac{8\pi G}{3H^2} \sum_{s=\oplus,\otimes} \frac{|h_k^{s\prime}|^2 + k^2|h_k^s|^2}{a^2}\,. \tag{17}$$

where $\rho_c = 3H^2/8\pi G$ is the critical density. The time evolution of $\Omega_{\rm GW}(\tau, k)$ is given by

$$\Omega_{\rm GW}(\tau, k) = \frac{k^2\Delta_h^2(\tau, k)}{12a^2H^2}\,, \tag{18}$$

where $\Delta_h^2(\tau, k) = T_h(\tau, k)\mathcal{P}_h(k)$. The transfer function is given by $T_h = \frac{1}{2}(a_k/a)^2$, where a_k is the value of $a(\tau)$ at the moment of horizon re-entry of the scale with momentum k. Today we have $T_h^0 = \frac{1}{2}(a_k/a_0)^2 = \Omega_R\left(\frac{a_0 H_0}{a_k H_k}\right)^2$, where $\Omega_R \simeq 10^{-4}$ is the density parameter of radiation at present and '0' denotes today.

Switching to frequency f, we employ the relation $f = \frac{k}{2\pi}\frac{a_k}{a_0}$. We end up with the expression [7]

$$\Omega_{\rm GW}(f) \propto f^{-2\left(\frac{1-3w}{1+3w}\right)}, \tag{19}$$

where w is the barotropic parameter of the Universe at the time of horizon reentry. Thus, we see that, for modes which re-enter the horizon during the radiation era, because $w = \frac{1}{3}$, we have $\Omega_{\rm GW}(f) =$ constant. That is, the spectrum is flat and unfortunately, its value is unobservable in the near future (see Fig. 1).

4 Kination

The inflationary paradigm suggests that the Universe inflates when dominated by the potential density of a scalar field, called the inflaton field. Non-oscillatory inflation considers a runaway inflaton scalar potential, with its minimum displaced at infinity [8]. In such models, after the inflaton field ϕ rolls away from the inflationary plateau, which is a relatively flat part of its scalar potential $V(\phi)$, it becomes dominated by its kinetic energy density $\frac{1}{2}\dot{\phi}^2$, because the potential reduces drastically and becomes negligible. The Universe is still dominated by the inflaton field, but the latter is kinetically dominated, with barotropic parameter $w = \frac{\frac{1}{2}\dot{\phi}^2 - V}{\frac{1}{2}\dot{\phi}^2 + V} \approx 1$. This phase is called *kination* [9]. Equation (19) suggests that, for the modes that re-enter the horizon during kination we have $\Omega_{\rm GW}(f) \propto f$.

Therefore, the GW spectrum features a peak. The corresponding frequencies, however, are unobservable at the moment because kination cannot be extended arbitrarily to later times, and therefore, lower frequencies. The reason is that, if kination is prolonged, the GW peak becomes too large and threatens to destabilise the delicate process of Big Bang Nucleosynthesis (BBN). The upper bound to $\Omega_{\rm GW}$ is obtained as follows.

At the time of BBN we require that $\Omega_{\rm GW}^{\rm BBN} < 10^{-2}$ so that BBN is not harmed [10]. Using that the density of GWs redshifts as radiation we can estimate the corresponding bound at present. We find

$$\Omega_{\rm GW}^0 = \frac{\rho_{\rm GW}^0}{\rho_c^0} = \left.\frac{\rho_{\rm GW}}{\rho_r}\right|_0 \Omega_R = \left.\frac{\rho_{\rm GW}}{\rho_r}\right|_{\rm BBN} \Omega_R = \Omega_{\rm GW}^{\rm BBN}\Omega_R < 10^{-6}, \tag{20}$$

where ρ_r is the density of radiation and we used that $\rho_{\rm GW}/\rho_r =$ constant. Thus, the sharp peak in $\Omega_{\rm GW}(f)$ of kination cannot be extended to observable frequencies (see Fig. 1).

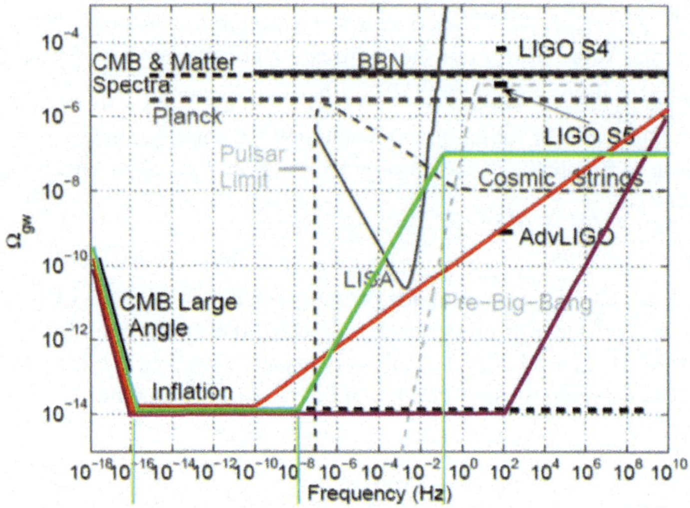

Fig. 1 Plot of $\Omega_{\rm GW}(f)$ superimposed with the observational expectations of LISA and Advanced LIGO (taken by Ref. [20]). The frequency range extends up to the scale of inflation, corresponding to the largest possible frequency, of GW which re-enter the horizon right at the end of inflation (assumed at the energy scale of grand unification). The uplift in the spectrum at low frequencies corresponds to the matter era of the hot Big Bang. The case when inflation is directly followed by the radiation era of the hot Big Bang (prompt reheating) is depicted by the low horizontal dashed thick line. As shown, the predicted spectrum is well beyond the observational capabilities of LISA and Advanced LIGO. The case when an early era of kination follows right after inflation corresponds to the purple line. As shown, there is a sharp peak in the spectrum ($\Omega_{\rm GW}(f) \propto f$), which however cannot be larger than the upper horizontal dashed line, which is the BBN constraint $\Omega_{\rm GW} < 10^{-6}$. Thus, the kination peak cannot extend to low frequencies and is not near the expected observations of LISA and Advanced LIGO. The case of a stiff period with $w \approx \frac{1}{2}$ following right after inflation is depicted with the red line. The peak of GW spectrum is milder ($\Omega_{\rm GW}(f) \propto f^{2/5}$), which means that it can be extended to lower frequencies without violating the BBN bound. As shown, the spectrum will be detectable by both LISA and Advanced LIGO. Finally, the case when a period of hyperkination (with a flat spectrum) and then regular kination follows the end of inflation is shown with a green line, which well overlaps with the expected observations of LISA and Advanced LIGO. In all cases, after reheating, the usual radiation era of the hot Big Bang begins and the GW spectra become flat as the frequency is lowered

5 Stiff Period

If the peak in $\Omega_{\rm GW}(f)$ is not so sharp then it might be extended to observable frequencies without disturbing BBN. This may be possible if $\frac{1}{3} < w < 1$. Indeed, In Ref. [11] it is shown that, when $0.46 \leq w \leq 0.56$ and $1\,{\rm MeV} < T_{\rm reh} < 150\,{\rm MeV}$, then the GW peak can be extended to frequencies low enough to be observable in the near future by Advanced LIGO and LISA, where $T_{\rm reh}$ is the reheating temperature, that is the temperature of the thermal bath at the onset of the usual radiation era of the hot Big Bang.

How can this possibility be realised? I have presented a concrete model to this end in Ref. [12]. Consider two flat directions φ and σ in field space, which meet at an Enhanced Symmetry Point (ESP) such that they are characterised by the standard hybrid potential [13]

$$V(\varphi,\sigma)=\frac{1}{2}g^2\sigma^2\varphi^2+\frac{1}{4}\lambda(\varphi^2-M^2)^2+V(\sigma)\,, \tag{21}$$

where $g<1$ is a perturbative interaction coupling, $\lambda<1$ is a perturbative self-coupling and $V(\sigma)$ is some unknown potential for the inflaton field σ, which forces it to vary (roll) to smaller values. In the above M is the vacuum expectation value (VEV) of the waterfall field φ. Below we consider that $M\sim m_P$, which is why φ can be called a flat direction (only lifted by Planck-suppressed interactions).

The waterfall field is non-canonical. Indeed, the the Lagrangian density is

$$\mathcal{L}=-\frac{1}{2}\partial_\mu\sigma\partial^\mu\sigma-\frac{\frac{1}{2}\partial_\mu\varphi\partial^\mu\varphi}{(1-\varphi^2/M^2)^2}-V(\varphi,\sigma)\,, \tag{22}$$

i.e. there are poles at the VEVs of φ due to non-trivial geometry in field space. This is the standard setup in α-attractors, where the poles could be due to a non-trivial Kähler metric in supergravity [14]. We can define a canonically normalised waterfall field ϕ using the transformation $\varphi=M\tanh(\phi/M)$. In terms of ϕ, the scalar potential in Eq. (21) assumes the form

$$V(\phi,\sigma)=\frac{1}{2}g^2M^2\sigma^2\tanh^2(\phi/M)+\frac{\frac{1}{4}\lambda M^4}{\cosh^4(\phi/M)}+V(\sigma)\,, \tag{23}$$

where the minima along the waterfall direction have been displaced at infinity.

When the inflaton expectation value is large, the waterfall field is heavy and is pushed towards the origin. At the origin, φ is canonical, so the hybrid mechanism operates normally. The waterfall transition occurs when the inflaton reaches the critical value $\sigma_c=(\sqrt{\lambda}/g)M$ [13]. Afterwards, the waterfall field finds itself on top of a potential hill and is released along its runaway direction towards large (absolute) values.

Near the origin, when $\phi\ll M$ (without loss of generality we assume $\phi>0$), the runaway waterfall potential is approximated as

$$V\simeq\frac{\frac{1}{4}\lambda M^4}{[1+\frac{1}{2}(\phi/M)^2]^4}\simeq\frac{1}{4}\lambda M^4\left[1-2\left(\frac{\phi}{M}\right)^2\right]. \tag{24}$$

Because $M\sim m_P$ (see below), the waterfall field undergoes a period of quadratic hilltop inflation, while ϕ dominates the Universe [15].

Eventually, $\phi \gg M$ and the waterfall potential is approximated as

$$V \simeq \frac{\frac{1}{4}\lambda M^4}{\left[\frac{1}{2}\exp(\phi/M)\right]^4} \simeq 4\lambda M^4 e^{-4\phi/M}\,. \tag{25}$$

During this roll, there is an attractor solution (power-law inflation [16]) in which the barotropic parameter of the rolling scalar field is

$$w = -1 + \frac{16}{3}\left(\frac{m_P}{M}\right)^2. \tag{26}$$

Thus, the value $M \approx 1.88\, m_P$ results in $w \simeq \frac{1}{2}$, which means that there is a stiff period of the Universe history when the GW modes re-entering the horizon correspond to a peak with $\Omega_{\rm GW}(f) \propto f^{2/5}$ [cf. Eq. (19)], which is not as sharp as the one due to kination and can be extended to observable frequencies (see Fig. 1).

A multiple of mechanisms can be responsible for reheating at the appropriate time. In Ref. [12], Ricci reheating is assumed as an example [17], where the Universe is reheated by the decay of a spectator field non-minimally coupled to gravity. It is shown that the appropriate $T_{\rm reh}$ is obtained with non-minimal coupling $\xi \simeq 30$.

6 Hyperkination

There is another example of creating enhanced primordial GWs by inflation, this time by truncating a peak in the GW spectrum generated by a stiff period. I have investigated this with collaborators in Ref. [18]. It can be done as follows.

In Palatini modified gravity we consider

$$\mathcal{L} = \frac{1}{2}m_P^2 R + \frac{1}{2}\alpha R^2 + \frac{1}{2}\xi\varphi^2 R - \frac{1}{2}\partial_\mu\varphi\,\partial^\mu\varphi - V(\varphi)\,, \tag{27}$$

where α and ξ are non-perturbative coefficients. Switching to the Einstein frame we obtain

$$\mathcal{L} = \frac{1}{2}m_P^2 R - \frac{1}{2}\partial_\mu\phi\,\partial^\mu\phi + \frac{1}{4}\alpha\frac{h^2+4\alpha V}{h^2 m_P^4}(\partial_\mu\phi\,\partial^\mu\phi)^2 - \frac{V m_P^4}{h^2+4\alpha V}\,, \tag{28}$$

where $h(\varphi) = m_P^2 + \xi\varphi^2$ and we employed the field redefinition

$$\frac{\mathrm{d}\phi}{\mathrm{d}\varphi} = \sqrt{\frac{h m_P^2}{h^2+4\alpha V}}\,. \tag{29}$$

In the above there is a strange quartic kinetic term. Such a term can be considered in general k-inflation models (no need for Palatini modified gravity) [19].

The EoM is

$$\left[1+3\alpha\left(1+\frac{4\alpha V}{h^2}\right)\frac{\dot{\phi}^2}{m_P^4}\right]\ddot{\phi}+3\left[1+\alpha\left(1+\frac{4\alpha V}{h^2}\right)\frac{\dot{\phi}^2}{m_P^4}\right]H\dot{\phi}$$
$$+3\alpha^2\frac{\dot{\phi}^4}{m_P^4}\frac{\mathrm{d}}{\mathrm{d}\phi}\left(\frac{V}{h^2}\right)+\frac{\mathrm{d}}{\mathrm{d}\phi}\frac{Vm_P^4}{h^2+4\alpha V}=0\,. \tag{30}$$

Then, from the energy-momentum tensor we can obtain the energy density and pressure of the field, which read

$$\rho_\phi=\frac{1}{2}\left[1+\frac{3}{2}\alpha\left(1+\frac{4\alpha V}{h^2}\right)\frac{\dot{\phi}^2}{m_P^4}\right]\dot{\phi}^2+\frac{Vm_P^4}{h^2+4\alpha V}\,, \tag{31}$$

$$p_\phi=\frac{1}{2}\left[1+\frac{1}{2}\alpha\left(1+\frac{4\alpha V}{h^2}\right)\frac{\dot{\phi}^2}{m_P^4}\right]\dot{\phi}^2-\frac{Vm_P^4}{h^2+4\alpha V}\,. \tag{32}$$

After exiting the inflationary plateau, the inflaton field ϕ becomes dominated by its kinetic energy density, i.e. it becomes oblivious to the potential V. Then the above reduce to

$$\left(1+3\alpha\frac{\dot{\phi}^2}{m_P^4}\right)\ddot{\phi}+3\left(1+\alpha\frac{\dot{\phi}^2}{m_P^4}\right)H\dot{\phi}=0 \tag{33}$$

and

$$\rho_\phi=\frac{1}{2}\left(1+\frac{3}{2}\alpha\frac{\dot{\phi}^2}{m_P^4}\right)\dot{\phi}^2 \quad \text{and} \quad p_\phi=\frac{1}{2}\left(1+\frac{1}{2}\alpha\frac{\dot{\phi}^2}{m_P^4}\right)\dot{\phi}^2\,. \tag{34}$$

Thus, when the quadratic kinetic term dominates one can effectively set $\alpha=0$ and we have regular kination with $w=1$. However, when the quartic kinetic term dominates we can effectively consider only the α-depended terms. Then, $w=p_\phi/\rho_\phi=\frac{1}{3}$. We call this period *hyperkination*. Equation (19) suggests that the corresponding part of the GW spectrum is flat as in the radiation era. This means that the peak generated by kination has been truncated, which implies that the kinetic regime can last longer without disturbing BBN. As such, the GW signal can be amply boosted at observable frequencies as shown in Fig. 1.

7 Conclusions

Cosmic Inflation resolves the fine-tunings of the hot Big Bang and provides seeds for structure formation. Inflation is spectacularly verified by CMB observations. Another generic prediction of inflation is a superhorizon spectrum of primordial gravitational waves (GWs) generated through particle production. The form of the resulting GW spectrum depends on the post-inflation history. However, when GW modes re-enter the horizon during radiation domination they form a flat spectrum, too faint to be observable at present.

A stiff period in the Universe history enhances primordial GWs forming a peak in their spectrum. Non-oscillatory inflation is followed by such a period, dominated by the inflaton's kinetic energy density, called kination, but the frequencies of the peak are too high. The GW peak can be extended to observable frequencies if the stiff period is milder than that of kination, with $w \approx 1/2$. A model realisation of this possibility considers two flat directions which intersect at an ESP and give rise to the hybrid mechanism with Planckian waterfall VEV, which is also a kinetic pole of the waterfall field, as in α-attractors.

Another possibility to obtain a boost in primordial GWs down to observable frequencies is by considering higher order kinetic terms, as with k-inflation. This is possible to realise in Palatini modified gravity. Considering $R + R^2$ gravity and a non-minimally coupled scalar field, results in additional quartic kinetic terms. When the quartic kinetic terms dominate, this gives rise to to hyperkination. Hyperkination is followed by regular kination, when the kinetic terms become canonical. The resulting truncated GW peak can be extended to observable frequencies without disturbing BBN.

Forthcoming observations of Advanced LIGO, LISA, DesiGO and BBO may well detect the primordial GWs generated by inflation. Detection of primordial GWs will not only confirm a prediction of inflation but offer tantalising evidence of the quantum nature of gravity, because the Bunch-Davis vacuum of virtual gravitons is assumed as an initial condition for the generation of GWs during inflation by particle production.

Acknowledgements This work was funded in part by STFC with the consolidated grant: ST/X000621/1.

References

1. A.H. Guth, The inflationary universe: a possible solution to the horizon and flatness problems. Phys. Rev. D **23**, 347 (1981). https://doi.org/10.1103/PhysRevD.23.347
2. A.A. Starobinsky, A new type of isotropic cosmological models without singularity. Phys. Lett. B **91**, 99 (1980). https://doi.org/10.1016/0370-2693(80)90670-X
3. Y. Akrami, F. Arroja, M. Ashdown, J. Aumont, C. Baccigalupi, M. Ballardini, A.J. Banday, R.B. Barreiro, N. Bartolo, S. Basak, et al., Planck 2018 results. X. Constraints on inflation. Astron. Astrophys. **641**, A10 (2020). https://doi.org/10.1051/0004-6361/201833887

4. M.S. Turner, Detectability of inflation produced gravitational waves. Phys. Rev. D **55**, R435–R439 (1997). https://doi.org/10.1103/PhysRevD.55.R435
5. T.S. Bunch, P.C.W. Davies, Quantum field theory in de sitter space: renormalization by point splitting. Proc. Roy. Soc. Lond. A **360**, 117–134 (1978). https://doi.org/10.1098/rspa.1978.0060
6. P.A.R. Ade, Z. Ahmed, M. Amiri, D. Barkats, R.B. Thakur, C.A. Bischoff, D. Beck, J.J. Bock, H. Boenish, E. Bullock et al., Improved constraints on primordial gravitational waves using Planck, WMAP, and BICEP/Keck observations through the 2018 observing season. Phys. Rev. Lett. **127**, 151301 (2021). https://doi.org/10.1103/PhysRevLett.127.151301
7. Y. Gouttenoire, G. Servant, P. Simakachorn, Kination cosmology from scalar fields and gravitational-wave signatures. [arXiv:2111.01150 [hep-ph]]
8. G.N. Felder, L. Kofman, A.D. Linde, Inflation and preheating in NO models. Phys. Rev. D **60**, 103505 (1999). https://doi.org/10.1103/PhysRevD.60.103505
9. M. Joyce, T. Prokopec, Turning around the Sphaleron bound: Electroweak baryogenesis in an alternative postinflationary cosmology. Phys. Rev. D **57**, 6022–6049 (1998). https://doi.org/10.1103/PhysRevD.57.6022
10. V. Sahni, M. Sami, T. Souradeep, Relic gravity waves from brane world inflation. Phys. Rev. D **65**, 023518 (2002). https://doi.org/10.1103/PhysRevD.65.023518
11. D.G. Figueroa, E.H. Tanin, Ability of LIGO and LISA to probe the equation of state of the early Universe. JCAP **08**, 011 (2019). https://doi.org/10.1088/1475-7516/2019/08/011
12. K. Dimopoulos, Waterfall stiff period can generate observable primordial gravitational waves. JCAP **10**, 027 (2022). https://doi.org/10.1088/1475-7516/2022/10/027
13. A.D. Linde, Hybrid inflation. Phys. Rev. D **49**, 748–754 (1994). https://doi.org/10.1103/PhysRevD.49.748
14. R. Kallosh, A.D. Linde, D. Roest, Superconformal inflationary α-attractors. JHEP **11**, 198 (2013). https://doi.org/10.1007/JHEP11(2013)198
15. L. Boubekeur, D.H. Lyth, Hilltop inflation. JCAP **07**, 010 (2005). https://doi.org/10.1088/1475-7516/2005/07/010
16. F. Lucchin, S. Matarrese, Power law inflation. Phys. Rev. D **32**, 1316 (1985). https://doi.org/10.1103/PhysRevD.32.1316
17. T. Opferkuch, P. Schwaller, B.A. Stefanek, Ricci reheating. JCAP **07**, 016 (2019). https://doi.org/10.1088/1475-7516/2019/07/016
18. S. Sánchez López, K. Dimopoulos, A. Karam, E. Tomberg, Observable gravitational waves from hyperkination in Palatini gravity and beyond. Eur. Phys. J. C, in press (2023). [arXiv:2305.01399 [gr-qc]], https://doi.org/10.48550/arXiv.2305.01399
19. C. Armendariz-Picon, T. Damour, V.F. Mukhanov, k-inflation. Phys. Lett. B **458**, 209–218 (1999). https://doi.org/10.1016/S0370-2693(99)00603-6
20. E. Thrane, J.D. Romano, Sensitivity curves for searches for gravitational wave backgrounds. Phys. Rev. D **88**(12), 124032 (2013). https://doi.org/10.1103/PhysRevD.88.124032

Modules and Assembly Status of the CMS HGCAL

Dimitra Tsionou

Abstract The CMS Collaboration is designing a new high granularity endcap calorimeter, HGCAL, to be installed as a replacement endcap calorimeter for the High Luminosity LHC era, later this decade. This sampling calorimeter will have approximately six million silicon sensor cells and four hundred thousand scintillator tiles readout with on-tile silicon-photomultipliers. After prototypes and past test beams, the module hardware design has been finalised and CMS is preparing for the mass assembly process. We present the status of the silicon modules for the HGCAL as well as the assembly process.

Keywords HGCAL · Cassette · Assembly

1 Introduction

The High Luminosity LHC (HL-LHC) is a major upgrade of the LHC which aims to collect 10 times more integrated luminosity than the pre HL-LHC era. It is planned to operate at an instantaneous luminosity of $5 - 7.5 \times 10^{34}\ \mathrm{s}^{-1}\mathrm{cm}^{-2}$, roughly 3 times higher than LHC, with a mean number of 140–200 events per bunch crossing (pile-up), roughly four times higher than LHC.

To cope in this new environment of higher radiation, pile-up and track density, the CMS experiment is undergoing an extensive upgrade program. In particular, the endcap calorimeter will need to be replaced to withstand the expected radiation of 1.5×10^{16} neq/cm^2 at the region close to the beam at the end of the HL-LHC operation. The endcap CMS High Granularity Calorimeter (HGCAL) will cover the pseudorapidity region $1.5 < |\eta| < 3$, will be radiation tolerant and will provide precise hit and cluster information adequate for particle flow algorithms [1].

Dimitra Tsionou—on behalf of the CMS collaboration.

D. Tsionou (✉)
National Taiwan University (NTU), Taipei, Taiwan
e-mail: dimitra.tsionou@cern.ch

J. Strologas and L. Perivolaropoulos (eds.), *HEP2023*, Springer Proceedings in Physics 333 https://doi.org/10.1007/978-3-031-76856-9_12

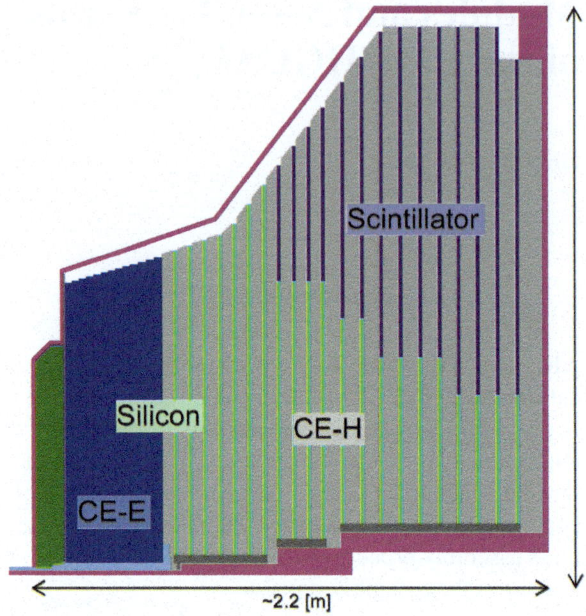

Fig. 1 Schematic view of the CMS high granularity calorimeter [2]

The CMS HGCAL shown in Fig. 1 is a sampling calorimeter consisting of 26 layers of silicon sensors for the electromagnetic part, and 21 layers for the hadronic part. The hadronic part will be equipped with silicon sensors in the higher radiation areas and with scintillating tiles read out by Silicon Photomultipliers (SiPM) in the lower radiation areas. The detector will be mounted on disks separated by the absorber material and will operate at $-30\,°C$ using CO_2 cooling.

2 Silicon Sensors and Modules

The silicon sensors that will be used for the HGCAL are produced in a hexagonal shape of 8″ wafers to maximise the coverage area and facilitate the tiling of the sensors. A number of different shapes of partial sensors will be used to provide further coverage on the inner and outer radii of the detector layers.

In addition, different variants also exist in terms of sensor thickness and number of cells. In particular, sensors of 120, 200 and 300 μm will be used in different areas of the detector. Close to the beam line where higher radiation is expected, 120 μm sensors will be used since it has been demonstrated that they can still achieve a good charge collection efficiency (>60%) even at 10^{16} neq/cm^2 fluences [3, 4]. These sensors will also have a smaller cell size of 0.5 cm^2, and a total of roughly 400 cells per sensor (High Density sensors). The 200 and 300 μm sensors will be placed further away from the interaction point e.g. in the middle and outer radii of the detector where

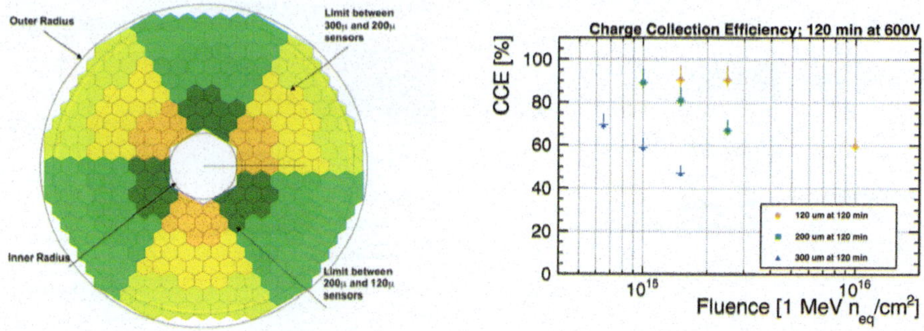

Fig. 2 *Left*: Placement of sensors of different thicknesses for one of the electromagnetic layers of HGCAL. *Right*: Charge collection efficiency of irradiated sensors of different thickness

lower fluences are expected. These sensors are expected to have a larger cell size and density, i.e. roughly 200 cells of 1 cm^2 size (Low Density sensors). An example of sensor placement of the different variants is shown in Fig. 2.

Sensors for the pre-production phase have been delivered and sensor testing is well advanced. Non-irradiated and irradiated sensor variants of different thickness have also been tested.

A silicon module is composed of a PCB (the hexaboard) where the readout chips are placed, the silicon sensor, a kapton sheet and a baseplate. The PCB is glued to the sensor and the input channels are wire bonded to the sensor through step holes that are then encapsulated. The PCB provides the bias voltage on the backside of the silicon sensor. A kapton sheet is placed between the sensor and the baseplate to provide electrical isolation. The baseplate is Cu-W or PCB for the electromagnetic and the hadronic modules respectively. CMS will build ~27000 such modules.

3 Scintillator and SiPM Modules

HGCAL regions with lower expected radiation levels ($<$ 3kGy and 8×10^{13} neq/cm^2) will be equipped with scintillator tiles [5]. Each scintillator tile will be wrapped in reflective foil and mounted on a silicon photomultiplier (SiPM) which will be located on a PCB to form a tile module. The size of the tiles depends on the radius where they will be placed and varies from 4 to 32 cm^2. A signal-to-noise ratio better than 3 is expected to be obtained at the end of the HL-LHC operation for the scintillators.

In Fig. 3 a zoom-in picture of a tile module is shown as well as the expected neutron equivalent fluence for the HGCAL at the end of the HL-LHC operation.

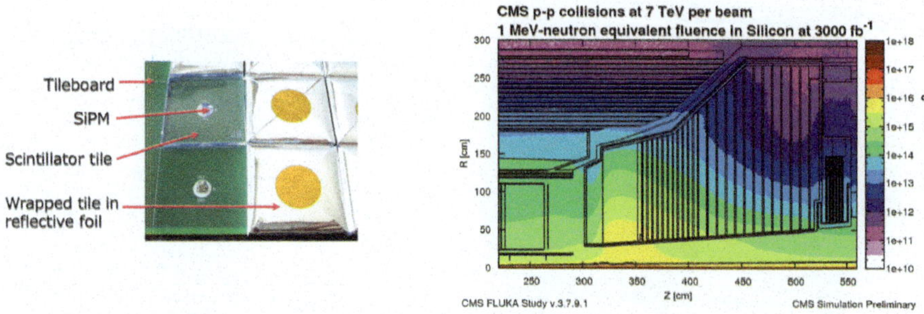

Fig. 3 *Left*: An example of a tile module showing also unwrapped tiles. *Right*: Fluence accumulated in HGCAL after an integrated luminosity of 3000 fb^{-1}. The purple and dark blue areas indicate where scintillators with SiPM-on-tile modules will be placed [1]

4 Cassettes and Assembly

The finished silicon and scintillator/SiPM-on-tile modules are mounted on cassettes for mechanical integration. The cassette is the assembly of many different components including the passive material, the detector modules, and the copper cooling plate. The modules are tiled together next to each other and are connected via wagon boards which are connected to engine boards. The wagon boards are purely passive and transmit data from the modules to the active engine boards which then transmit the data to the off-detector electronics through optical fibers.

For the electromagnetic part, the silicon modules will be mounted on both sides of a copper cooling plate. These double-sided layers will then be sandwiched between lead absorbers. The azimuthal angle of a double layer electromagnetic cassette is 60°. For the hadronic part, the silicon and scintillator SiPM-on-tile modules will be mounted on one side of the cassette which will then be mounted between steel absorbers. The azimuthal angle of a hadronic cassette is 30° to keep them at a manageable size. In total ~700 cassettes will be built.

Figure 4 shows an example of an electromagnetic cassette with some of its components.

5 Electronics

There are several different boards used for the electronics sequence for the HGCAL. The one that takes part in the module building is the hexaboard. Different sensor variants also require different hexaboards for the modules. There are therefore the Low Density and High Density hexaboard variations and a number of partials as well. The HGCROC readout chip [6] is housed on the hexaboard (3 chips for the Low Density and 6 for the High Density boards). The chip is used both for the silicon and the SiPM-on-tile parts.

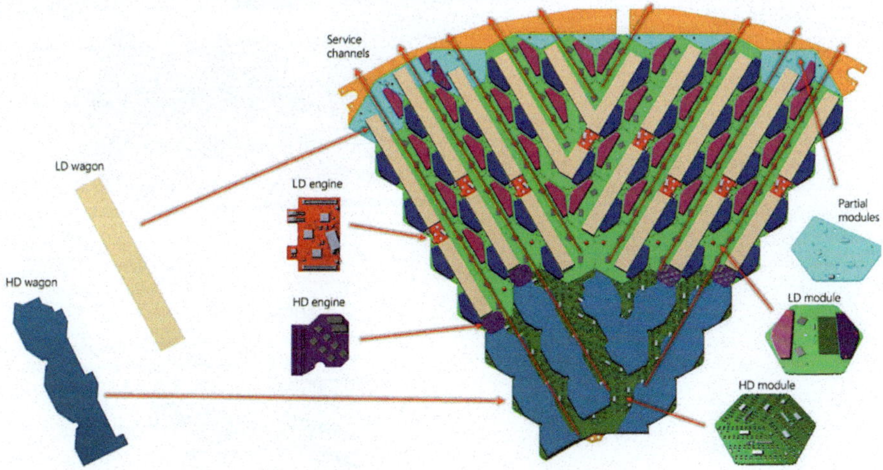

Fig. 4 Design of a full silicon cassette with some of its main components

A high dynamic energy range is required in order to be sensitive to single minimum ionising particles as well as high energy deposits. The HGCROC features a dynamic range from 0.2 fC to 10 pC. Depending on the range, the energy is calculated either through ADC (10 bits) or with time over threshold (ToT) measurement (12 bits). The time of arrival (ToA) is measured in 10 bits with a timing resolution at the level of $\sim$30 ps.

Two data flows are in place: one is the DAQ path for the ADC, ToT and ToA data and the other is the trigger path that transmits data of trigger sums of 4 or 9 cells. ECON-D and ECON-T boards are connected to the hexaboard to perform this part. Additionally, the boards need to be radiation tolerant and have low power consumption ($<$20 mW per channel) [7].

6 Test Beam Data

During the past years, several test beams have taken place. In 2018, a test beam was performed at CERN using positron, pion, and muon beams [8, 9]. The setup was composed of 28 electromagnetic layers, 10 hadronic layers with silicon modules and 39 layers of the AHCAL CALICE calorimeter with scintillator and SiPM-on-tile modules, as shown in the top figure in Fig. 5. The bottom figure shows an event display of the hadronic shower of a 300 GeV pion at the test beam detector setup. The shower starts in the last layers of the electromagnetic (CE-E) calorimeter and deposits energy throughout the silicon (CE-H) and scintillator (AHCAL) hadronic parts. An excellent performance agreement between data and simulation has been

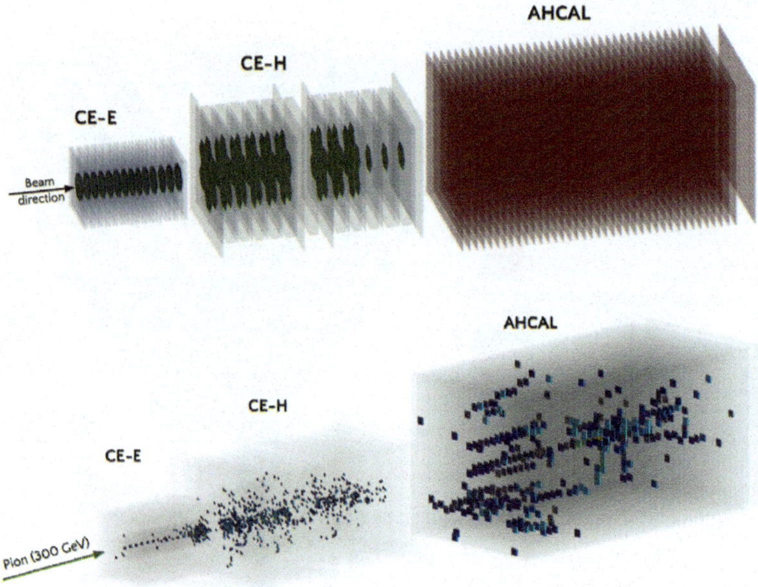

Fig. 5 *Top*: Simulated geometry of the test beam detector layout. *Bottom*: Event display of the hadronic shower of a 300 GeV pion [8]

achieved. Two test beams are planned for 2023 to test the system with the latest electronics version and a chain as close as possible to the final system.

7 Reconstruction

The small cell size and the large number of layers offer great opportunities for reconstruction algorithms including particle flow and machine learning. New computing technologies are available and GPUs can be used to increase the computational efficiency. One difficulty in the reconstruction of events is the large expected number of recorded hits ($\sim 10^5$) per event. Therefore, this information needs to be reduced to clusters of energy. The Clustering of Energy (CLUE) algorithm calculates the energy density around hits and defines seeds, followers, and outliers [10]. The found energy clusters can then be passed on to the next reconstruction step. Its advantage is that it can be parallelised and can run on GPU. The Iterative Clustering (TICL) algorithm uses the 2D layer clusters from CLUE and based on angular requirements it links them together to form tracksters which represent particle showers. It is an iterative procedure where simpler objects are reconstructed first, and the more complex objects are reconstructed in following iterations.

8 Conclusion

The CMS Collaboration is developing a high granularity endcap calorimeter for the HL-LHC era. It is a sampling calorimeter using silicon sensors in the high radiation areas and scintillator with SiPM-on-tile technology in the lower radiation areas. It will provide precise energy, spatial and timing information. The sensors for the pre-production stage have started arriving, the electronics are close to final, the module pre-series will start being built soon and will be followed by the pre-series cassette production. Testing procedures at all the various production stages are being put in place.

References

1. CMS Collaboration, The Phase-2 Upgrade of the CMS Endcap Calorimeter, CERN-LHCC-2017-023, CMS-TDR-019 (2017)
2. D. Barney, Overview slide of CE with main parameters. https://cms-docdb.cern.ch/cgi-bin/PublicDocDB/ShowDocument?docid=13251
3. N. Akchurin et al., Charge collection and electrical characterization of neutron irradiated silicon pad detectors for the CMS high granularity calorimeter. JINST **15**(09), P09031 (2020)
4. M. Wiehe, The CMS high granularity calorimeter for the high luminosity LHC. Nucl. Instrum. Methods Phys. Res. A **1041**, 1673123 (2022)
5. M. Reinecke, The CMS high granularity calorimeter scintillator/SiPM tileboards. J. Phys. : Conf. Ser. **2374**, 012019 (2022)
6. F. Bouyjou et al., HGCROC2: the front-end readout ASICs for the CMS high granularity calorimeter. J. Phys. : Conf. Ser. **2374**, 012070 (2022)
7. A. Lobanov, Electronics and triggering challenges for the CMS high granularity calorimeter. JINST **13**, C02056 (2018)
8. CALICE Collaboration, Performance of the CMS high granularity calorimeter prototype to charged pion beams of 20-300 GeV/c. arXiv:2211.04740
9. CMS Hgcal Collaboration, Response of a CMS HGCAL silicon-pad electromagnetic calorimeter prototype to 20–300 GeV positrons. JINST **17**, P05022 (2022)
10. A. Di Pilato et al., Reconstruction in an imaging calorimeter for HL-LHC. JINST **15**, C06023 (2020)

Dedicated Experiments for Feebly Interacting Particles at the LHC

Vasiliki A. Mitsou

Abstract The LHC experiments have been optimised for the discovery of hypothetical particles with prompt decays. In parallel, (meta)stable particles are constrained by cosmological observations. Between these two extremes, there is a gap with particles with moderate lifetimes, connecting hidden sectors and the Standard Model through feeble interactions in "portal" models of dark matter, neutrino masses and other scenarios. The detector design and expected sensitivity of dedicated experiments targeting such particles is reviewed.

Keywords Long-lived particles · Millicharged particles · HNLs · Supersymmetry · MAPP · FASER · CODEX-b · milliQan · ANUBIS · MATHUSLA

1 Introduction

Long-lived particles (LLPs) [1, 2] travel a macroscopic distance before either giving rise to anomalous ionisation or decaying within a detector volume. From the theory point-of-view, long lifetimes can occur because of a symmetry, tiny couplings, tight mass splittings, or a heavy mediator. A type of LLPs denominated feebly interacting particles (FIPs) is characterised by indirect interactions of FIPs with Standard Model (SM) states through low-dimensional operators. Such interactions, referred to as *portals*, may give rise to a dark Higgs boson (scalar portal), to dark photons (vector portal), to heavy neutral leptons (fermion portal), or to axion-like particles (pseudoscalar portal).

Several experiments at the Large Hadron Collider (LHC), the Super Proton Synchrotron (SPS) at CERN and in other laboratories elsewhere [3, 4] are sensitive to searches for FIPs. This mini-review discusses the possibility to explore such theories with LHC LLP detectors, in a way complementary to searches performed by the general-purpose LHC experiments [5].

V. A. Mitsou (✉)
Instituto de Física Corpuscular (IFIC), CSIC—Universitat de València, C/ Catedrático José Beltrán 2, 46980 Paterna, Valencia, Spain
e-mail: vasiliki.mitsou@ific.uv.es

J. Strologas and L. Perivolaropoulos (eds.), *HEP2023*, Springer Proceedings in Physics 333 https://doi.org/10.1007/978-3-031-76856-9_13

2 The Experiments

Several LLP experiments are proposed, planned, approved or are currently taking data at the LHC. They span different detector designs and cover a range of time scales [3]. Besides the detectors highlighted below, A Laboratory for Long-Lived eXotics (AL3X) has been proposed [6] to move the ALICE interaction point (IP) outside the magnet to allow LLPs to travel before decaying, in case ALICE ends its physics program before the end of High-Luminosity LHC (HL-LHC).

2.1 *MAPP MoEDAL Apparatus for Penetrating Particles*

The Monopole and Exotics Detector at the LHC (MoEDAL) [7], the first dedicated *search* LHC experiment, is specialised in highly ionising particles (HIPs). Its main objective is the search for magnetic monopoles [8], dyons, and any massive, detector-stable, slow-moving particle with single or multiple electric charge arising in various scenarios beyond the SM (BSM) [9]. The detector is deployed around IP8 in the LHCb cavern [10].

MoEDAL has pioneered the quest for high magnetic charges [11–14], by being first in considering photon fusion as production process [14, 15], and in constraining dyons [16] in colliders. By investigating the Schwinger thermal production of monopoles [17, 18] and considering resummation techniques [19, 20], it has bypassed the non-perturbativity of the large HIP–photon coupling. It has recently probed LLPs in the high-electric-charge regime [21, 22], while prospects for discovering electric charges lower than $\sim 10e$ look promising [23–26].

The MAPP detector, an extension of MoEDAL covering FIPs, is shielded by approximately 100 m of rock from cosmic rays [27, 28]. The Phase-1 MAPP (MAPP-1) is composed of a detector sensitive to particles with fractional charge as small as $0.001e$, the so-called millicharged particles (mCPs). It is made of plastic scintillation bars, as depicted in Fig. 1, and it is installed in the UA83 gallery around 100 m from IP8. An *Outrigger* detector, made of scintillator planks, is planned for deployment in three ducts connecting the UA83 gallery and the beam tunnel at angles from 2°–6° from the beam axis. It is expected to improve the acceptance of MAPP-1 for heavy mCPs. MAPP-1 is expected to start taking data in 2025.

MAPP-LLP is proposed to be installed in the Phase-2 MAPP (MAPP-2) as several nested boxes of scintillator hodoscope detectors following the shape of the UGC1 cavern. It is optimised for long-livedneutral particles emerging in new-physics scenar-

Fig. 1 Diagram of the MAPP-1 detector. The main bar-scintillator detector is surrounded by the scintillator veto and the aluminium flame shield. The electronic boards are installed in the crate on the left. From [29]

ios via their decay in flight in a zone of approximately 5 m (wide) × 10 m (deep) × 3 m (high). It will be deployed during the Long Shutdown 3 to be operated in the HL-LHC [27, 28].

2.2 *FASER ForwArd Search ExpeRiment*

FASER [30, 31] is looking for new particles produced in decays of light mesons abundantly present at zero angle at the LHC Run-3 and beyond. Such copiously produced particles may travel for hundreds of metres without interacting, before decaying to SM particles. FASER is located 480 m away from the ATLAS IP in the service tunnel TI12 at $\eta > 9.1$.

A schematic view of the different detector components can be seen in Fig. 2. The 0.55 T permanent dipole magnets have a radius of 10 cm. Three scintillator stations offer timing, trigger and veto of incoming charged particles. The electromagnetic calorimeter comprises four spare outer ECAL modules from LHCb. The three tracking stations, each containing three layers, is instrumented with spare ATLAS semiconductor-tracker barrel modules.

Moreover, FASERν [33] is designed to directly detect collider neutrinos and study their cross sections at TeV energies by employing emulsion films, also visible in Fig. 2. Candidate vertices consistent with neutrino interactions at the LHC were observed above muon-induced backgrounds at 13-TeV [34] and 13.6-TeV [35] *pp* collision data. The SND@LHC experiment [36, 37] is also designed to observe and study LHC-produced neutrinos and recently reported first neutrino observations [38].

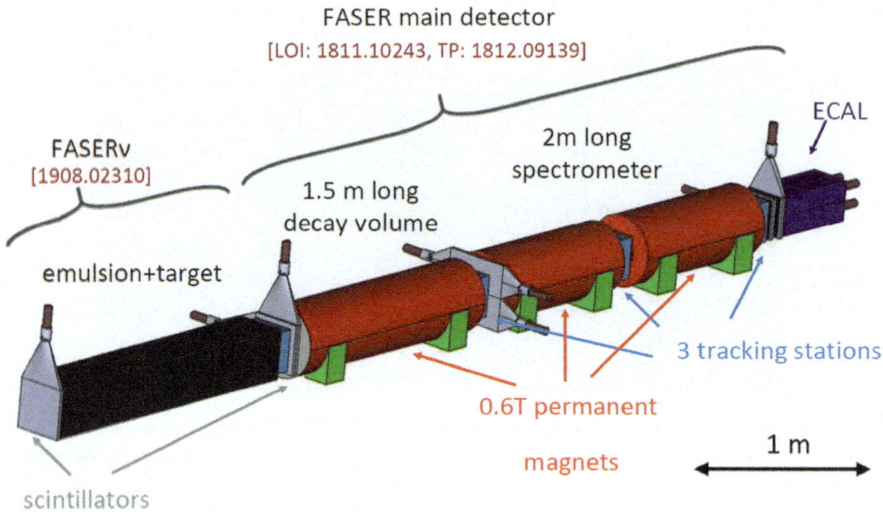

Fig. 2 Sketch of the FASER detector design. From [32, 33]

2.3 *Other Forward Detectors*

The Forward Physics Facility (FPF) is proposed to operate during the HL-LHC aiming at improving the LHC potential in BSM physics searches, neutrino physics and QCD measurements [39, 40]. It will be a 65 m-long and 9 m-wide purpose-built facility located 620–685 m west of the ATLAS IP. It will house FASER2 [31], a larger version of FASER with $R \sim 1$ m and $L \sim 5$ m. Other experiments proposed to be housed in FPF include FASERν2, AdvSND, the Forward Liquid Argon Experiment (FLArE) and FORMOSA.

FORMOSA [41] is designed to discover mCPs in the far-forward direction, where it can benefit from an enhanced mCP production cross section compared to the transverse direction. It is a scintillator-based detector to be eventually hosted in the FPF. FORMOSA is proposed to start in Run 3 by moving the milliQan demonstrator to a gallery near the LHC tunnel, as a phase called FORMOSA-I, with the full milliQan-type detector, FORMOSA-II, to be deployed at a later stage.

The Forward-Aperture CMS ExTension (FACET) [42] is a multi-particle spectrometer to be situated at $z \sim +100$ m from the CMS IP5 and take data during HL-LHC. The detector will occupy a radius of $\sim$50 cm and coverage $6 < \eta < 8$, so it will be closer to the IP and with a larger decay volume than FASER. It will be fully integrated in CMS and can be used either a forward part of CMS or a standalone detector.

Fig. 3 The milliQan bar detector [46]

2.4 *MilliQan*

Just like MAPP-1, milliQan [43, 44] is searching for mCPs. It is operational in an underground tunnel at a distance of 33 m from the CMS IP5, as shown in Fig. 3, with 17 m of rock shielding it from most particles produced in LHC collisions. A large active area of scintillator is needed to make it sensitive to particles with charges as low as $0.001e$. The bar detector is made of $0.2 \times 0.2 \times 3$ m plastic scintillator bars surrounded by an active muon veto shield. The slab detector will increase the reach for heavier mCPs through $40 \times 60 \times 5$ cm scintillator slabs [45].

A small portion of $\sim$1% of the full detector, the *milliQan* demonstrator* was installed and operated to measure backgrounds and provide proof of principle. It was the first search for mCPs at the LHC [46], where no excess over the background expectation was observed. The bounds set on mCPs are discussed in Sect. 3.1.

2.5 *ANUBIS AN Underground Belayed In-Shaft*

ANUBIS [48] is designed for neutral LLPs with $c\tau \gtrsim 5$ m, proposed to cover the shaft bottom and the cavern ceiling of the ATLAS cavern, as shown in Fig. 4. This geometry offers a wider solid angle with respect to the, initially conceived, occupancy of the shaft volume, also visible in Fig. 4. The tracking stations will use the same resistive plate chamber (RPC) technology as the new ATLAS layers, while ATLAS

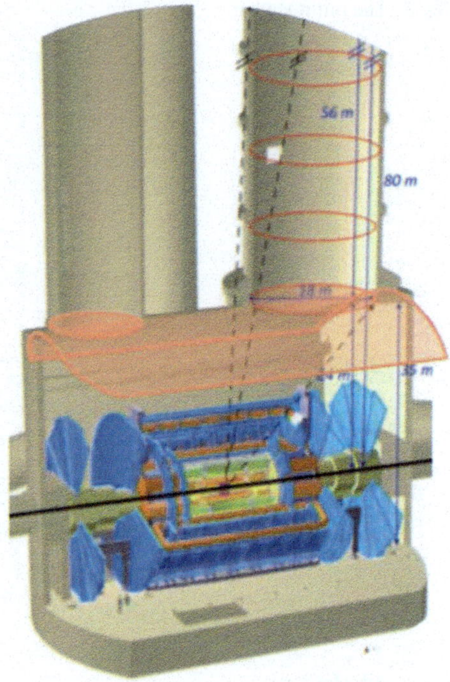

Fig. 4 Schematic view of ANUBIS (in orange) in the ATLAS cavern, proposed to be installed in shaft bottoms and the cavern ceiling. From [47]

can be used as an active veto of SM activity. The proANUBIS prototype, installed in ATLAS cavern in March 2023 and in commissioning, is expected to validate the ANUBIS proof of concept.

2.6 CODEX-b COmpact Detector for EXotics at LHCb

CODEX-b [49, 50] is proposed as a cubic detector with a nominal fiducial volume of $10\ \mathrm{m} \times 10\ \mathrm{m} \times 10\ \mathrm{m}$ to be situated in a transverse location ~ 25 m from the LHCb IP8, corresponding to the pseudorapidity range $0.2 < \eta < 0.6$. It will be composed of six RPC layers—being fast, precise and cheap for a large-area tracker—at 4 cm intervals on each box face with 1 cm granularity. An additional passive Pb shielding of $25\lambda_{\mathrm{int}}$ with an active scintillator veto is planned to reduce IP and secondary backgrounds.

The CODEX-β demonstrator, with dimensions $2 \times 2 \times 2\ \mathrm{m}^3$, will be operated during Run 3 to prove the detector concept. Its main goal will be the online integration with LHCb, the K_L^0 reconstruction and background measurements. It will offer competitive sensitivity to $b \to s\chi(\to$ hadrons), too.

2.7 MATHUSLA MAsive Timing Hodoscope for Ultra Stable NeutraL pArticles

MATHUSLA [51, 52] has been proposed as a large decay volume (25 m high) and a large footprint ($100 \times 100\ \text{m}^2$ wide) to be situated in the surface above CMS. The $\sim$90 m of rock between the IP5 and the detector decay volume is expected to provide enough shielding to work in a relatively clean environment. The decay volume will be occupied by several detector layers for tracking in a modular way, which provides a high level of scalability. Plastic scintillators and RPCs are proven technologies that fulfil the set specifications.

3 Physics Sensitivity

3.1 Stable Millicharged Particles

Millicharged particles are connected to the mechanism of electric charge quantisation and possible non-conservation of electric charge [53]. There are three LHC experiments sensitive to the detection of low ionisation coming from an mCP: MAPP-1 sub-detector (Sect. 2.1), milliQan (Sect. 2.4) and FORMOSA (Sect. 2.3).

One of the mechanisms that predict mCPs is the presence of a new massless $U'(1)$ gauge field, the dark photon, A'. A new massive dark fermion ψ with mass M_{mCP} occurres, which has charge e' [54–56].

The milliQan* demonstrator, discussed in Sect. 2.4, provided constraints on mCPs [46], as shown in Fig. 5, where pair-production of mCPs at 13 TeV is considered. MAPP-1 can extend these limits especially towards higher masses, while the reach in low charges depends on the integrated luminosity. Charges as low as $10^{-3}e$ will require the HL-LHC [28, 54]. Similar sensitivity is expected to be achieved by the full milliQan detector during the HL-LHC run [45]. We recall the MoEDAL-MAPP IP8 receives a factor $\sim$10–50 less instantaneous luminosity than the milliQan IP5.

When considering DM relic density, mCPs can account for a fraction of it undetectable by direct-detection experiments due to attenuation effects, when the ambient DM has non-negligible cross section with SM particles making it to lose most of its kinetic energy. For some model parameters [57, 58], the DM particles, generally referred to as *strongly interacting DM*, cannot be detected by ground-based direct detection experiments. mCP-hunting experiments at the LHC can constrain significantly such models [41].

In addition to DM, heavy neutrinos with a large electric dipole moment may induce ionisation in mCP-sensitive detectors. It has been shown that MAPP-1 and milliQan have good sensitivity in such a scenario [59, 60].

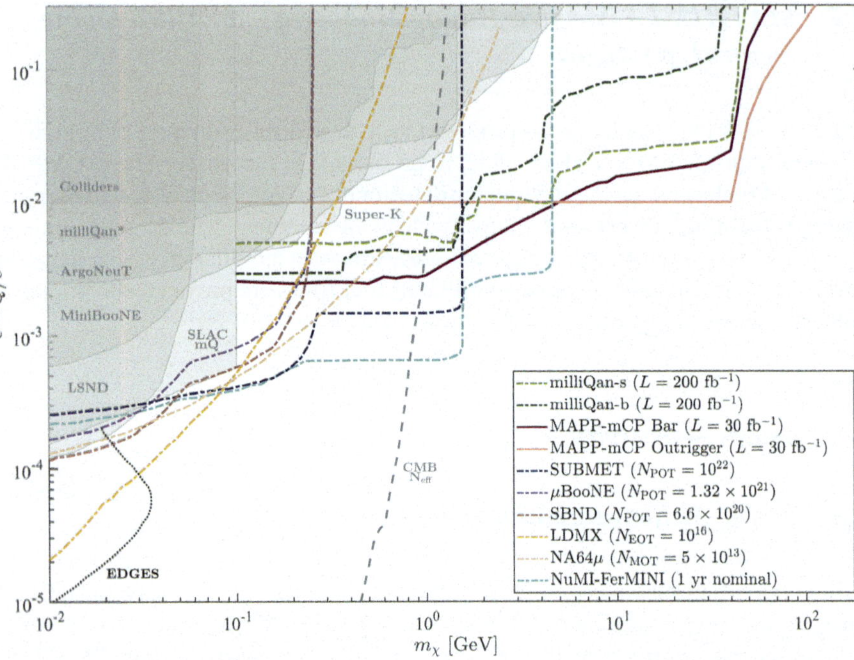

Fig. 5 Direct limits from accelerator-based searches and indirect limits from Planck (grey regions). Projected sensitivity for mCPs for the milliQan slab (s) and bar (b) detectors and for MAPP-1 (for the bar (B) and Outrigger (O) detectors) in pp collisions at $\sqrt{s} = 14$ TeV (coloured lines). From [28]

3.2 *Dark Photons*

A large class of BSM models includes interactions with light new *vector* particles. Such particles can result from additional gauge symmetries of BSM physics. New vector states can mediate interactions between the SM fields and extra dark-sector fields that may eventually play the role of the DM states. A minimal vector portal interaction can include a new $U'(1)$ gauge group, the field strength of which couples to the hypercharge field through a *kinetic mixing* parameter ϵ [61]. New matter fields χ, charged under $U'(1)$, can be incorporated, which if they are (meta)stable, they may constitute a fraction or entirety of dark matter. In the minimal dark photon model, DM is assumed to be either heavy or contained in a different sector. LHC experiments may be sensitive to such scenarios when the dark photon decays into visible states. Extensive studies on the discovery reach of LHC main [62] and LLP experiments [1, 63–65] have been performed for various species of dark photon models and signatures.

3.3 Dark Higgs

The *scalar portal* involves a new dark scalar, which mixes slightly with the SM-like Higgs, and provides a simple target for new physics searches [66]. The dark Higgs mixing portal leads to exotic inclusive $B \to X\phi$ decays, where ϕ is a light CP-even scalar, the *dark Higgs*, that mixes with the SM Higgs h via a mixing angle $\theta \ll 1$. The particle lifetime depends on the degree of mixing. In addition, there can appear a non-negligible trilinear interaction term between ϕ and h with the corresponding coupling denoted by λ, i.e. $\propto \lambda h\phi\phi$. Thus, the signal sought after in LLP experiments is two charged lepton tracks originating from dark Higgs decays $\phi \to \ell^+\ell^-$ in their fiducial volume [1, 63, 64].

3.4 Extended Neutrino Sector

In the *fermion portal*, new heavy neutral leptons (HNLs) are added to the SM to provide an elegant way to generate non-zero neutrino masses via the seesaw mechanism [67–69]. In type-I seesaw models, one extends the SM by adding neutral right-handed fermions (the HNLs), that couple to the SM neutrinos similarly to the coupling between left- and right-handed components of the charged leptons. In a large class of these models, the HNLs become long lived and the LLP experiments have good prospects to detect them [63, 70–75].

To highlight an example, in the case of right-handed neutrinos being produced in the decay of an additional Z' boson in the gauged $B - L$ model, which also contains a singlet Higgs that spontaneously breaks the extra $U(1)_{B-L}$ gauge symmetry, MAPP-2 will fill the gap left by CMS, LHCb, MATHUSLA, FASER2 and CODEX-b [76, 77], as shown in Fig. 6.

Light sterile neutrinos can account for dark matter [78–80], while sterile neutrinos with a broad range of masses can account for the baryon asymmetry of the Universe through leptogenesis [81]. Sterile neutrinos may be long lived in simplified models where the SM is extended with one sterile neutrino [82] or in neutrino-extended SM Effective Field Theories, νSMEFT [83]. Intermediate-mass sterile neutrinos can be produced in leptonic and semi-leptonic decays of charmed and bottomed mesons, decaying to leptons via neutral and charged weak currents, thus becoming detectable in LLP experiments [84].

3.5 Axions and Axion-Like Particles (ALPs)

Axions, and more generally ALPs, are pseudo-Nambu-Goldstone bosons arising in approximate BSM Abelian global symmetries, broken spontaneously at a scale much greater than the electroweak scale [86]. Specifically, axions were postulated by the

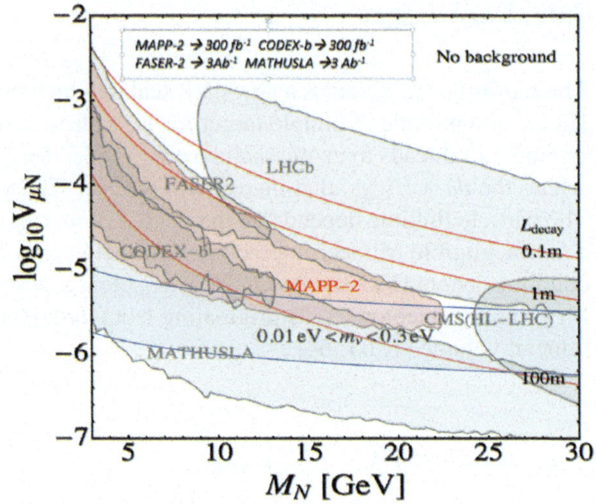

Fig. 6 95% C.L. exclusion bounds for MAPP-2 and other LLP experiments for HNLs in the minimal $Z'(B-L)$ model at $\sqrt{s} = 14$ TeV. The $U(1)_{B-L}$ gauge coupling and the Z' mass are $g'_1 = 10^{-3}$ and $m_{Z'} = 3.33\, m_N$, respectively. From [28]

Peccei-Quinn theory to resolve the strong CP problem in QCD. ALPs also offer an interesting connection to the puzzle of dark matter, since they can mediate interactions between the DM particle and SM states and allow for additional annihilation channels necessary for the thermal freeze-out of DM, serving as *pseudoscalar portal*, relevant for LLP experiments [1, 31, 50, 63, 75, 87–89].

3.6 Supersymmetry

Supersymmetry (SUSY), a theoretical framework that, among other attractive features, provides in a natural way a DM candidate, predicts the existence of charged or neutral LLPs in several models. For instance, in R-parity violating (RPV) SUSY, LLPs such as light long-lived neutralinos, $\tilde{\chi}_1^0$, decay via λ'_{ijk} couplings to visible particles. Benchmark scenarios related to either charm or bottom mesons decaying into $\tilde{\chi}_1^0$ have been considered, in similar fashion as in sterile neutrinos, demonstrating that these experiments can probe various meson production and decay modes and $\tilde{\chi}_1^0$ lifetimes [50, 72, 85, 90–92]. As an example, if the RPV couplings λ'_{131} and λ'_{112} for production and decay, respectively, are considered, MAPP-1, MAPP-2 and ANUBIS are sensitive to a wide range of $\tilde{\chi}_1^0$ lifetimes [85], as becomes evident from Fig. 7.

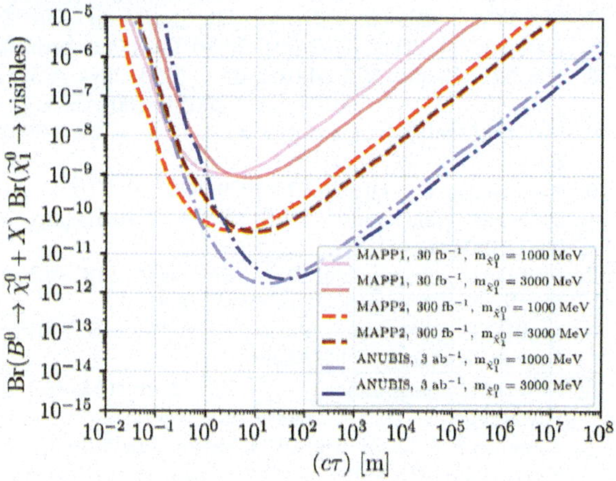

Fig. 7 Sensitivity estimate of MAPP-1 (pink solid), MAPP-2 (red dashed) and ANUBIS (blue dot-dashed) for an RPV SUSY model with a neutral long-lived fermion produced in bottom-meson decays. Estimates for two light $\tilde{\chi}_1^0$ masses are depicted: 1 GeV and 3 GeV. From [85]

4 Summary

There is a growing interest in searches for long-lived particles at the LHC, other CERN facilies and other laboratories to explore open questions in Particle Physics. Besides the continuous efforts in the large experiments ATLAS, CMS and LHCb, small specialised experiments take data in Run 3 or have been proposed for the HL-LHC. Several of these detectors have already yielded important physics results.

Acknowledgements The author thanks the HEP2023 organisers for the kind invitation and acknowledges support by the Generalitat Valenciana via the Excellence Grant CIPROM/2021/073, by the Spanish MCIN/AEI/10.13039/501100011033/ and the European Union/FEDER via the grant PID2021-122134NB-C21, and by the Spanish MCIU/AEI via the Severo Ochoa project CEX2023-001292-S.

References

1. J. Alimena et al., J. Phys. G **47**, 090501 (2020). https://doi.org/10.1088/1361-6471/ab4574 [arxiv:1903.04497]
2. T. Bose et al., *Report of the Topical Group on Physics Beyond the Standard Model at Energy Frontier for Snowmass 2021* [arxiv:2209.13128]
3. V.A. Mitsou, LHC experiments for long-lived particles of the dark sector, in *16th Marcel Grossmann Meeting on Recent Developments in Theoretical and Experimental General Relativity, Astrophysics and Relativistic Field Theories* (2021). https://doi.org/10.1142/9789811269776_0160 [arxiv:2111.03036]
4. V.A. Mitsou, Hidden sectors meet the lifetime frontier in specialised experiments, to appear (2025)
5. V.A. Mitsou, Int. J. Mod. Phys. A **28**, 1330052 (2013). https://doi.org/10.1142/S0217751X13300524 [arxiv:1310.1072]
6. V.V. Gligorov, S. Knapen, B. Nachman, M. Papucci, D. J. Robinson, Phys. Rev. D **99**, 015023 (2019). https://doi.org/10.1103/PhysRevD.99.015023 [arxiv:1810.03636]

7. MoEDAL collaboration, *Technical Design Report of the MoEDAL Experiment*. CERN-LHCC-2009-006, MoEDAL-TDR-001 (2009). https://cds.cern.ch/record/1181486
8. N.E. Mavromatos, V.A. Mitsou, Int. J. Mod. Phys. A **35**, 2030012 (2020). https://doi.org/10.1142/S0217751X20300124 [arxiv:2005.05100]
9. MoEDAL collaboration, Int. J. Mod. Phys. A **29**, 1430050 (2014). https://doi.org/10.1142/S0217751X14300506 [arxiv:1405.7662]
10. V.A. Mitsou, PoS **DISCRETE2020-2021**, 017 (2022). https://doi.org/10.22323/1.405.0017
11. MoEDAL collaboration, JHEP **08**, 067 (2016). https://doi.org/10.1007/JHEP08(2016)067 [arxiv:1604.06645]
12. MoEDAL collaboration, Phys. Rev. Lett. **118**, 061801 (2017). https://doi.org/10.1103/PhysRevLett.118.061801 [arxiv:1611.06817]
13. MoEDAL collaboration, Phys. Lett. B **782**, 510 (2018). https://doi.org/10.1016/j.physletb.2018.05.069 [arxiv:1712.09849]
14. MoEDAL collaboration, Phys. Rev. Lett. **123**, 021802 (2019). https://doi.org/10.1103/PhysRevLett.123.021802 [arxiv:1903.08491]
15. S. Baines, N.E. Mavromatos, V.A. Mitsou, J.L. Pinfold, A. Santra, Eur. Phys. J. C **78**, 966 (2018). https://doi.org/10.1140/epjc/s10052-018-6440-6 [arxiv:1808.08942]. [Erratum: Eur.Phys.J.C 79, 166 (2019)]
16. MoEDAL collaboration, Phys. Rev. Lett. **126**, 071801 (2021). https://doi.org/10.1103/PhysRevLett.126.071801 [arxiv:2002.00861]
17. MoEDAL collaboration, Nature **602**, 63 (2022). https://doi.org/10.1038/s41586-021-04298-1 [arxiv:2106.11933]
18. MoEDAL collaboration, Phys. Rev. Lett. **133**, 071803 (2024). https://doi.org/10.1103/PhysRevLett.133.071803 [arxiv:2402.15682]
19. J. Alexandre, N.E. Mavromatos, Phys. Rev. D **100**, 096005 (2019). https://doi.org/10.1103/PhysRevD.100.096005 [arxiv:1906.08738]
20. J. Alexandre, N.E. Mavromatos, V.A. Mitsou, E. Musumeci, Phys. Rev. D **109**, 036026 (2024). https://doi.org/10.1103/PhysRevD.109.036026 [arxiv:2310.17452]
21. MoEDAL collaboration, Eur. Phys. J. C **82**, 694 (2022). https://doi.org/10.1140/epjc/s10052-022-10608-2 [arxiv:2112.05806]
22. MoEDAL collaboration, *Search for Highly-Ionizing Particles in pp Collisions During LHC Run-2 Using the Full MoEDAL Detector* [arxiv:2311.06509]
23. K. Sakurai, D. Felea, J. Mamuzic, N.E. Mavromatos, V.A. Mitsou, J.L. Pinfold et al., J. Phys. Conf. Ser. **1586**, 012018 (2020). https://doi.org/10.1088/1742-6596/1586/1/012018 [arxiv:1903.11022]
24. D. Felea, J. Mamuzic, R. Masełek, N.E. Mavromatos, V.A. Mitsou, J.L. Pinfold et al., Eur. Phys. J. C **80**, 431 (2020). https://doi.org/10.1140/epjc/s10052-020-7994-7 [arxiv:2001.05980]
25. M. Hirsch, R. Masełek, K. Sakurai, Eur. Phys. J. C **81**, 697 (2021). https://doi.org/10.1140/epjc/s10052-021-09507-9 [arxiv:2103.05644]. [Erratum: Eur. Phys. J. C 82, 774 (2022)]
26. M.M. Altakach, P. Lamba, R. Masełek, V.A. Mitsou, K. Sakurai, Eur. Phys. J. C **82**, 848 (2022). https://doi.org/10.1140/epjc/s10052-022-10805-z [arxiv:2204.03667]
27. J.L. Pinfold, Phil. Trans. Roy. Soc. Lond. A **377**, 20190382 (2019). https://doi.org/10.1098/rsta.2019.0382
28. MoEDAL- MAPP collaboration, MoEDAL-MAPP, an LHC dedicated detector search facility, in *Snowmass 2021* (2022) [arxiv:2209.03988]
29. J. Pinfold, PoS **ICHEP2022**, 168 (2022). https://doi.org/10.22323/1.414.0168
30. J.L. Feng, I. Galon, F. Kling, S. Trojanowski, Phys. Rev. D **97**, 035001 (2018). https://doi.org/10.1103/PhysRevD.97.035001 [arxiv:1708.09389]
31. FASER collaboration, *FASER: ForwArd Search ExpeRiment at the LHC*. arxiv:1901.04468
32. FASER collaboration, *Technical Proposal for FASER: ForwArd Search ExpeRiment at the LHC*. arxiv:1812.09139
33. FASER collaboration, Eur. Phys. J. C **80**, 61 (2020). https://doi.org/10.1140/epjc/s10052-020-7631-5 [arxiv:1908.02310]

34. FASER collaboration, Phys. Rev. D **104**, L091101 (2021). https://doi.org/10.1103/PhysRevD.104.L091101 [arxiv:2105.06197]
35. FASER collaboration, Phys. Rev. Lett. **131**, 031801 (2023). https://doi.org/10.1103/PhysRevLett.131.031801 [arxiv:2303.14185]
36. SHiP collaboration, *SND@LHC* [arxiv:2002.08722]
37. SND@LHC collaboration, *SND@LHC: The Scattering and Neutrino Detector at the LHC* [arxiv:2210.02784]
38. SND@LHC collaboration, Phys. Rev. Lett. **131**, 031802 (2023). https://doi.org/10.1103/PhysRevLett.131.031802 [arxiv:2305.09383]
39. L.A. Anchordoqui et al., Phys. Rept. **968**, 1 (2022). https://doi.org/10.1016/j.physrep.2022.04.004 [arxiv:2109.10905]
40. J.L. Feng et al., J. Phys. G **50**, 030501 (2023). https://doi.org/10.1088/1361-6471/ac865e [arxiv:2203.05090]
41. S. Foroughi-Abari, F. Kling, Y.-D. Tsai, Phys. Rev. D **104**, 035014 (2021). https://doi.org/10.1103/PhysRevD.104.035014 [arxiv:2010.07941]
42. S. Cerci et al., JHEP **06**, 110 (2022). https://doi.org/10.1007/JHEP06(2022)110 [arxiv:2201.00019]
43. A. Haas, C.S. Hill, E. Izaguirre, I. Yavin, Phys. Lett. B **746**, 117 (2015). https://doi.org/10.1016/j.physletb.2015.04.062 [arxiv:1410.6816]
44. A. Ball et al., *A Letter of Intent to Install a milli-charged Particle Detector at LHC P5* [arxiv:1607.04669]
45. MILLIQAN collaboration, Phys. Rev. D **104**, 032002 (2021). https://doi.org/10.1103/PhysRevD.104.032002 [arxiv:2104.07151]
46. A. Ball et al., Phys. Rev. D **102**, 032002 (2020). https://doi.org/10.1103/PhysRevD.102.032002 [arxiv:2005.06518]
47. ANUBIS collaboration, PoS **EPS-HEP2023**, 051 (2024). https://doi.org/10.22323/1.449.0051 [arxiv:2401.11604]
48. M. Bauer, O. Brandt, L. Lee, C. Ohm, *ANUBIS: Proposal to Search for Long-Lived Neutral Particles in CERN Service Shafts* [arxiv:1909.13022]
49. V.V. Gligorov, S. Knapen, M. Papucci, D.J. Robinson,Phys. Rev. D **97**, 015023 (2018). https://doi.org/10.1103/PhysRevD.97.015023 [arxiv:1708.09395]
50. G. Aielli et al., Eur. Phys. J. C **80**, 1177 (2020). https://doi.org/10.1140/epjc/s10052-020-08711-3 [arxiv:1911.00481]
51. MATHUSLA collaboration, JINST **15**, C06026 (2020). https://doi.org/10.1088/1748-0221/15/06/C06026 [arxiv:1901.04040]
52. MATHUSLA collaboration, *An Update to the Letter of Intent for MATHUSLA: Search for Long-Lived Particles at the HL-LHC* [arxiv:2009.01693]
53. A.Y. Ignatiev, V.A. Kuzmin, M.E. Shaposhnikov, Phys. Lett. B **84**, 315 (1979). https://doi.org/10.1016/0370-2693(79)90048-0
54. M.A. Staelens, *Physics From Beyond the Standard Model: Exotic Matter Searches at the LHC with the MoEDAL-MAPP Experiment*, Ph.D. thesis, Alberta U., (2021). 10.7939/r3-g8yh-hv16
55. M. de Montigny, P.-P. A. Ouimet, J. Pinfold, A. Shaa and M. Staelens, *Minicharged Particles at Accelerators: Progress and Prospects* [arxiv:2307.07855]
56. M. Kalliokoski, V.A. Mitsou, M. de Montigny, A. Mukhopadhyay, P.-P. A. Ouimet, J. Pinfold et al., JHEP **04**, 137 (2024). https://doi.org/10.1007/JHEP04(2024)137 [arxiv:2311.02185]
57. T. Emken, R. Essig, C. Kouvaris, M. Sholapurkar, JCAP **09**, 070 (2019). https://doi.org/10.1088/1475-7516/2019/09/070 [arxiv:1905.06348]
58. R. Plestid, V. Takhistov, Y.-D. Tsai, T. Bringmann, A. Kusenko, M. Pospelov, Phys. Rev. D **102**, 115032 (2020). https://doi.org/10.1103/PhysRevD.102.115032 [arxiv:2002.11732]
59. M. Sher, J.R. Stevens, Phys. Lett. B **777**, 246 (2018). https://doi.org/10.1016/j.physletb.2017.12.022 [arxiv:1710.06894]
60. M. Frank, M. de Montigny, P.-P. A. Ouimet, J. Pinfold, A. Shaa, M. Staelens, Phys. Lett. B **802**, 135204 (2020). https://doi.org/10.1016/j.physletb.2020.135204 [arxiv:1909.05216]
61. B. Holdom, Phys. Lett. B **166**, 196 (1986). https://doi.org/10.1016/0370-2693(86)91377-8

62. T. Lagouri, Symmetry **14**, 1299 (2022). https://doi.org/10.3390/sym14071299
63. J. Beacham et al., J. Phys. G **47**, 010501 (2020). https://doi.org/10.1088/1361-6471/ab4cd2 [arxiv:1901.09966]
64. C. Antel et al., Eur. Phys. J. C **83**, 1122 (2023). https://doi.org/10.1140/epjc/s10052-023-12168-5 [arxiv:2305.01715]
65. FASER collaboration, Phys. Lett. B **848**, 138378 (2024). https://doi.org/10.1016/j.physletb.2023.138378 [arxiv:2308.05587]
66. B. Patt, F. Wilczek, *Higgs-Field Portal into Hidden Sectors* [arxiv:hep-ph/0605188]
67. R.N. Mohapatra, G. Senjanovic, Phys. Rev. Lett. **44**, 912 (1980). https://doi.org/10.1103/PhysRevLett.44.912D
68. T. Yanagida, https://doi.org/10.1143/PTP.64.1103Prog. Theor. Phys. **64**, 1103 (1980)
69. J. Schechter, J.W.F. Valle, Phys. Rev. D **22**, 2227 (1980). https://doi.org/10.1103/PhysRevD.22.2227
70. F. Kling, S. Trojanowski, Phys. Rev. D **97**, 095016 (2018). https://doi.org/10.1103/PhysRevD.97.095016 [arxiv:1801.08947]
71. M. Hirsch, Z.S. Wang, Phys. Rev. D **101**, 055034 (2020). https://doi.org/10.1103/PhysRevD.101.055034 [arxiv:2001.04750]
72. D. Dercks, H.K. Dreiner, M. Hirsch, Z.S. Wang, Phys. Rev. D **99**, 055020 (2019). https://doi.org/10.1103/PhysRevD.99.055020 [arxiv:1811.01995]
73. C. Arbeláez, J.C. Helo, M. Hirsch, Phys. Rev. D **100**, 055001 (2019). https://doi.org/10.1103/PhysRevD.100.055001 [arxiv:1906.03030]
74. G. Cottin, J.C. Helo, M. Hirsch, A. Titov, Z.S. Wang, JHEP **09**, 039 (2021). https://doi.org/10.1007/JHEP09(2021)039 [arxiv:2105.13851]
75. R. Beltrán, G. Cottin, M. Hirsch, A. Titov, Z.S. Wang, JHEP **05**, 031 (2023). https://doi.org/10.1007/JHEP05(2023)031[arxiv:2302.03216]
76. F. Deppisch, S. Kulkarni, W. Liu, Phys. Rev. D **100**, 035005 (2019). https://doi.org/10.1103/PhysRevD.100.035005 [arxiv:1905.11889]
77. F.F. Deppisch, S. Kulkarni, W. Liu, *Sterile Neutrinos at MAPP in the B-L Model* [arxiv:2311.01719]
78. M. Drewes, Int. J. Mod. Phys. E **22**, 1330019 (2013). https://doi.org/10.1142/S0218301313300191 [arxiv:1303.6912]
79. A. Kusenko, Phys. Rept. **481**, 1 (2009). https://doi.org/10.1016/j.physrep.2009.07.004 [arxiv:0906.2968]
80. M. Drewes et al., JCAP **01**, 025 (2017). https://doi.org/10.1088/1475-7516/2017/01/025 [arxiv:1602.04816]
81. F.F. Deppisch, P.S. Bhupal Dev, A. Pilaftsis, New J. Phys. **17**, 075019 (2015). https://doi.org/10.1088/1367-2630/17/7/075019 [arxiv:1502.06541]
82. G. Cottin, J.C. Helo, M. Hirsch, Phys. Rev. D **98**, 035012 (2018). https://doi.org/10.1103/PhysRevD.98.035012 [arxiv:1806.05191]
83. G. Zhou, J.Y. Günther, Z.S. Wang, J. de Vries, H.K. Dreiner, JHEP **04**, 057 (2022). https://doi.org/10.1007/JHEP04(2022)057 [arxiv:2111.04403]
84. J. De Vries, H.K. Dreiner, J.Y. Günther, Z.S. Wang, G. Zhou, JHEP **03**, 148 (2021). https://doi.org/10.1007/JHEP03(2021)148[arxiv:2010.07305]
85. H.K. Dreiner, J.Y. Günther, Z.S. Wang, Phys. Rev. D **103**, 075013 (2021). https://doi.org/10.1103/PhysRevD.103.075013 [arxiv:2008.07539]
86. I.G. Irastorza, J. Redondo, Prog. Part. Nucl. Phys. **102**, 89 (2018). https://doi.org/10.1016/j.ppnp.2018.05.003 [arxiv:1801.08127]
87. S. Alekhin et al., Rept. Prog. Phys. **79**, 124201 (2016). https://doi.org/10.1088/0034-4885/79/12/124201 [arxiv:1504.04855]
88. J.L. Feng, I. Galon, F. Kling, S. Trojanowski, Phys. Rev. D **98**, 055021 (2018). https://doi.org/10.1103/PhysRevD.98.055021 [arxiv:1806.02348]
89. D. Curtin et al., Rept. Prog. Phys. **82**, 116201 (2019). https://doi.org/10.1088/1361-6633/ab28d6 [arxiv:1806.07396]

90. D. Dercks, J. De Vries, H.K. Dreiner, Z.S. Wang, Phys. Rev. D **99**, 055039 (2019). https://doi.org/10.1103/PhysRevD.99.055039 [arxiv:1810.03617]
91. H.K. Dreiner, D. Köhler, S. Nangia, Z.S. Wang, JHEP **02**, 120 (2023). https://doi.org/10.1007/JHEP02(2023)120 [arxiv:2207.05100]
92. H.K. Dreiner, D. Köhler, S. Nangia, M. Schürmann, Z.S. Wang, JHEP **08**, 058 (2023). https://doi.org/10.1007/JHEP08(2023)058 [arxiv:2306.14700]

The microNet (μNet) Project: An Extended Outreach Program for Astroparticle Physics

Leonidas Xiros, Apostolos Tsirigotis, Stergios Kyanidis, and Antonios Leisos

Abstract The μNet project is an extended educational program aiming for the deployment and operation of an extensive school network of educational cosmic-ray telescopes throughout Greece. In the first year of operation (school term 2022–2023), 20 telescopes were installed in the region of Peloponnese, while more than 50 schools and 500 students from all over Greece attended remotely the extended educational program of μNet. The experimental devices, the educational activities as well as the developed software tools are briefly covered in this report, along with the main findings and outcomes of the first year of operation of the network.

Keywords Cosmic ray telescopes · Educational outreach · Astroparticle physics

1 Introduction

Science educators believe that science education is much more effective when it is done using hands-on experience, than when using book/classroom learning [1, 2]. Astronomy and Astrophysics are fields that have always fascinated not only students but also the general public. However, the incorporation of these research fields into the school curriculum is not an easy task for various reasons: pedagogical methods, complexity and cost of the equipment, as well as the use of advanced scientific methodologies which the high school teachers are not familiar with.

The Physics Laboratory of the Hellenic Open University (HOU) initiated the μNet project [7], as an attempt to engage Greek high school students with the experimental methodology of particle and astroparticle physics. In this framework, students build, test and operate their own telescope in order to observe high energy cosmic rays.

In this work we present the latest developments of the μNet project and some results of the second phase of the project that took place during the 2022–2023 school year, engaging 500 students in 60 schools over Greece. The paper is organized as

L. Xiros (✉) · A. Tsirigotis · S. Kyanidis · A. Leisos
Physics Laboratory, School of Science and Technology, Hellenic Open University, Perivola Patras, Greece
e-mail: xiros.leonidas@ac.eap.gr

J. Strologas and L. Perivolaropoulos (eds.), *HEP2023*, Springer Proceedings in Physics 333 https://doi.org/10.1007/978-3-031-76856-9_14

follows: In the rest of the present section, we present some general principles of the construction and operation of an educational cosmic-ray telescope. In Sect. 2, the design of the μCosmics telescope is presented, while in Sect. 3 we discuss the main aspects of the μNet project. In Sect. 4 we discuss the main results of the pilot runs, as well as the first year of full deployment and operation, while in Sect. 5 we summarize our work and discuss our future plans.

1.1 Educational Cosmic-Ray Telescopes

An educational cosmic ray telescope like those constructed and operated in the framework of the μNet project, consists of three or four particle detectors (almost always scintillator detectors), that are usually placed on the roof of a (high school) building. The telescope detects extensive air showers by detecting mainly the secondary particles of the shower that reach the detector level [6]. The telescopes operate under the students' supervision, while GPS timing can be used to combine data from different schools and form larger telescope networks. The telescope operates by measuring the time that a particle shower front passes through the different detection units, separated by adequate distance; assuming that the particle front is planar, one can calculate the shower axis direction with an accuracy of a few degrees [3–5].

The project of a Greek educational cosmic-ray telescope was started by the Hellenic Open University [7]. By 2014, twelve large ($\approx$1 m^2) particle detectors (Scintillation Detector Moules—SDM) were constructed and tested at the HOU physics laboratory. Even though the SDM proved to be very efficient for cosmic-ray research [8], their large dimensions, big mass and large cost made them inappropriate for use in an educational cosmic-ray telescope. For that reason the Physics Laboratory of HOU constructed the μCosmics detector which is presented in the next paragraph.

2 μCosmics Detector

For the development of a Greek network of educational cosmic-ray telescopes, the Physics Laboratory of HOU first designed and developed a small, cheap and portable telescope called the μCosmics detector. Its main feature is that it can be assembled by students and operated within the classroom or school laboratory (see Figs. 1, 2).

The HOU educational telescope consists of three detection units, a PC-based digital oscilloscope and a computer for controlling its operation. The active area of the detector unit consists of three rows of plastic scintillator tiles, with dimensions of 10×12 cm^2. Each tile includes a groove where an optical fiber is attached in order to collect the scintillation light, with 12 tiles making up one row.

Hence, a total of 36 optical fibers monitor each detection unit and their edges are coupled to a Silicon Photomultiplier tube in order to convert the collected light into an electrical signal. The SiPMT is supplied with a voltage of less than 40 V. In order

Fig. 1 The μCosmics detector in "compact" form. The three scintillation detector units, the DAQ electronics board (under the screen on the right), and the control and communication computer are shown

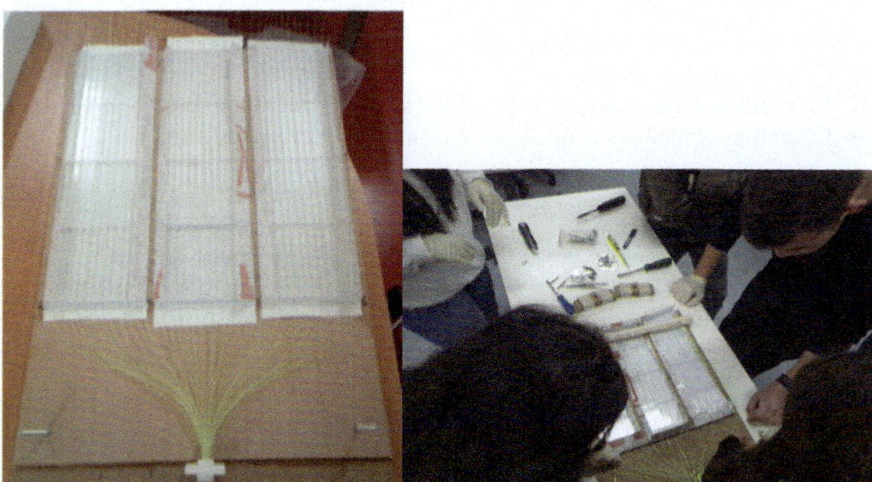

Fig. 2 The μCosmics detector components on the left; Students assembling the detector units on the right

to reduce light losses, we wrap the tiles with reflective paper (Tyvek). The whole construction is easy to assemble and disassemble. The final telescope at each site consists of combinations of detector units, as discussed in the next section.

The final assembled detector unit is very small, easy to carry, weighs about 6 kg and it is much cheaper than the large SDM. The Data Acquisition System consists of a digital PC-based oscilloscope that acquires pulses with a sampling rate of 250 MHz.

Simulation studies and experimental data show that the resolution of the μCosmics telescope is about 5° while the recording rate is about 10 showers per hour, which is sufficient even for the short duration of a high school class period [10].

3 The μNet Project

The μNet project aims at the active involvement of high school students in modern experimental methods in astroparticle physics and especially in cosmic-ray physics. In the framework of μNet, high school laboratories are equipped with cosmic-ray telescopes, while remote operated detectors deployed at the HOU campus are also available to high school students for distant experimentation [2].

In the framework of the μNet project, 20 educational cosmic-ray telescopes have been installed at high school laboratories in the area of Peloponnese, while remote operated devices have been deployed at the HOU campus that are available to schools for educational activities by distance. The schools equipped with μCosmics detectors as well as the schools participating in distant education activities, constitute the μNet network, the 1st Greek school network of educational cosmic-ray telescopes.

Specifically, the μNet educational program includes a series of educational activities that are carried out by students under the guidance of their teachers.

Specifically, there are 7 main educational activities designed:

(a) Construction of the detection unit
(b) Calibration of the telescope
(c) Estimation of the characteristics of the atmospheric muon flux.
(d) Operation of the data acquisition system and online monitoring
(e) Detector geometry studies
(f) Telescope operation.

When the collection of data is over (at the end of the school year) the students estimate the detection rate as well as the arrival distribution of air showers and compare with other measurements from other schools.

4 Pilot Runs—First Year of Full Operation

The μNet project has been evolving continuously over the last years, despite some setbacks during the pandemic years. In 2018 and 2019 two summer schools for high school students were organized at the HOU Physics Laboratory [10]. In 2020 a small network consisting of five stations was installed and run in the prefecture of Achaia (Peloponnese), but due to pandemic restrictions at the time, that pilot run was carried out completely remotely.

In the academic year 2021–22 an extensive pilot was run with the participation of 15 schools and 150 students from all over Greece. Twelve schools operated the telescopes remotely while three of them operated the telescope in their school laboratory. More information about these pilot runs has been presented in Ref. [5].

For the remote users 16 detection units were arranged at the Physics Laboratory as shown in Fig. 3. Four stacks of detectors were placed at the tips of a square of side 8–10 m. In each stack there were 4 detectors one on top of each other. That way 16

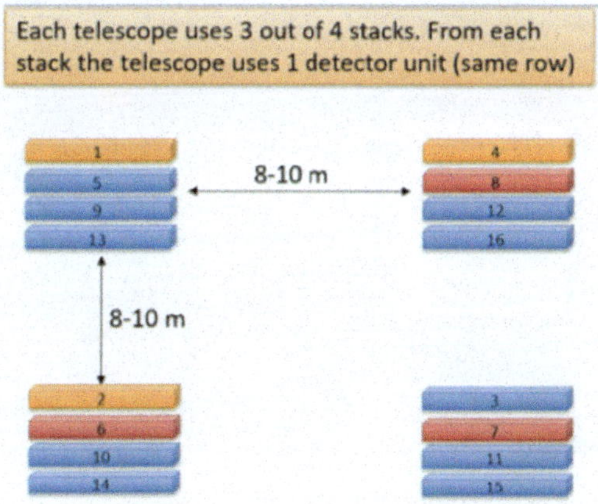

Telescope	Detector Units
1	2,3,4
2	1,3,4
3	1,2,4
4	1,2,3
5	6,7,8
6	5,7,8
7	5,6,8
8	5,6,7
9	10,11,12
10	9,11,12
11	9,10,12
12	9,10,11
13	14,15,16
14	13,15,16
15	13,14,16
16	13,14,15

Fig. 3 The detector units operated in the HOU campus. 16 different combinations of detector units triplets can be selected to constitute 16 different telescopes

different telescopes could be formed following 4 different geometries, for example, telescope 1 consisted of detectors 2, 3 and 4 which is actually the same with telescope 5 which consists of detectors 6, 7 and 8. The signals of the 16 detection units were acquired by 4 digital oscilloscopes.

The latest academic year (2022–23) was the first year of full operation of the μNet project with 60 participating schools. Twenty school units from Peloponnese were hosting a μCosmic detector in their laboratory, while 40 school units from the rest of Greece operated telescopes located at the Physics Laboratory of HOU.

For that reason, eight detection units were deployed at the HOU campus, offering 56 different μCosmics telescopes (3 out of 8 combinations). The 56 telescopes were available for high school students that participated to μNet as remote users, by assigning one telescope for each school. Each combination had different geometry offering the opportunity to study the effect of the geometry on shower reconstruction rate and angular resolution.

The architecture of the Data Acquisition System of the detector array is shown in Fig. 4: The signals from the 8 detectors are split in order to form a coincidence logic. When three out of 8 signals are within a narrow time window (around 200 ns), a trigger is formed and the 8 signals are digitized by three four-channel oscilloscopes (the 4th channel of the oscilloscope is used for the trigger signal).

The main aspect of the educational program is that the school students construct and operate the detection units by themselves. Students from various schools worked to construct the detection unit, getting a more in-depth understanding of its principle of operation. The operation of the stations is performed online using a web-based application. Participating students entered the application and either operated on their local station or selected the remote telescope (located in HOU) to operate on. The tasks of data operation included the calibration of the units as well as the extraction of the shower characteristics from the collected data (Figs. 5 and 6).

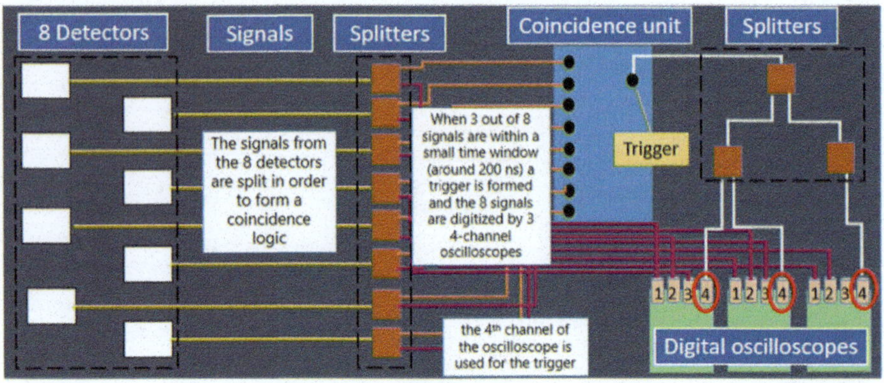

Fig. 4 The architecture of the data acquisition system of the detector array

Fig. 5 The μNet online web application. The operation of the stations is performed online using a web-based application at https://mnet-online.eap.gr/

In the calibration setup, three detection units are arranged on top of each other in order to be penetrated by the same atmospheric muon. The mean pulse height and the pulse height distribution of each detector unit are calculated (to determine the optimum pulse voltage threshold), while the time differences of the pulses from any two pairs of detectors are used for the time synchronization of the detectors. The web application during a timing calibration run is shown in Fig. 7.

After the calibration procedure, the detection units are ready to be used for the detection of air showers. During data taking each school monitored the detection and reconstruction rate of showers, the number of acquired pulses for each detected shower, along with some other characteristics, i.e., timing and peak values of the pulses. Then, the reconstructed direction of the shower was also recorded, in terms of the zenith and azimuth angles and shown in a 3D animation.

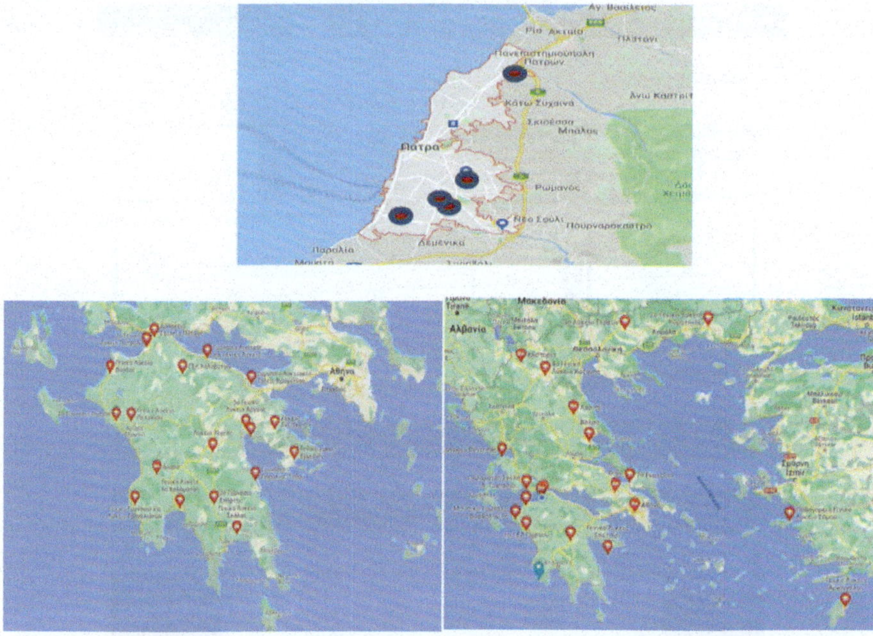

Fig. 6 The evolution of the μNet project from 2018 (top) to 2023 (bottom). In 2018, just 5 telescopes were operated by high schools at the prefecture where HOU is located (Achaia). In the first year of operation (2022–23), detectors were installed in 20 schools all over the Peloponnese, while more than 40 schools participated in remote data taking, calibration and data analysis

Finally, the online monitoring page was used to present histograms that summarized the data quality of the telescope. These histograms included information from previously recorded corresponding data, in order to compare directly the acquired data with data from previous periods of operation (reference period), as shown in Fig. 7.

A dedicated online educational platform was developed for the training of high school students and teachers. The entire educational program is organized using Moodle as a Learning Management System [11], where all the developed educational material and technological tools are placed (Fig. 8). The Moodle platform included "gamification" methods [12], assignments, fora, etc. in order to enhance students' attendance. The performance of the students is monitored through a series of activities that also include competitive "games", with point-collection and badges[13].

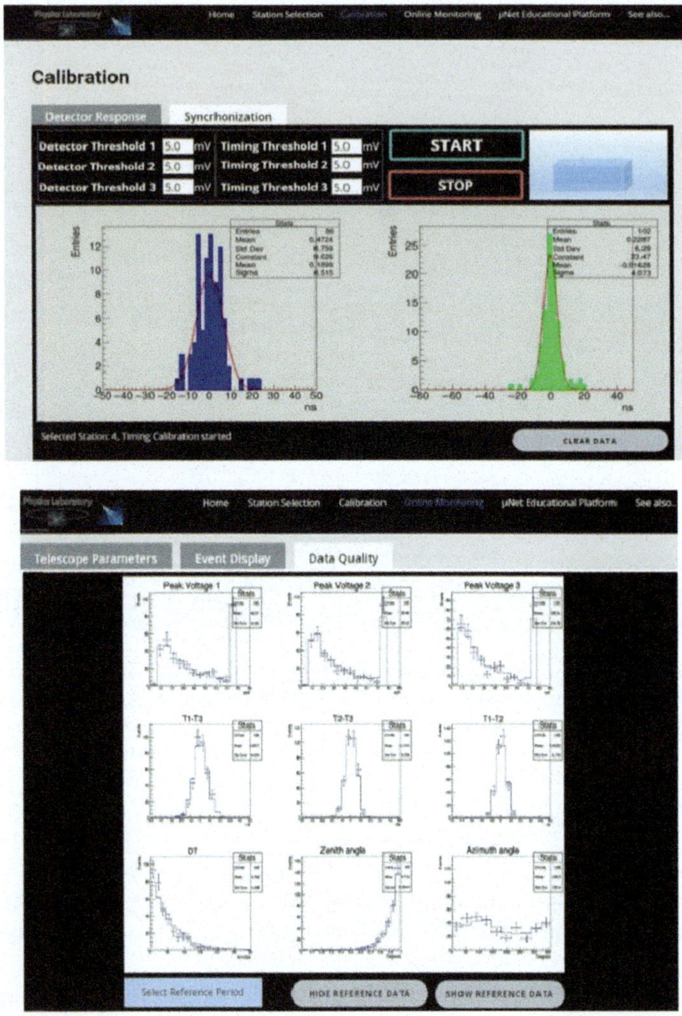

Fig. 7 The web application window during a timing calibration run (top). The web application window of a data quality analysis: Historic data from the same telescope appear as continuous lines, current data as points with errors (bottom)

5 Conclusions—Future Work

We applied the lessons we learned from summer schools and test runs between 2018 and 2021 during the first year of the μNet project in the 2022–23 school year. These lessons covered teaching students and teachers, making telescope parts efficiently, and operating telescopes both in situ as well as by distant. In this first year, 500 students from 60 schools across Greece took part. We collected videos, presentations, worksheets, and online games that we had created earlier, and organized them into

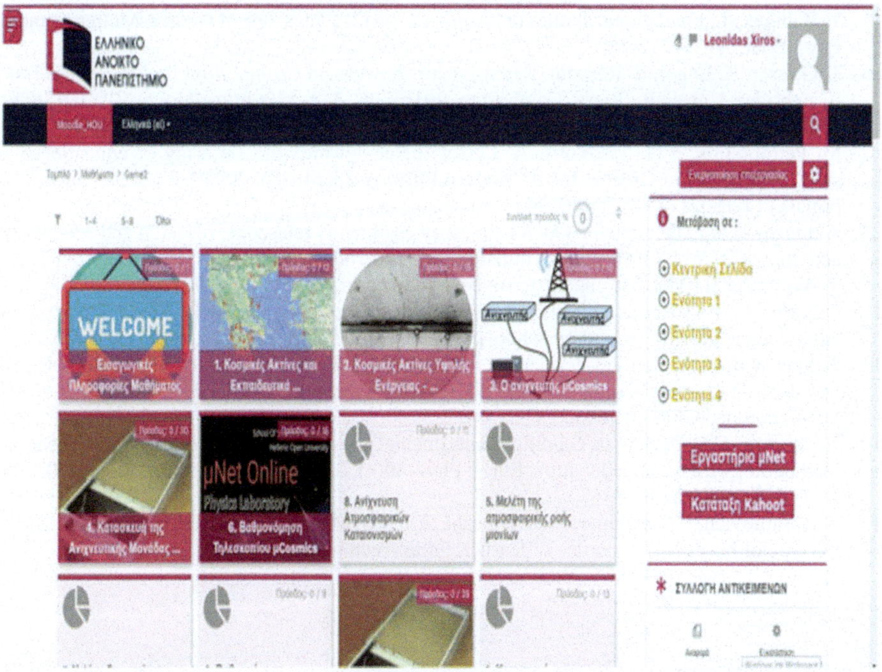

Fig. 8 The Moodle application developed for the training and monitoring of the operation of HOU educational cosmic-ray telescope network, μNet

an online class similar to Moodle. The tools we used for setting up and running the telescopes worked really well, even with lots of users.

Because our project is designed to be flexible, we can easily expand to more schools all over Greece. Looking ahead, our main focus will be on analyzing the data we've gathered. This includes how the project ran, how students and teachers behaved, and how different teaching methods, like using games, performed. We've automatically collected all this data from the Moodle web application.

References

1. I. Abrahams, M.J. Reiss, Practical work: its effectiveness in primary and secondary schools in England. J. Res. Sci. Teach. **49**(8), 1035–1055 (2012). https://doi.org/10.1002/tea.21036
2. D. Hodson, Re-thinking old ways: towards a more critical approach to practical work in school science. Stud. Sci. Educ. **22**(1), 85–142 (1993). https://doi.org/10.1080/03057269308560022
3. B. Bruant Gulejova and the IPPOG Collaboration, IPPOG Global cosmic rays portal: making cosmic rays studies available to schools worldwide. J. Phys.: Conf. Ser. **2156**, 012161 (2021)
4. F. Blanco et al., Cosmic ray measurements by scintillators with metal resistor semiconductor avalanche photo diodes. Phys. Educ. **43**, 536 (2008)

5. T. Wibig et al., Educational studies of cosmic rays with a telescope of Geiger–Müller counters. Phys. Educ. **41**, 542 (2006)
6. T. Gaisser, R. Engel, E. Resconi, Extensive air showers, in *Cosmic Rays and Particle Physics* (Cambridge University Press, Cambridge, 2016), pp. 313–340. https://doi.org/10.1017/CBO9781139192194.018
7. M. Petropoulos, A.G. Tsirigotis, A. Leisos, μNet: towards the 1st array of educational air shower detectors in Greece. Int. J. Modern Phys. A **25**(34n35), 2044022 (2020). https://doi.org/10.1142/S0217751X20440224
8. A. Leisos et al., The Hellenic Open University cosmic ray telescope: research and educational activities. EPJ Web Conf. (2018). https://doi.org/10.1051/epjconf/201818202072
9. T. Avgitas et al., The Astroneu extensive air shower array. JINST **15**, T03003
10. L. Xiros, A. G. Tsirigotis, A. Leisos, Using of distant education approaches to introduce cutting edge science into the secondary school classroom. Int. J. Inf. Educ. Technol. **13**(4) (2023)
11. T. Martín-Blas, A. Serrano-Fernández, The role of new technologies in the learning process: moodle as a teaching tool in physics. Comput. Educ. **52**(1), 35–44 (2009). https://doi.org/10.1016/j.compedu.2008.06.005
12. F. Forndran, C.R. Zacharias, Gamified experimental physics classes: a promising active learning methodology for higher education. Eur. J. Phys. **40**(4), 045702 (2019). https://doi.org/10.1088/1361-6404/ab215e
13. S. Abramovich, S.C. Schunn, R.M. Higashi, Are badges useful in education?: it depends upon the type of badge and expertise of learner. Educ. Technol. Res. Dev. **61**, 217–232

Simulations Studies for the RF Performance of the Astroneu II Array

Stavros Nonis, Antonios Leisos, Apostolos Tsirigotis, Kostantinos Papageorgiou, Ioannis Gkialas, and Spyros Tzamarias

Abstract Astroneu is an array for extensive air shower detection which is developed and operated at the Hellenic Open University campus near the city of Patras, Greece. In the first stage of operation (Astroneu I) it consisted of 9 scintillators detectors and 6 antennas, which were arranged in 3 stations. The Astroneu array is entering the next phase of operation, where 48 particle detectors and 16 RF antennas are about to be installed. In this work we estimate the efficiency of the new antenna array which depends on the topology as well as on various parameters of the air shower, such as the energy, direction of arrival and mass of the primary particle.

Keywords Cosmic rays · Astroneu I · Astroneu II · Extensive air showers · Radio detection

1 Introduction

Cosmic rays (CR) is the term regularly used to describe highly energetic atomic nuclei that bombard the Earth and are originating from sources beyond our planet. When these particles penetrate the Earth's atmosphere, they generate extensive air showers (EAS) [1] composed of secondary particles. A detailed investigation of these EAS offers valuable insights in the properties of the primary cosmic rays, such as their energy, incoming direction, and composition. As the air showers develop, the relativistic motion of secondary electrons and positrons leads to the production of

---

WWW home page:https://physicslab.eap.gr.

S. Nonis (✉) · A. Leisos · A. Tsirigotis
Physics Laboratory, School of Science and Technology, Hellenic Open University, Patras 26222, Greece
e-mail: snonis@eap.gr

K. Papageorgiou · I. Gkialas
School of Engineering, Department of Financial and Management Engineering, University of the Aegean, Chios 82100, Greece

S. Tzamarias
Department of Physics, Aristotle University of Thessaloniki, Thessaloniki 54124, Greece

J. Strologas and L. Perivolaropoulos (eds.), *HEP2023*, Springer Proceedings in Physics 333 https://doi.org/10.1007/978-3-031-76856-9_15

coherent electromagnetic pulses in the radio frequency (RF) range, typically from a few to hundreds of MHz [2]. Antennas have been able to detect this RF radiation, as evidenced by a series of experiments conducted over the past two decades [3–7]. With notable advancements in theoretical understanding and experimental techniques, the detection of RF signals is now recognized as an effective approach for identifying EAS and reconstructing important parameters of the primary cosmic rays (for a summary, please refer to [8, 9]).

During the development of an EAS, two physical processes play a significant role in the RF emission. As electrons and positrons traverse the geomagnetic field, they experience systematic deflection in opposite directions due to the influence of the Lorentz force. This force acts perpendicular to both the geomagnetic field and the direction of propagation of the EAS [10]. The electric field generated by the geomagnetic mechanism aligns with the direction of the Lorentz force and its amplitude is proportional to $\sin a$, where a represents the geomagnetic angle. Another contribution to the RF signal originates from an excess of electrons in the shower front caused by medium ionization and positron annihilation [11]. The corresponding electric field is polarized radially with respect to the shower axis. Unlike the geomagnetic contribution, the electric field associated with the charge-excess mechanism depends on the observer's location.

The Astroneu [12], a hybrid instrument for observing cosmic rays, is comprised of scintillator detector modules (SDM) and RF antennas. It is positioned within the campus of the Hellenic Open University (HOU) situated on the outskirts of Patras in Greece. Throughout the 8-year operation, studies have been conducted pertaining to the identification and reconstruction of EAS parameters utilizing both the particle and RF detectors. Despite the telescope's operation in an area characterized by elevated levels of electromagnetic interference, the RF signal emitted during the development of EAS was effectively discerned from the background noise through the application of noise filters, signal rising time and signal polarization [13]. Analyses concerning the timing and amplitude characteristics of the RF signals were carried out by combining RF and SDM data alongside simulation predictions [14]. Additionally, the response of the RF system was established by considering the antenna's effective length while the measured RF spectrum enabled the determination of the EAS axis direction [15]. Moreover, a technique involving the RF signal and Monte Carlo (MC) simulations was devised to determine the core of the EAS, while measurements of the electric field direction emitted during the expansion of EAS facilitated the computation of the Charge excess to Geomagnetic Ratio (CGR) [16].

In this study, we present the forthcoming phase of operation for the Astroneu II array, wherein 48 particle detectors and 16 RF antennas are scheduled for installation. Our focus lies in evaluating the efficiency of the upgraded antenna array, which depends on both the configuration of the system and diverse characteristics of the atmospheric shower, including the energy, the arrival direction, and the mass of the primary particle. In Sect. 2, we provide a brief report on the instrumental setup of Astroneu II array. Section 3 discusses the simulation sample generated for this study. The event selection algorithm implemented in the sample is presented in Sect. 4. Moving on to Sect. 5, we present the main results which include the reconstruction probability of the RF system for various EAS axis directions, distance to shower core, and energy bins, as well as the resolution in estimating the shower direction and core. Finally, Sect. 6 outlines the conclusions and ongoing work.

2 Instrumental Setup of Astroneu II Array

The Astroneu II array will comprise 16 stations, with distances between stations ranging from 80 to 100 m, as shown in Fig. 1a. A dense array will provide the necessary statistics to study the radio emission of low energy showers and develop techniques for determining the EAS parameters (direction, energy and primary mass) in areas with strong electromagnetic background. Each station will be equipped with 3 small SDMs, a large muon counter,[1] and 1 RF antenna. The new SDM unit is smaller compared to the one used in Astroneu and covers an area of $0.15\ m^2$ [18]. It consists of 24 scintillator tiles, and the light produced is transmitted to the photomultiplier (PMT) using 36 wavelength shifting optical fibers (Fig. 1b). Instead of the traditional photomultiplier tube, a silicon PMT has been introduced. The muon counter follows a similar design but has larger dimensions, covering an area of $1\ m^2$. To optimize its response in the frequency range of 30–80 MHz and at higher zenith angles (up to 75°), a new antenna has been developed with a suitable low noise amplifier (LNA).

The antenna is composed of two bow-tie-shaped dipoles oriented in the east-west and north-south directions (Fig. 1c), which are connected to the LNA (Fig. 1d). Figure 2 illustrates the antenna gain (in dB) in relation to the azimuth angle (0°-east, 90°-north) for different zenith angles (upper) and the antenna's effective length at the central frequency of 55 MHz as a function of the zenith and azimuth angles (bottom). In addition to the GPS timing information for each station, multiple stations can be operated in combination utilizing a trigger decision system realized with VME and NIM electronics (amplifier, coincidence and digitization units, as shown in Fig. 1e).

3 Simulation Sample

The simulation sample was generated using the Corsika simulation package (ver. 77000) [20] while the Coreas extension (ver. 1.4) [21] is employed for the electric field calculations in the antenna positions. The model SIBYLL 2.3c [22] was used for the hadronic interactions while GHEISHA 2002d [23] package was applied for electromagnetic interactions. The sample consists of 10^5 events, using protons or iron nuclei as primary particles, in the energy range 10^{17}–$5 \cdot 10^{18}$ eV. The primary energy distributions (Fig. 3, upper left) have been selected according to the flux arising from

[1] The muonic and electromagnetic components of air showers are sensitive to the mass of the primary cosmic particle. The combination of radio and muon detection increase the mass sensitivity of the array [17].

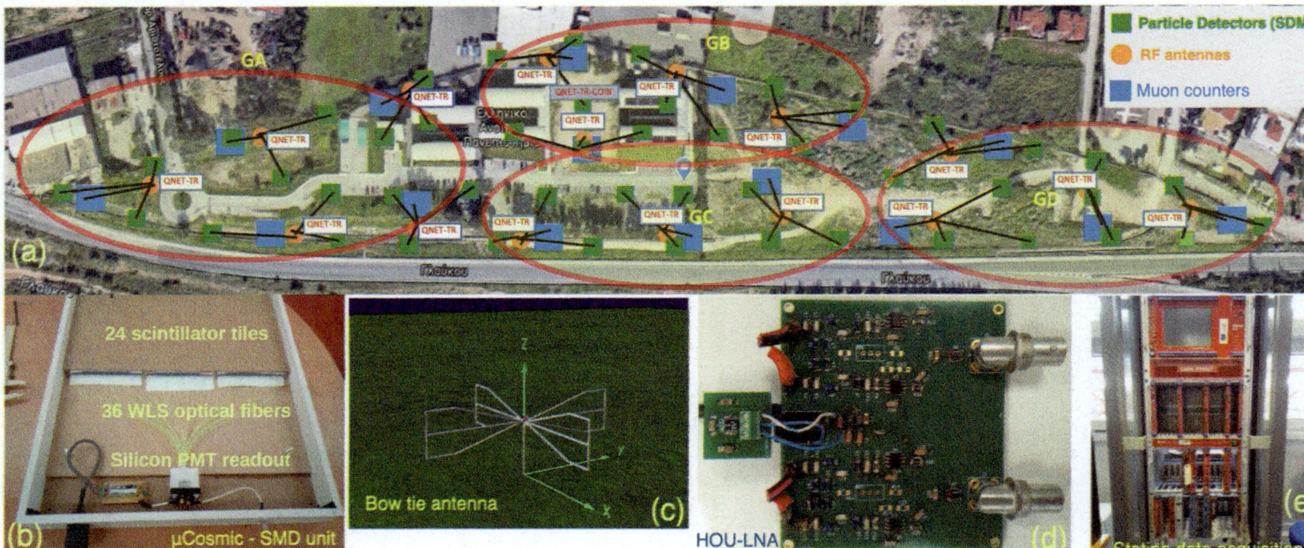

Fig. 1 (**a**) The 16 stations of the Astroneu II array at the site of the Hellenic Open University campus. The SDMs are indicated with green squares, the RF detectors (antennas) are indicated with orange circles while the muon counter with blue square. The SDM unit (**b**), the RF antenna (**c**), the LNA (**d**), and the DAQ system (**e**) of each station are displayed in a left-to-right sequence

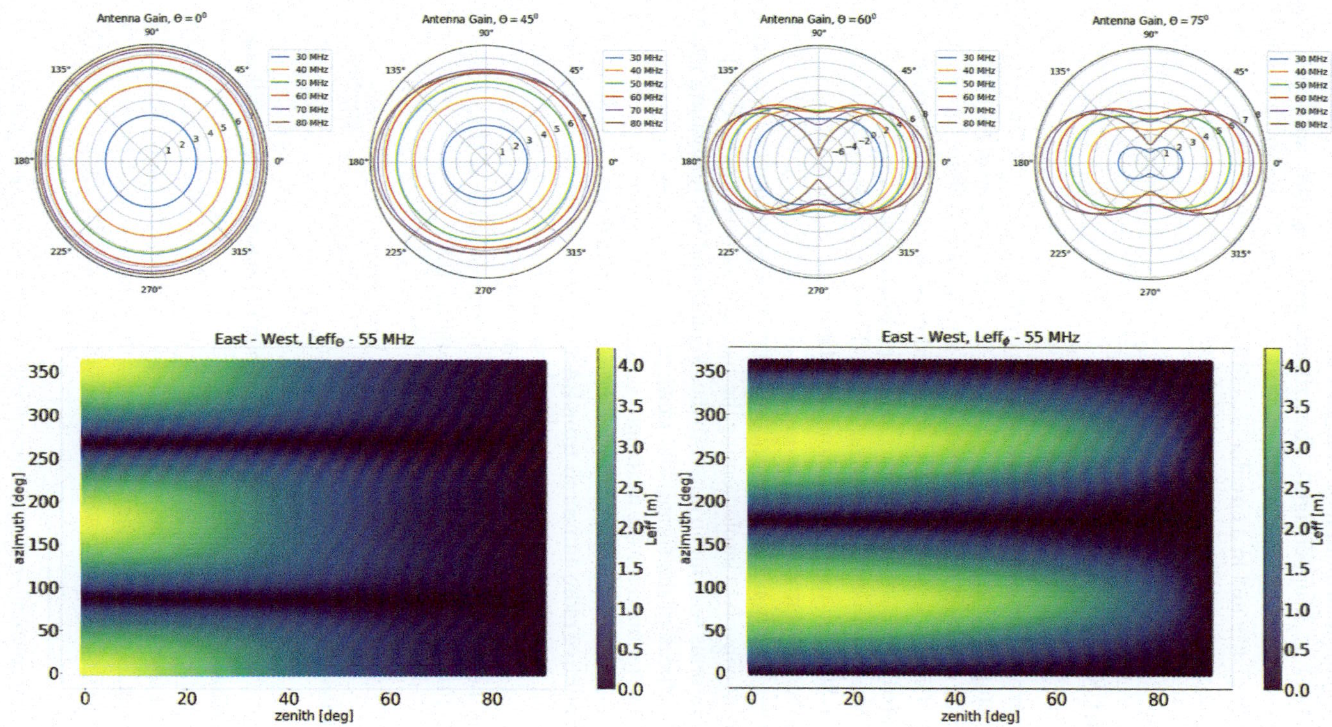

Fig. 2 (upper) The antenna's gain (in dB) as varies with the azimuth angle (0°-east, 90°-north) at different zenith angles. (bottom) The effective length of the antenna at frequency 55 MHz in relation to the zenith and azimuth angles. Both plots produced using NEC [19] antenna simulation package

cosmic rays spectrum in this energy range [24]. The zenith angle (θ) varies from 0 to 75° following the distribution obtained for flat detectors (proportional to $\sin\theta\cos\theta$) as depicted in Fig. 3 (upper right). The azimuth angles (ϕ) are uniformly distributed over a range from 0 to 360° (Fig. 3, bottom left). Finally the shower core position is uniformly distributed in a circular area of approximately 500 m radius around the center of the array covering an area of 1.1 km^2 (Fig. 3 bottom right).

4 Event Selection Algorithm

This section provides a brief overview of the event setup process and event selection algorithm employed in this study. The event setup process involves converting electric field data obtained from the simulations into induced voltages using the antenna response (effective length) while background noise measurements from Astroneu I are integrated in the voltage traces. Finally the filtering process, which uses a Tukey window filter and a Fast Fourier Transform (FFT), rejects signal's frequencies outside the 30–80 MHz range, optimizing the dataset for subsequent analysis. Figure 4 illustrates the signal traces both before (left) and after (right) the event setup process.

The event selection algorithm incorporates three subroutines for event filtering. The initial step involves implementing the signal-to-noise ratio (SNR) rejection algorithm. Waveforms with mean power in the "signal window" (extending 100 ns before and after the peak of the signal) six times greater than the mean power of the noise window were retained for the next analysis stage. Subsequently, a subroutine within the noise rejection algorithm examines the signal's rise time (R_T) values. Due to the transient nature of radio emissions from EAS, a short R_T (less than 28 ns) is expected for the signal. Consequently, events with a rise time R_T >28 ns were excluded. In the final stage of the noise rejection algorithm, the degree of polarization (d.o.p.)[2] estimation for the RF transient is performed. A threshold value of 0.85 is set, resulting in the rejection of events with a d.o.p. value below 0.85.

5 Results

Initially we estimate the reconstruction probability based on the zenith and azimuth angles. In each bin, the detection probability is defined as the ratio of the successfully reconstructed events to the total number of events in the bin. Our analysis defines a successful event reconstruction as meeting the following two criteria: a minimum of five antennas exhibiting signals that pass the event selection criteria (described in Sect. 4), along with the reconstruction of both the angular direction and the shower

[2] The degree of polarization (d.o.p) is the ratio of the polarized part of the electric field intensity to the total intensity and can be expressed with the Stokes parameters, according to the equation $d.o.p = \left(Q^2 + U^2 + V^2\right)^{1/2} / I = I_{pol}/I$.

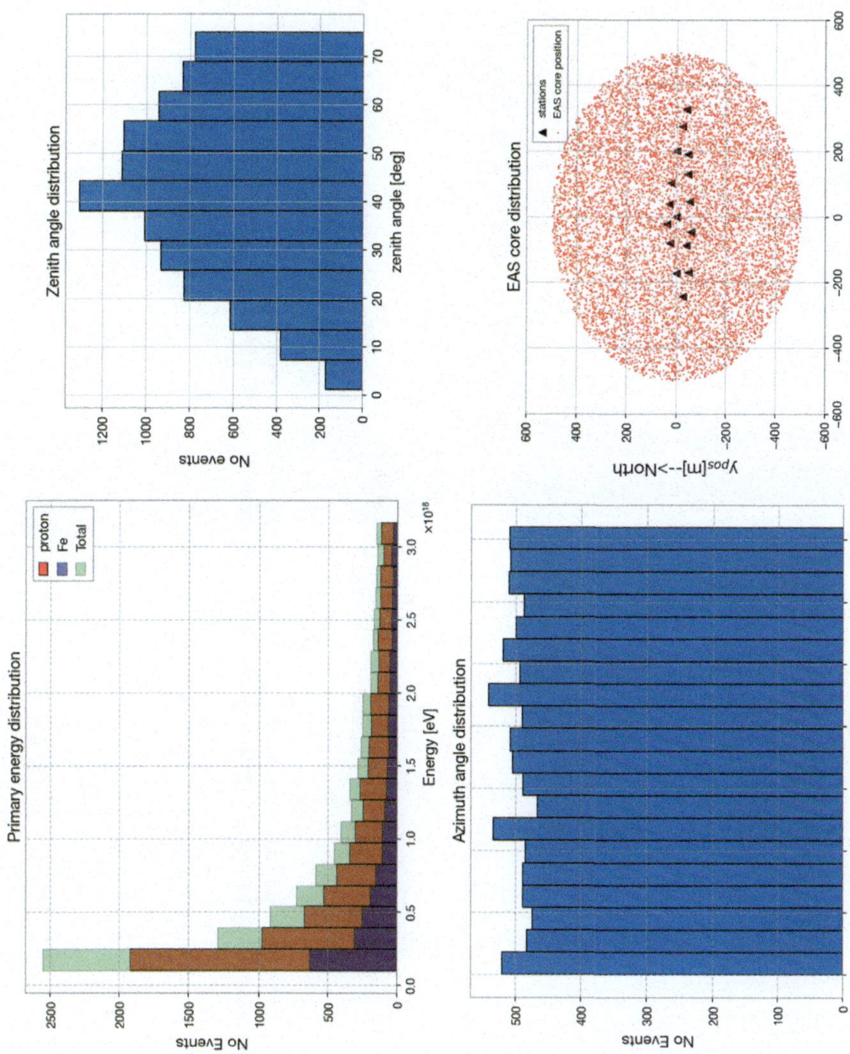

Fig. 3 (upper) The primary energy (left) and zenith angle (right) distributions for the simulation sample. (bottom) The azimuth angle distribution (left) and the EAS core distribution (right) of the simulation sample (the stations are denoted with black triangles)

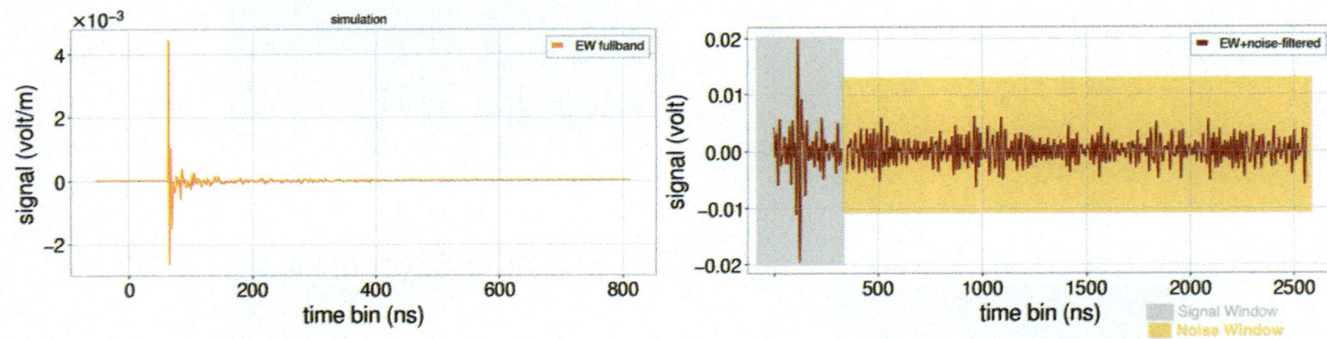

Fig. 4 (left) The RF signal traces (in Volts/m) as derived from the simulation for the EW polarization. (right) The signal traces (in Volts) after adding background noise and filtering. The noise (orange) and signal (grey) window, used in SNR rejection algorithm is clearly visible in the plot

core position. The reconstruction probability as a function of the zenith and azimuth angles of the shower axis is illustrated in Fig. 5 (left). By examining the diagram, it becomes evident that the probability remains relatively constant and sufficiently high (50–55%) for zenith angles up to 50°. However, for azimuth angles falling within the range of 225–315°, the probability significantly decreases. Within this azimuth angle range, the $\sin\alpha$ (α is the geomagnetic angle) reaches lower values, leading to induced voltages in the antennas that are below the noise level. This is clearly illustrated in the diagram presented in Fig. 6, where the variation of $\sin\alpha$ (at the site of Astroneu II array) is represented as a function of the zenith and azimuth angles.

It is important to highlight that the discussed detection probability applies under the assumption of a 100% SDM trigger efficiency. The reconstruction probability concerning the primary energy and the distance to the shower core is shown in Fig. 5 (right). Consistent with expectations, the reconstruction probability is reduced as the distance from the core increases and the primary energy decreases.

It should be noted that in the above study the assumed environmental noise is the one that has been measured in Astroneu I and consequently the shower energy threshold for a single RF antenna is the same. However in Astroneu II there are 16 RF antennas which means that more events are reconstructed in much larger area. Therefore we don't expect to have a lower energy threshold but a substantial growth of reconstructed events in all energies.

The resolution in estimating the zenith or azimuth angle of the primary particle is defined as the square root of the variance of the difference between the true angle (as used in the input simulation file) of the EAS primary particle and the respective reconstructed angle of the EAS using the simulated events. To determine the direction of the shower axis (zenith and azimuth angles), we consider the positions of the RF detectors and the time at which the signal's envelope reaches its maximum value assuming a plane wavefront for the signal (details provided in reference [25]). In the upper left and right plots of Fig. 7 are depicted the distributions of the difference between the true and the estimated zenith ($\theta_{true} - \theta_{recon}$) and azimuth ($\phi_{true} - \phi_{recon}$) angles respectively. Both distributions are fitted using Gaussian functions, with mean values centered around zero. The zenith angle distribution has a standard deviation (sigma) of 1.4°, while the azimuth angle distribution has a standard deviation of 2.85°. Likewise, the resolution for estimating the X and Y coordinates of the shower core is determined by taking the square root of the variance between the true X or Y coordinate (as provided in the input simulation file) of the shower core and the corresponding reconstructed core of the EAS obtained from simulated events. The novel approach described in [16] is employed to reconstruct the shower core position by comparing the expected electric field map (as derived from simulations for the measured EAS direction) and the reconstructed field map employing data. The distribution of the difference between the true and reconstructed X core coordinate ($X_{true} - X_{recon}$) is shown in the bottom left plot of Fig. 7, while the right plot displays the corresponding distribution for the Y core coordinate ($Y_{true} - Y_{recon}$). Both distributions are modeled using Gaussian functions, with a mean centered around zero. The sigma values are 13.02 m for the X coordinate and 13.82 m for the Y coordinate.

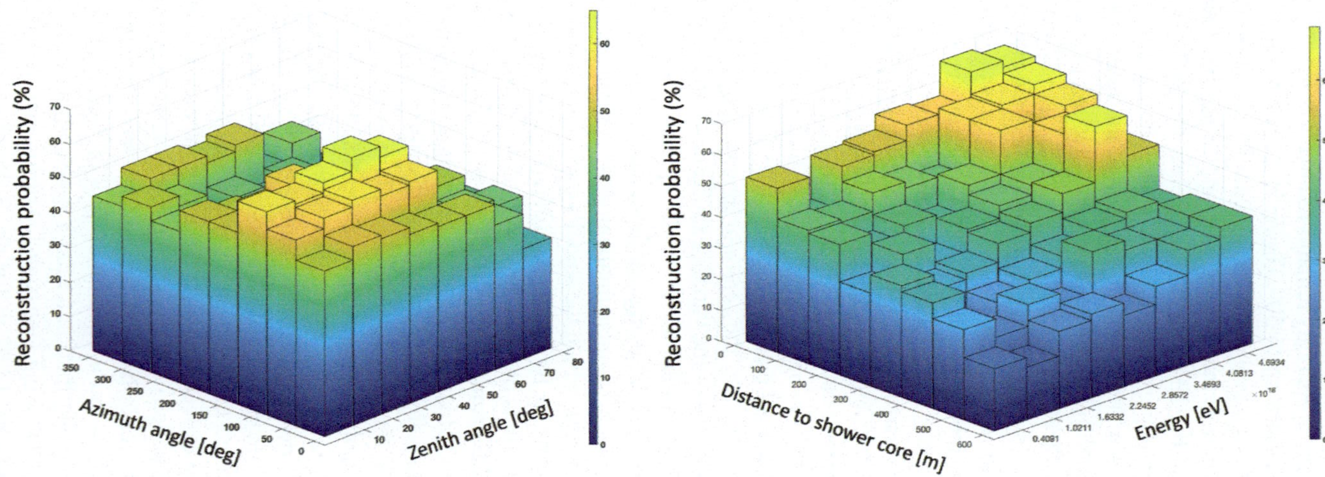

Fig. 5 (left) The probability of reconstruction is evaluated as a function of the zenith and azimuth angles. (right) The reconstruction probability as a function of the primary energy and the distance to the shower core

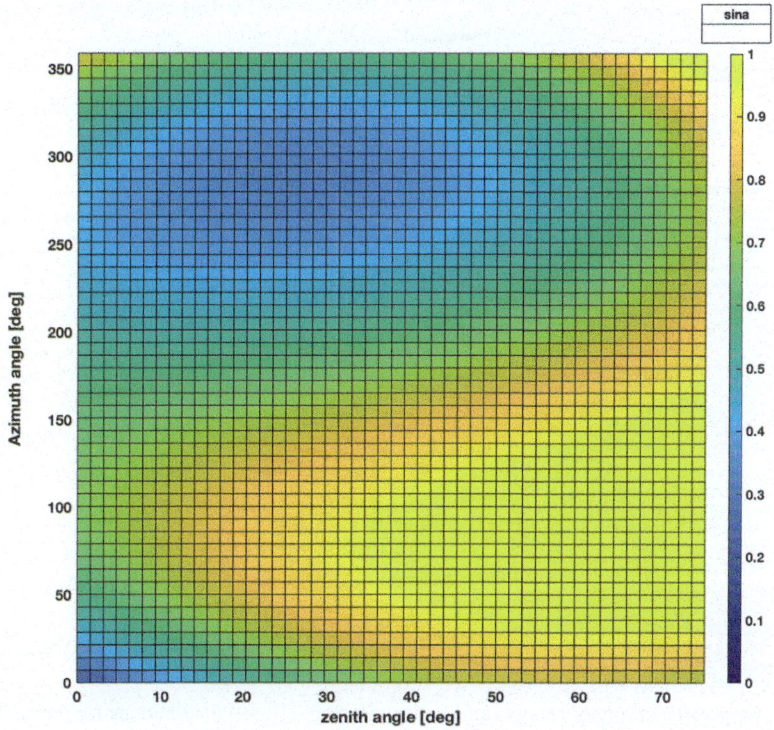

Fig. 6 The dependence of the geomagnetic angle (α), in the area of the Astroneu II array, on the zenith and azimuth angles of the shower

6 Summary-Conclusions

Details regarding the forthcoming expansion of the Astroneu array (Astroneu II), consisting of 16 stations, have been provided. Each station will consist of three particle detectors, one RF antenna, and one large muon counter. An ongoing simulation study is being conducted to evaluate the RF performance of the new Astroneu II array in estimating various shower parameters, including direction, core position, energy, and X_{max}.[3] The preliminary findings demonstrate a significant detection probability, taking into account the strong electromagnetic background of the region, across a wide range of shower directions. This includes inclined showers of up to 75°. According to simulation studies, Astroneu II is expected to encompass an active detection area of approximately 1 Km^2 around the center of the array. This active detection area will be effective in the energy range of 10^{17}–5×10^{18} eV. Additionally, it demonstrates a very good resolution in estimating both the shower direction and core, with the potential for further improvement by leveraging the particle detectors as well.

[3] The atmospheric depth at which the maximum number of secondary particles appears during the EAS development.

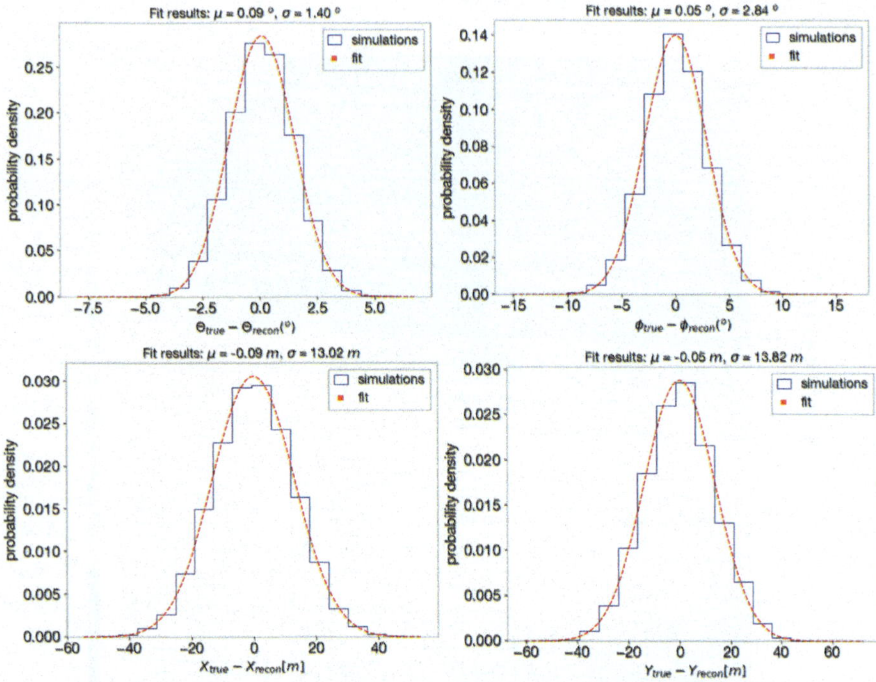

Fig. 7 (upper) The primary energy (left) and zenith angle (right) distributions for the simulation sample. (bottom) The azimuth angle distribution (left) and the EAS core distribution (right) of the simulation sample

Specifically for the direction resolution the combination of the RF signals with the independent measurements from the particle detectors is expected to improve the direction resolution by 30% (i.e. $\sigma_{combined} = \sigma_{RF}/\sqrt{2}$). On the other hand for the shower core estimation the improvement is expected to be lower since it mainly depends on the lever arm of the shower footprint. However, detailed studies are being contacted in order to quantify these improvements and also to determine the resolution in estimating the primary energy and X_{max} of the shower.

References

1. K.H. Kampert, A.A. Watson, Extensive air showers and ultra high-energy cosmic rays: a historical review. EPJ H **37**, 359–412 (2012). https://doi.org/10.1140/epjh/e2012-30013-x
2. J.V. Jelley, J.H. Fruin, N.A. Porter, T.C. Weekes, F.G. Smith, R.A. Porter, Radio pulses from extensive cosmic-ray air showers. Nature **205**, 327–328 (1965). https://doi.org/10.1038/205327a0
3. H. Falcke et al., LOPES collaboration, Detection and imaging of atmospheric radio flashes from cosmic ray air showers. Nature **435**, 313–316 (2005). https://doi.org/10.1038/nature03614

4. D. Ardouin et al., CODALEMA collaboration, Radio-detection signature of high-energy cosmic rays by the CODALEMA experiment. Nucl. Instrum. Methods A **555**, 148–163 (2005). https://doi.org/10.1016/j.nima.2005.08.096
5. P. Schellart et al., LOFAR collaboration, Detecting cosmic rays with the LOFAR radio telescope. A&A **560**, A98 (2013). https://doi.org/10.1051/0004-6361/201322683
6. F.G. Schröder, for the Pierre Auger Collaboration, Radio detection of high-energy cosmic rays with the Auger engineering radio array. Nucl. Instrum. Methods A **824**, 648–651 (2016). https://doi.org/10.1016/j.nima.2015.08.047
7. P.A. Bezyazeekov et al., Tunka-Rex collaboration, Measurement of cosmic-ray air showers with the Tunka radio extension (Tunka-Rex). Nucl. Instrum. Methods A **802**, 89–96 (2015). https://doi.org/10.1016/j.nima.2015.08.061
8. T. Huege, Radio detection of cosmic ray air showers in the digital era. Phys. Rep. **620**, 1–52 (2016). https://doi.org/10.1016/j.physrep.2016.02.001
9. F.G. Schröder, Radio detection of cosmic-ray air showers and high-energy neutrinos. Prog. Part. Nucl. Phys. **93**, 1–68 (2017). https://doi.org/10.1016/j.ppnp.2016.12.002
10. F.D. Kahn, I. Lerche, Radiation from cosmic ray air showers. R. Soc. Lond. Proc. Ser. A **289**, 206–213 (1966). https://doi.org/10.1098/rspa.1966.0007
11. G. Askaryan, Excess negative charge of an electron-photon shower and its coherent radio emission. Soviet Phys. JETP-USSR **14**, 441–443 (1962)
12. T. Avgitas, G. Bourlis, G.K. Fanourakis, I. Gkialas, A. Leisos, I. Manthos, A.G. Tsirigotis, S.E. Tzamarias, The Astroneu extensive air shower array. JINST **15**, T03003 (2020). https://doi.org/10.1088/1748-0221/15/03/T03003
13. I. Manthos, G. Bourlis, I. Gkialas, A. Leisos, A. Papaikonomou, A.G. Tsirigotis, S.E. Tzamarias, Cosmic ray RF detection with the Astroneu array. New Astron. **81**, 101443 (2020). https://doi.org/10.1016/j.newast.2020.101443
14. A. Leisos, S. Nonis, A.G. Tsirigotis, G. Bourlis, K. Papageorgiou, I. Gkialas, I. Manthos, S.E. Tzamarias, Hybrid detection of high energy showers in urban environments. Universe **5**(1), 3 (2019). https://doi.org/10.3390/universe5010003
15. S. Nonis, A. Leisos, A.G. Tsirigotis, G. Bourlis, K. Papageorgiou, I. Gkialas, I. Manthos, S.E. Tzamarias, Angular reconstruction of high energy air showers using the radio signal spectrum. Phys. Scr. **95**, 084007 (2020). https://doi.org/10.1088/1402-4896/ab9f79
16. S. Nonis, A. Leisos, A.G. Tsirigotis, G. Bourlis, K. Papageorgiou, I. Gkialas, I. Manthos, S.E. Tzamarias, Performance of the RF detectors of the Astroneu Array. Universe **9**(1), 17 (2023). https://doi.org/10.3390/universe9010017
17. E.M. Holt, F.G. Schröder, A. Haungs, Enhancing the cosmic-ray mass sensitivity of air-shower arrays by combining radio and muon detectors. Eur. Phys. J. C **79**, 371 (2019). https://doi.org/10.1140/epjc/s10052-019-6859-4
18. A.G. Tsirigotis, A. Leisos, S. Nonis, M. Petropoulos, G. Georgis, K. Papageorgiou, I. Gkialas, I. Manthos, S.E. Tzamarias, A low cost hybrid detection system of high energy air showers. Eng. Res. Express **2**, 025027 (2020). https://doi.org/10.1088/2631-8695/ab9126
19. G. Burke, A. Poggio, *Numerical Electromagnetics Code (NEC) Method of Moments, Parts I, II, III* (Lawrence Livermore National Laboratory Livermore, CA, USA, 1983), pp.89–223
20. D. Heck, J. Knapp, J. Capdevielle, G. Schatz, T. Thouw, Corsika: a Monte Carlo code to simulate extensive air showers. FZKA Report, 6019, Forschungszentrum Karlsruhe (1998)
21. T. Huege, M. Ludwig, C.W. James, Simulating radio emission from air showers with CoREAS. AIP Conf. Proc. **1535**, 128–132 (2013). https://doi.org/10.1063/1.4807534
22. R.S. Fletcher, T.K. Gaisser, L. Paolo, T. Stanev, Sibyll: an event generator for simulation of high energy cosmic ray cascades. Phys. Rev. D **50**, 5710 (1994). https://doi.org/10.1103/PhysRevD.50.5710
23. S.A. Bass et al., Microscopic models for ultrarelativistic heavy ion collisions. Prog. Part. Nucl. Phys. **41**, 225 (1998). https://doi.org/10.1016/S0146-6410(98)00058-1

24. T.K. Gaisser, T. Stanev, S. Tilav, Cosmic ray energy spectrum from measurements of air showers. Front. Phys. **8**, 748–758 (2013). https://doi.org/10.1007/s11467-013-0319-7
25. S. Nonis, A. Leisos, A.G. Tsirigotis, G. Bourlis, K. Papageorgiou, I. Gkialas, I. Manthos, S.E. Tzamarias, Extensive air shower reconstruction using the timing information from the RF-system of the Astroneu array. J. Phys.: Conf. Ser. **2105**, 012018 (2021). https://doi.org/10.1088/1742-6596/2105/1/012018

Ultraviolet Freeze-in Baryogenesis Scenarios

Ioannis Dalianis, Andreas Goudelis, Dimitrios Karamitros, Pantelis Papachristou, and Vassilis C. Spanos

Abstract In this talkwe will present various models that can explain simultaneously the production of dark matter and the observed baryon asymmetry of the Universe. In particular, we will discuss scenarios, where the relevant processes are due to higher-dimensional, non-renormalizable operators with dim = 5 or 6. These operators are effective close to the reheating temperatures. Moreover, we see that additional cosmic component has a significant impact on the predictions of concrete microscopic models, allowing for reheating temperatures which are much lower than those required in the simplest cosmological scenario.

Keywords Cosmology · Dark matter · Baryogenesis

1 Introduction

The nature of dark[1] matter (DM) and the origin of the baryon asymmetry of the Universe are still two of the most fundamental open questions in cosmology and particle physics. In [1] it was suggested that a common framework for the simultaneous explanation of both could be found within the context of freeze-in DM [2, 3]. In freeze-in scenarios, DM particles interact so weakly with the Standard Model (SM) thermal bath particles, that the relevant interactions never reach thermal equilibrium. This fact satisfies the third Sakharov condition [4] required for successful baryogenesis. This idea was exploited in [1] in the context of asymmetric frozen-in

[1] Talk given by V. C. Spanos.

I. Dalianis · P. Papachristou · V. C. Spanos (✉)
Section of Nuclear and Particle Physics, Department of Physics, National and Kapodistrian University of Athens, Athens, Greece
e-mail: vspanos@phys.uoa.gr

A. Goudelis
Laboratoire de Physique de Clermont (UMR 6533), CNRS/IN2P3, University of Clermont Auvergne, Aubière Cedex, France

D. Karamitros
School of Physics and Astronomy, The University of Manchester, Manchester, UK

J. Strologas and L. Perivolaropoulos (eds.), *HEP2023*, Springer Proceedings in Physics 333 https://doi.org/10.1007/978-3-031-76856-9_16

DM and, later, in [5, 6] in the context of symmetric DM candidates. In this talk we will discuss scenarios assuming that after the end of inflation, the energy density of the universe was dominated by a fluid which subsequently decays, resulting to a dilution of all abundances. We will see that indeed, the predictions of the ultraviolet freeze-in baryogenesis scenario presented in [8] is modified substantially [9] and that the mechanism can be efficient for a quite heavier dark matter mass and assuming a much lower reheating temperature.

The talk is organised as follows: In Sect. 2 we review the main features of symmetric DM freeze-in baryogenesis in its UV-dominated version and briefly introduce the models that will be studied in the following. In Sect. 3 we show how the predicted abundance is modified in the presence of a decaying fluid. Finally, in Sect. 4 we summarize our main findings and conclude.

2 Ultraviolet Freeze-in Baryogenesis

Here we will summarise the main features of the UV-dominated freeze-in baryogenesis mechanism, as presented in [8]. The starting point assumption is that the DM particle species only interacts feebly with the rest of thermal bath. This interaction is governed by the operator $\hat{\mathcal{O}}_{(n)}$ of mass dimension $(n+4)$

$$\mathcal{L} \supset \frac{1}{\Lambda^n} \hat{\mathcal{O}}_{(n)} \,, \tag{1}$$

where $n = 1, 2, \ldots$ and Λ is the energy scale of the effective field theory (EFT). This operator is responsible for the simultaneous generation of the freeze-in DM abundance and of the baryon asymmetry of the universe. Due to the non-renormalizable nature of the underlying interactions, both are mostly produced in the ultraviolet regime, i.e. close to the reheating temperature $T_{\rm RH}$.

In the models that we are discussing here, the DM is produced via pair-annihilations of bath particles. In the limit that Λ is much larger than the masses involved, the DM yield obtained by solving the corresponding Boltzmann equation can be written in a fairly compact form [10]

$$Y_{\rm DM}(T) \approx 2\,A\, \frac{4^{n+1} n!(n+1)!}{2n-1} \frac{45}{1024 \times 1.66\,\pi^7 g_{*s}\sqrt{g_{*\rho}}} \frac{M_{\rm Pl}\,(T_{\rm RH}^{2n-1} - T^{2n-1})}{\Lambda^{2n}} \,, \tag{2}$$

where A is a model-dependent constant, $M_{\rm Pl}$ is the reduced Planck mass and g_{*s}, $g_{*\rho}$ are the relativistic degrees of freedom of the plasma associated with the entropy and energy densities, respectively.

The motivation for studying the simultaneous generation of freeze-in DM and baryon asymmetry is that the relevant interactions are always out-of-equilibrium, so that the third Sakharov is satisfied. On the other hand, the CP symmetry is explicitly broken through the complex couplings appearing in the effective operators. The most non-trivial part of the proposed baryogenesis mechanism is the violation of the baryon number. In this model here the effective interactions conserve the lepton and baryon number, resulting to lepton asymmetries generated in the SM, $L_{\rm SM}$, and an exotic sector, $L_{\rm ex}$, while the total lepton asymmetry remains zero, $L = L_{\rm SM} + L_{\rm ex} = 0$. However, if the exotic particles are taken to be $SU(2)_{\rm L}$-singlets the electroweak sphalerons, which violate $B + L$ but conserve $B - L$, only affect the non-zero lepton asymmetry stored in the SM sector. Hence, they "see" a non-vanishing lepton asymmetry and convert it into a baryon one, which is proportional to $Y_B \propto Y_{B-L_{\rm SM}}$. Once they depart from equilibrium at $T = T_{\rm sph}$ the baryon and lepton asymmetries are separately conserved and the baryon asymmetry remains frozen at the value $Y_B \propto Y_{B-L_{\rm SM}}|_{T_{\rm sph}}$, which, in principle, is non-zero.

In order to present the corresponding predictions for the baryon asymmetry we will introduce two specific models: one based on a scalar DM candidate produced through a dimension-5 operator, and that of a fermionic candidate produced through a dimension-6 one.

2.1 Dimension-5 Operator

The first case study that we consider is an extension of the SM by a complex gauge-singlet scalar field φ, which is our DM candidate, along with two heavy vector-like Dirac fermions $F_{1,2}$, which are singlets under $SU(3)_c \times SU(2)_{\rm L}$ but carry hypercharge. The interaction Lagrangian reads

$$\mathcal{L}_{\rm int} = \mathcal{L}_{\rm gauge} + \frac{\lambda_1}{2\Lambda}\left(\bar{e}\,{\rm P_L}\,F_1\right)\varphi^*\varphi^* + \frac{\lambda_2}{2\Lambda}\left(\bar{e}\,{\rm P_L}\,F_2\right)\varphi^*\varphi^* + \frac{\kappa}{\Lambda^2}\left(\bar{e}\,{\rm P_L}\,F_1\right)\left(\bar{F}_2\,{\rm P_R}\,e\right) + \text{h.c.}\,, \tag{3}$$

where e can be any of the Standard Model charged leptons and $\mathcal{L}_{\rm gauge}$ contains the gauge interactions of the heavy fermions F_i. The latter keep the F_i in equilibrium with the SM bath particles. Besides, the heavy fermions carry the same lepton number as the SM leptons, and so total lepton number is conserved. A $\mathbb{Z}_3$ symmetry is imposed, under which both the F_i's and φ are charged, to have DM stability (Fig. 1).

The predicted DM abundance is given by Eq. (2) upon substituting $n = 1$. The asymmetries generated in the F and Standard Model sectors, on the other hand, are described by a set of three differential equations

$$-sHT\frac{{\rm d}Y_{\Delta F_i}}{{\rm d}T} = -\left[F_i\bar{e} \leftrightarrow \varphi\varphi\right] - \left[F_i\bar{e} \leftrightarrow F_j\bar{e}\right] + (-1)^i\left[F_i\bar{F}_j \leftrightarrow e\bar{e}\right], \tag{4a}$$

$$-sHT\frac{{\rm d}Y_{B-L_{\rm SM}}}{{\rm d}T} = -\sum_i\left[F_i\bar{e} \leftrightarrow \varphi\varphi\right], \tag{4b}$$

where $Y_{\Delta F_i} \equiv Y_{F_i} - Y_{\bar{F}_i}$ and we have introduced the shorthand notations

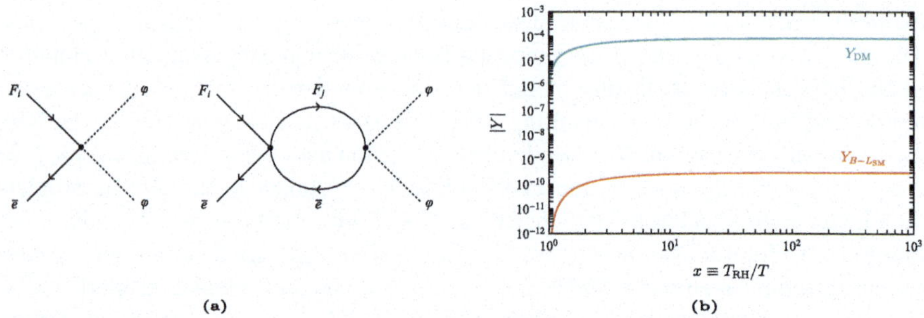

Fig. 1 **a** Tree-level and 1-loop Feynman diagrams of the dominant scattering processes that contribute to the generation of the DM abundance and the baryon asymmetry. **b** The yields $Y_{B-L_{\rm SM}}$ and $Y_{\rm DM}$ as functions of the dimensionless parameter $x \equiv T_{\rm RH}/T$ for a benchmark set of parameter values that reproduce the correct DM relic density and baryon asymmetry

$$[a\,b \leftrightarrow c\,d] \equiv (a\,b \leftrightarrow c\,d) - (\bar{a}\,\bar{b} \leftrightarrow \bar{c}\,\bar{d})\,, \tag{5a}$$

$$(a\,b \leftrightarrow c\,d) \equiv \frac{1}{S_{\rm in}S_{\rm out}} \int d\Pi_a\, d\Pi_b\, d\Pi_c\, d\Pi_d\; (2\pi)^4\, \delta^{(4)} \Big[|\mathcal{M}|^2_{ab\to cd}\, f_a\, f_b\, (1 \pm f_c)\,(1 \pm f_d) - |\mathcal{M}|^2_{cd\to ab}\, f_c\, f_d\, (1 \pm f_a)\,(1 \pm f_b) \Big]\,, \tag{5b}$$

where $\delta^{(4)}$ denotes the $\delta^{(4)}\,(p_a + p_b - p_c - p_d)$ and $S_{\rm in}$, $S_{\rm out}$ are the symmetry factors for the incoming and outgoing states respectively, which are equal to $N!$ in case of identical particles and 1 otherwise. The solution of these equations allows us to compute the predicted baryon asymmetry.

As shown in [8], Eq. (4) result in $Y_{\rm B-L_{SM}} \sim T_{\rm RH}^4/\Lambda^6$. From Eq. (2), on the other hand, setting $n = 1$ one finds that the DM yield scales as $Y_{\rm DM} \sim T_{\rm RH}/\Lambda^2$. This difference in the scaling behaviour between $Y_{\rm B-L_{SM}}$ and $Y_{\rm DM}$ is of critical importance for what follows.

2.2 *Dimension-6 Operator*

The dimension-6 case is quite similar to the dimension-5. In this case, the SM is extended by a pair of vector-like fermions F_i which are charged under $U(1)_Y$, along with a Dirac fermion χ which acts as a DM candidate pair-created through a dimension-6 operator. The interaction Lagrangian that we consider is

$$\mathcal{L} \supset \frac{\lambda_1}{2\Lambda^2}\,(\bar{e}\,{\rm P_L}\,F_1)\,(\bar{\chi}\,{\rm P_R}\,\chi^c) + \frac{\lambda_2}{2\Lambda^2}\,(\bar{e}\,{\rm P_L}\,F_2)\,(\bar{\chi}\,{\rm P_R}\,\chi^c) + \frac{\kappa}{\Lambda^2}\,(\bar{e}\,{\rm P_L}\,F_1)\,(\bar{F}_2\,{\rm P_R}\,e) + \text{h.c.}\,, \tag{6}$$

where for simplicity we ommited the F_i gauge interaction terms.

The DM abundance can, again, be computed through Eq. (2) by substituting $n = 2$, whereas the evolution of the sector-wise lepton asymmetries is described by a system of equations which is identical with the one governing the dimension-5 case, with the substitution $\varphi \leftrightarrow \chi$.

In [8] it was found that the yields of $B - L_{SM}$ and DM scale as $Y_{B-L_{SM}} \sim T_{RH}^8/\Lambda^{10}$ and $Y_{DM} \sim T_{RH}^3/\Lambda^4$, respectively. The viable parameter ranges are found to be $T_{RH} > 3 \times 10^{13}$ GeV, $\Lambda > 3 \times 10^{14}$ GeV and $4.6 < m_\chi < 18.1$ keV, i.e. it is possible to satisfy all DM and baryogenesis constraints with relatively lower values of the reheating temperature compared to the dimension-5 case.

3 Ultraviolet Freeze-in Baryogenesis in the Presence of a Decaying Fluid

From the previous discussion it becomes clear that the mechanism of UV freeze-in baryogenesis is viable if the DM mass is small and the reheating temperature extremely large. In the inflationary framework, thermalization occurs approximately when $\Gamma_{inf} \sim 3H$, where Γ_{inf} is the decay rate of the inflation, and the corresponding reheating temperature of the universe is

$$T_{RH} = \left(\frac{\pi^2}{10} g_{*RH}\right)^{-1/4} \sqrt{\Gamma_{inf} M_{Pl}}\,. \tag{7}$$

Let us call T_{max} the maximum possible reheating temperature of the universe after the decay of the inflaton field. We can write $T_{RH} = T_{max} \exp[-\frac{3}{4}(1 + \bar{w}_{RH})\tilde{N}_{RH}]$, where $\bar{w}_{RH}$ and $\tilde{N}_{RH}$ are the mean equation of state and the e-folds during the reheating stage, respectively. The maximum temperature can be written in terms of the energy density at the end of inflation as $T_{max} = \rho_{end}^{1/4}(30/\pi^2 g_{*RH})^{1/4}$ and is achieved in the instant reheating scenario, i.e. when $\tilde{N}_{RH} = 0$. From the hierarchy of the energy densities $\rho_{end} \lesssim V_* \simeq 3H_*^2 M_{Pl}^2 = (3\pi^2/2) r_* \mathcal{P}_\mathcal{R}(k_*) M_{Pl}^4$ we can obtain an upper bound on the reheating temperatures after inflation,

$$T_{RH} \lesssim 5 \times 10^{15}\text{GeV} \left(\frac{106.75}{g_{*\rho}(T_{RH})}\right)^{1/4} \left(\frac{r_*}{0.036}\right)^{1/4} \left(\frac{\mathcal{P}_\mathcal{R}(k_*)}{2 \times 10^{-9}}\right)^{1/4}, \tag{8}$$

where $g_{*\rho}(T_{RH})$ is the effective number of relativistic species upon thermalization, r_* the tensor-to-scalar ratio and $\mathcal{P}_R(k_*)$ the curvature power spectrum CMB scales, and we considered the BICEP2 bound $r_* < 0.036$ [20]. This upper bound on the temperature restricts considerably the viable parameter space of our particle physics models of DM and baryon asymmetry generation.

3.1 The Fluid System of Radiation and a Scalar Condensate

In many theories beyond the SM there generically exist scalar fields with rather flat potentials and very weak or gravitationally suppressed interactions. These fields might never be in thermal contact with the plasma and decay slowly. Well-known examples are given by the moduli fields, predicted by stringy setup.

In particular, the energy density of a coherently oscillating scalar about the minimum of an effectively quadratic potential scales in the same way as pressureless matter, $\rho_X \propto a^{-3}$, and redshifts slower than the background radiation plasma. We suppose that at the cosmic time $t_{\mathrm{dom},X} \gtrsim m_X^{-1}$ this scalar temporarily dominates the energy density of the early universe until it decays, at $t_{\mathrm{dec},X}$, reheating the universe for a second time at the temperature $T_{\mathrm{dec},X}$. This transit EMD era dilutes any pre-existing abundances of the relativistic degrees of freedom by the amount

$$\Delta_{\mathrm{EMD}} \equiv \frac{S_{\mathrm{final}}}{S_{\mathrm{initial}}} \approx \frac{T_{\mathrm{dom},X}}{T_{\mathrm{dec},X}}, \tag{9}$$

where S_{initial} and S_{final} denote the comoving entropy of the universe at times well before ($t \ll t_{\mathrm{dec},X}$) and after ($t \gg t_{\mathrm{dec},X}$) the decay of the X field. Note that $\Delta_{\mathrm{EMD}} \approx (\rho_{\mathrm{dom},X}/\rho_{\mathrm{dec},X}))^{1/4}$.

In order to incorporate this interfering process in our analysis, we assume a universe initially dominated by radiation where freeze-in DM and baryon asymmetry are produced. We also assume the presence of the aforementioned scalar field X. The X-decay enriches the radiation fluid and the evolution of the cosmological background is described by the system of equations,

$$\frac{d\rho_X}{d\tilde{N}} = -3\rho_X - \frac{\Gamma_X}{H}\rho_X \tag{10a}$$

$$\frac{d\rho_{\mathrm{rad}}}{d\tilde{N}} = -4\rho_{\mathrm{rad}} + (1 - \mathrm{B_{DM}})\frac{\Gamma_X}{H}\rho_X \tag{10b}$$

$$\frac{d\rho_{\mathrm{DM}}}{d\tilde{N}} = -4\rho_{\mathrm{DM}} + \mathrm{B_{DM}}\frac{\Gamma_X}{H}\rho_X \tag{10c}$$

$$\frac{dH}{d\tilde{N}} = -\frac{1}{2HM_{\mathrm{Pl}}^2}\left(\rho_X + \frac{4}{3}\rho_{\mathrm{DM}} + \frac{4}{3}\rho_{\mathrm{rad}}\right), \tag{10d}$$

where $d\tilde{N} = d(\ln a) = Hdt$ is the differential of the e-folds number. We assume that X decays symmetrically into SM particles and that the branching ratio of X into DM particles, $\mathrm{B_{DM}} \equiv \mathrm{Br}(X \to \mathrm{DM})$, is vanishing. For $\mathrm{B_{DM}} = 0$ the above system can be effectively reduced to a system of two interacting fluids, the scalar field X and the radiation. Our DM particle is relativistic, albeit not in thermal equilibrium, and its energy density only redshifts with time and is not sourced by the X decay, $\mathrm{B_{DM}}\Gamma_X\rho_X/H = 0$. The result of the interchange in the energy density among the different components is that the early produced freeze-in DM abundance and lepton asymmetry are diluted by an amount Δ_{EMD} at later cosmic times $t > \Gamma_X^{-1}$.

3.2 The Viable Parameter Space in the Scenario with Late Entropy Production

The viable parameter space of our microscopic models is, first of all, constrained by the inferred present-day baryon asymmetry $Y_B = (8.71 \pm 0.06) \times 10^{-11}$ and the DM cosmic abundance $\Omega_{\rm DM} h^2 = 0.1200 \pm 0.0012$ [29]. Furthermore, there are cosmological constraints on the DM mass value arising either from the possibility of DM thermalization or from large-scale structure formation considerations. Concretely, throughout our analysis we assume that the DM particles never reach thermal equilibrium with the SM thermal bath, whereas they are found to decouple from the plasma while they are still relativistic.

Given these remarks, the non-thermalization requirement is satisfied if the number density of DM lies below its would-be equilibrium value during the period in which its interactions were sizeable. That is, for the period close to the reheating temperature we impose the constraint

$$Y_{\rm DM} < Y_{\rm DM}^{\rm eq}, \tag{11}$$

otherwise the freeze-in assumption breaks down. Assuming that the DM particles saturate the entire DM content of the universe, i.e., $\Omega h^2 \approx 2.8 \times 10^8\, Y_{\rm DM}\, (m_{\rm DM}/{\rm GeV})$, Eq. (11) yields

$$m_{\rm DM} > 0.17\, \Delta_{\rm EMD}\, {\rm keV}\,. \tag{12}$$

Structure formation bounds are related to the fact that the DM particles have a mass at the keV range and decouple when relativistic. Observations favour DM candidates that assist the formation of structures in the universe, which would be hampered if the free-streaming length of the DM particles is too large. The strongest constraints come from Lyman-α forest observations, which impose restrictions on the momentum of the particles created and on their redshift at relevant time-scales. Assuming a pure radiation-dominated universe, this imposes a lower bound on the freeze-in DM mass $m_{\rm DM} \gtrsim m_{\rm Ly-\alpha}$, where $m_{\rm Ly-\alpha} \approx 4(4.6)$ keV in the $d = 5(6)$ case [30]. If, however, entropy is injected in the plasma, the momentum of the DM particles redshifts faster than the temperature of the plasma by a factor $\Delta_{\rm EMD}^{1/3}$.

$$m_{\rm DM} \gtrsim m_{\rm Ly-\alpha}\, \Delta_{\rm EMD}^{-1/3}\,. \tag{13}$$

Note that by increasing the dilution size $\Delta_{\rm EMD}$ the Lyman-α lower bound (13) on the DM mass becomes weaker, whereas the thermalization bound (12) becomes stronger. For $\Delta_{\rm EMD} \gtrsim 10\,(m_{\rm Ly-\alpha}/4\,{\rm keV})^{3/4}$ the thermalization bound is the one that becomes the most restrictive one, both for the $d = 5$ and for the $d = 6$ case.

Lastly, let us briefly comment on the fact that the decays of the surviving (post-freeze-out) F_i particles should not disrupt the formation of light elements in the universe. Such BBN constraints, which essentially impose restrictions on the allowed combinations of (m_F, Λ), were discussed in some detail in [8]. Given the fact that, as we will see in what follows, an early matter domination phase allows for lower values of Λ, i.e. faster F_i decay rates, this constraint is easily satisfied within the context discussed in this talk. Due to the lack of space we will discuss the effect of the extra cosmic component to the Dimension-5 operator case.

Dimension-5 operator and late entropy production In [8] it was shown that the viable parameter space found after a full numerical solution of Eq. (4) could be determined to a very good approximation by a few fairly simple analytical formulae. In the scenario that we consider here, in which entropy is injected in the plasma *after* the generation of the DM and lepton asymmetry, these expressions are reformulated as

$$\Lambda \approx 2 \times 10^{16}\,\text{GeV}\, \frac{\left(|\lambda_1|^2 + |\lambda_2|^2\right)^2}{\sqrt{\left||\kappa||\lambda_1||\lambda_2|^2(|\lambda_1|^2 - |\lambda_2|^2)\sin(\Delta\phi)\right|}} \left(\frac{m_\varphi}{10\,\text{keV}}\right)^2 \Delta_{\text{EMD}}^{-3/2} \tag{14a}$$

$$T_{\text{RH}} \approx 3 \times 10^{15}\,\text{GeV}\, \frac{\left(|\lambda_1|^2 + |\lambda_2|^2\right)^3}{\left||\kappa||\lambda_1||\lambda_2|^2(|\lambda_1|^2 - |\lambda_2|^2)\sin(\Delta\phi)\right|} \left(\frac{m_\varphi}{10\,\text{keV}}\right)^3 \Delta_{\text{EMD}}^{-2}, \tag{14b}$$

where $\Delta\phi \equiv \phi_1 - \phi_2 - \phi_3$ and ϕ_1, ϕ_2, ϕ_3 are the phases of the couplings λ_1, λ_2 and κ, respectively. Since the parameter Δ_{EMD} can take values within a range that spans many orders of magnitude, we expect a significant modification of the viable parameter space compared to the case of pure radiation domination. Alternatively, we can treat T_{RH} as free and constrain m_{DM}, using $m_\varphi = \Omega_{\text{DM}}\rho_c/(Y_\varphi(T_0)s_0)$ and Eq. (2) with $n = 1$, as

$$m_\varphi \approx 10\,\text{keV} \left(\frac{\Lambda}{2 \times 10^{16}\,\text{GeV}}\right)^2 \left(\frac{3 \times 10^{15}\,\text{GeV}}{T_{\text{RH}}}\right) \frac{\Delta_{\text{EMD}}}{\left(|\lambda_1|^2 + |\lambda_2|^2\right)}. \tag{15}$$

The results for this case are presented in Fig. 2. In the left-hand side panel, we show the viable parameter space in the (Λ, T_{RH}) plane for the benchmark choice of parameters $|\lambda_2| = 1$, $|\kappa| = 1$ and $|\Delta\phi| = \pi/2$, with all other model parameters being varied freely. The different colour regions correspond to different values for the dilution factor Δ_{EMD}, whereas the solid line depicts the limit $T_{\text{RH}} < \Lambda$, above which the use of Effective Field Theory becomes unjustified and knowledge of the full UV completion of our theory would be required. In the right-hand side panel, our results are projected on the (Λ, m_φ) plane.

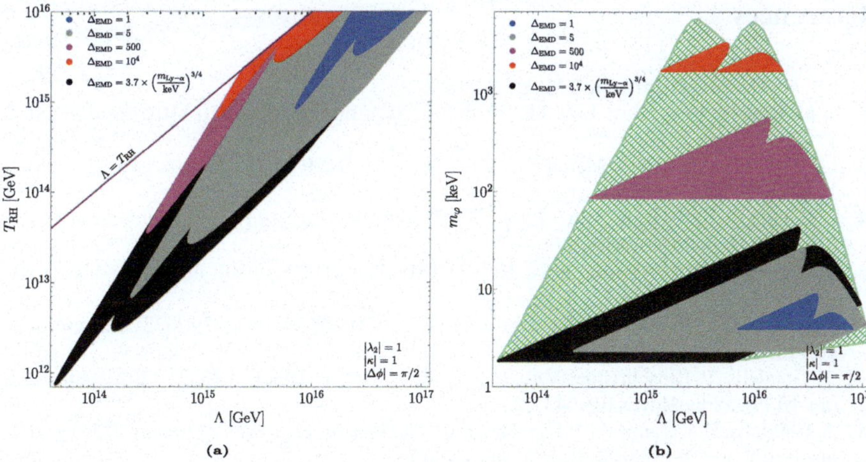

Fig. 2 **a** The allowed parameter space for the $d = 5$ model in the $(\Lambda, T_{\rm RH})$ plane. The parameter space is saturated for $\Delta_{\rm EMD} \approx 3.7\,(m_{\rm Ly-\alpha}/\rm keV)^{3/4}$ due to the thermalization of DM. **b** The allowed parameter space in the (Λ, m_φ) plane

4 Summary and Conclusions

In this talk we discussed the implications of an early matter-dominated cosmological era for a mechanism that allows for the simultaneous generation of dark matter and a viable matter-antimatter asymmetry in the universe through the same microscopic out-of-equilibrium (freeze-in) processes. We considered two examples, in which these processes are described by simple non-renormalizable operators: in the first case the DM is a scalar produced through a dimension-5 operator, while in the second case it is a fermion produced through a dimension-6 one. In both scenarios the freeze-in processes occur in the ultraviolet, i.e. at very high temperatures. In our freeze-in scenario the relic DM abundance and the size of baryon asymmetry have a different scaling with temperature, therefore the additional component has a non-trivial impact on the values of the favoured reheating temperature and on the allowed DM particle mass.

Acknowledgements The work of I.D. and V.C.S. was supported by the Hellenic Foundation for Research and Innovation (H.F.R.I.) under the "First Call for H.F.R.I. Research Projects to support Faculty members and Researchers and the procurement of high-cost research equipment grant" (Project Number: 824). This research is co-financed by Greece and the European Union (European Social Fund- ESF) through the Operational Programme "Human Resources Development, Education and Lifelong Learning" in the context of the project "Strengthening Human Resources Research Potential via Doctorate Research - 2nd Cycle" (MIS-5000432), implemented by the State Scholarships Foundation (IKY). The work of D.K. is supported by the Lancaster-Manchester-Sheffield Consortium for Fundamental Physics, under STFC Research Grant ST/T001038/1.

References

1. L.J. Hall, J. March-Russell, S.M. West (2010)
2. J. McDonald, Phys. Rev. Lett. **88**, 091304 (2002). https://doi.org/10.1103/PhysRevLett.88.091304
3. L.J. Hall, K. Jedamzik, J. March-Russell, S.M. West, JHEP **03**, 080 (2010). https://doi.org/10.1007/JHEP03(2010)080
4. A.D. Sakharov, Pisma. Zh. Eksp. Teor. Fiz. **5**, 32 (1967). https://doi.org/10.1070/PU1991v034n05ABEH002497
5. B. Shuve, D. Tucker-Smith, Phys. Rev. D **101**(11), 115023 (2020). https://doi.org/10.1103/PhysRevD.101.115023
6. J. Berman, B. Shuve, D. Tucker-Smith, Phys. Rev. D **105**(9), 095027 (2022). https://doi.org/10.1103/PhysRevD.105.095027
7. A. Goudelis, P. Papachristou, V.C. Spanos, Phys. Rev. D **105**(4), 043521 (2022). https://doi.org/10.1103/PhysRevD.105.043521
8. A. Goudelis, D. Karamitros, P. Papachristou, V.C. Spanos, Phys. Rev. D **106**(2), 023515 (2022). https://doi.org/10.1103/PhysRevD.106.023515
9. I. Dalianis, A. Goudelis, D. Karamitros, P. Papachristou, V.C. Spanos, arXiv:2304.07345 [hep-ph]
10. F. Elahi, C. Kolda, J. Unwin, JHEP **03**, 048 (2015). https://doi.org/10.1007/JHEP03(2015)048
11. I. Baldes, N.F. Bell, A. Millar, K. Petraki, R.R. Volkas, JCAP **11**, 041 (2014). https://doi.org/10.1088/1475-7516/2014/11/041
12. I. Baldes, N.F. Bell, K. Petraki, R.R. Volkas, Phys. Rev. Lett. **113**, 181601 (2014). https://doi.org/10.1103/PhysRevLett.113.181601
13. I. Baldes, N.F. Bell, A.J. Millar, R.R. Volkas, JCAP **10**, 048 (2015). https://doi.org/10.1088/1475-7516/2015/10/048
14. M. Kawasaki, K. Kohri, N. Sugiyama, Phys. Rev. Lett. **82**, 4168 (1999). https://doi.org/10.1103/PhysRevLett.82.4168
15. A.R. Liddle, S.M. Leach, Phys. Rev. D **68**, 103503 (2003). https://doi.org/10.1103/PhysRevD.68.103503
16. W.H. Kinney, A. Riotto, JCAP **03**, 011 (2006). https://doi.org/10.1088/1475-7516/2006/03/011
17. J. Martin, C. Ringeval, Phys. Rev. D **82**, 023511 (2010). https://doi.org/10.1103/PhysRevD.82.023511
18. R. Easther, R. Galvez, O. Ozsoy, S. Watson, Phys. Rev. D **89**(2), 023522 (2014). https://doi.org/10.1103/PhysRevD.89.023522
19. I. Dalianis, Y. Watanabe, JHEP **02**, 118 (2018). https://doi.org/10.1007/JHEP02(2018)118
20. P.A.R. Ade et al., Phys. Rev. Lett. **127**(15), 151301 (2021). https://doi.org/10.1103/PhysRevLett.127.151301
21. G.D. Coughlan, W. Fischler, E.W. Kolb, S. Raby, G.G. Ross, Phys. Lett. B **131**, 59 (1983). https://doi.org/10.1016/0370-2693(83)91091-2
22. J.R. Ellis, D.V. Nanopoulos, M. Quiros, Phys. Lett. B **174**, 176 (1986). https://doi.org/10.1016/0370-2693(86)90736-7
23. G. German, G.G. Ross, Phys. Lett. B **172**, 305 (1986). https://doi.org/10.1016/0370-2693(86)90257-1
24. B. de Carlos, J.A. Casas, F. Quevedo, E. Roulet, Phys. Lett. B **318**, 447 (1993). https://doi.org/10.1016/0370-2693(93)91538-X
25. T. Banks, M. Berkooz, P.J. Steinhardt, Phys. Rev. D **52**, 705 (1995). https://doi.org/10.1103/PhysRevD.52.705
26. M. Fujii, T. Yanagida, Phys. Lett. B **549**, 273 (2002). https://doi.org/10.1016/S0370-2693(02)02958-1
27. A.L. Erickcek, K. Sigurdson, Phys. Rev. D **84**, 083503 (2011). https://doi.org/10.1103/PhysRevD.84.083503

28. R. Allahverdi, J.K. Osiński, Phys. Rev. D **105**(2), 023502 (2022). https://doi.org/10.1103/PhysRevD.105.023502
29. N. Aghanim et al., Astron. Astrophys. **641**, A6 (2020). https://doi.org/10.1051/0004-6361/201833910. [Erratum: Astron. Astrophys. **652**, C4 (2021)]
30. G. Ballesteros, M.A.G. Garcia, M. Pierre, JCAP **03**, 101 (2021). https://doi.org/10.1088/1475-7516/2021/03/101

Higgs Vacuum Metastability in $R + R^2$ Gravity

Andreas Mantziris

Abstract Experimental data suggest that the Higgs potential has a lower ground state at high field values. Consequently, decaying from the electroweak to the true vacuum nucleates bubbles that expand rapidly and can have dire consequences for our Universe. This overview of our last study [1] regarding the cosmological implications of vacuum metastability during Starobinsky inflation was presented at the HEP2023 conference. Following the framework established in [2], we showcased our state-of-the-art lower bounds on the non-minimal coupling ξ, which resulted from a dedicated treatment of the effective Higgs potential in $R + R^2$ gravity. The effects of this consideration involved the generation of destabilising terms in the potential and the sensitive dependence of bubble nucleation on the last moments of inflation. In this regime, spacetime deviates increasingly from de Sitter and thus the validity of our approach reaches its limit, but at earlier times, we recover our result hinting against eternal inflation.

Keywords Standard model of particle physics · Inflationary cosmology · Quantum field theory in curved spacetime

1 Introduction

1.1 *The Metastability of the Higgs Vacuum*

The data from collider experiments regarding the parameters of the Standard Model (SM) of particle physics [3] indicate that the quartic coupling λ of the Higgs field h changes sign at high energies. The evolution of the self-coupling with the energy scale μ is formally computed by its β-function, where bosons and fermions contribute with opposing signs [4]. The Higgs boson and the top quark, being the heaviest ones,

A. Mantziris (✉)
Department of Physics, Imperial College London, London, UK
e-mail: andreas.mantziris@fuw.edu.pl

Faculty of Physics, University of Warsaw, ul. Pasteura 5, 02-093 Warsaw, Poland

J. Strologas and L. Perivolaropoulos (eds.), *HEP2023*, Springer Proceedings in Physics 333 https://doi.org/10.1007/978-3-031-76856-9_17

dominate and seem to balance each other, resulting in the smooth function shown in Fig. 1. The sign change of λ leads to the development of a lower ground state than the electroweak (EW) vacuum in the Higgs potential

$$V_{\mathrm{H}}(h, \mu) = \frac{\lambda(\mu)}{4} h^4 + \dots . \tag{1}$$

This realisation renders the EW vacuum metastable, since the Higgs field does not reside currently in the lowest energy state and is therefore prone to decay to its true vacuum [5, 6]. This process can take place quantum mechanically by tunnelling through the potential barrier, classically by fluctuating over it, or a combination of both, as shown in Fig. 2.

The survival of the Higgs vacuum during our cosmological evolution can be utilised to impose significant constraints on fundamental physics in a manner that would enable this long-lasting metastability [7]. In our studies [1, 2], we adopted a minimal model of the early universe consisting of the SM and cosmological inflation in order to obtain vacuum decay bounds on the Higgs curvature coupling ξ. This coupling is the last missing renormalizable parameter of the SM, since it is impossible to probe it with accelerators in the flat spacetime of our late Universe, and thus can

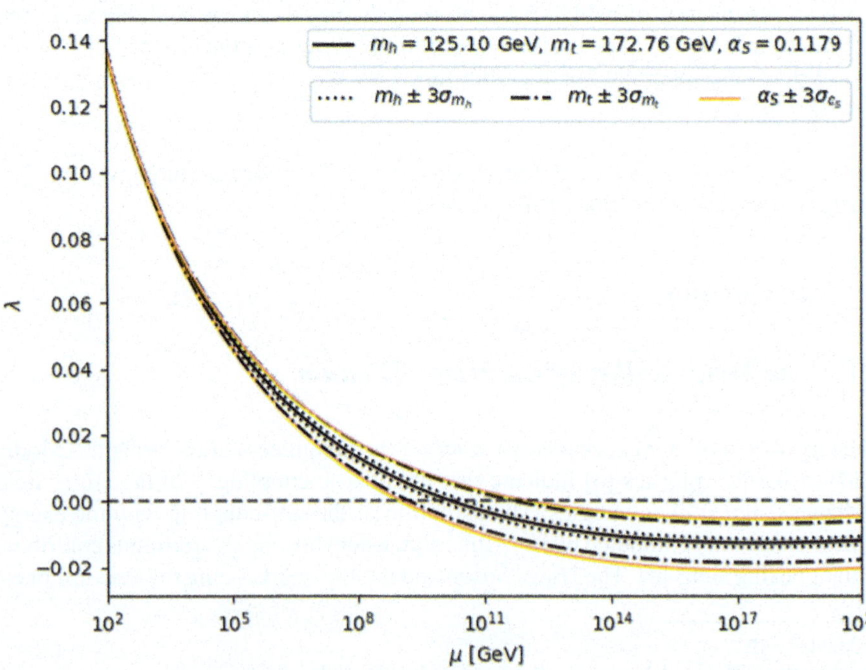

Fig. 1 The "running" of the Higgs self-interaction λ with the renormalization scale μ according to the experimentally measured values of the Higgs mass m_h, top quark mass m_t, and the strong coupling α_S to 3σ variance [3]

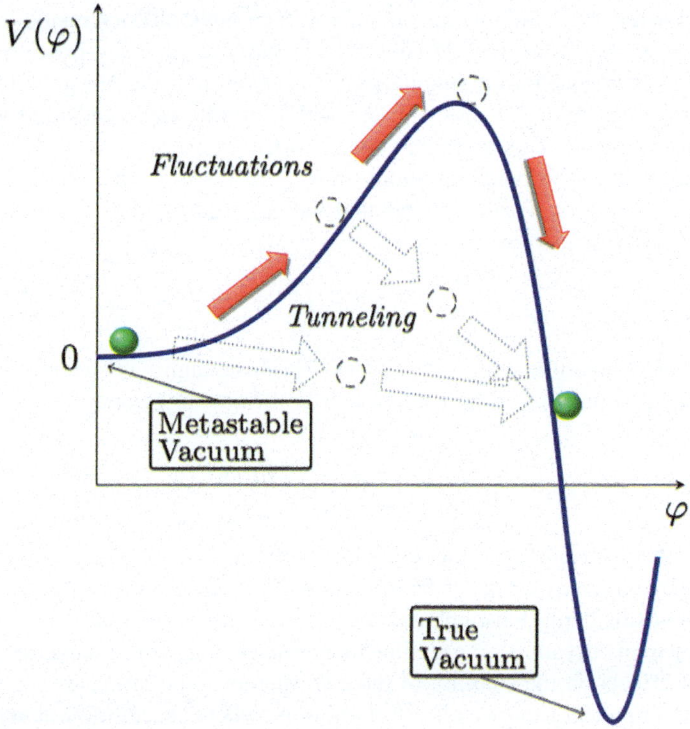

Fig. 2 The metastability of the vacuum state of a scalar field φ in a potential $V(\varphi)$, where it can decay to the true vacuum by fluctuating over, tunnelling through, or a combination of both mechanisms [7]

only be constrained by cosmological studies. The non-minimal coupling appears in the Higgs potential at tree-level

$$V_{\mathrm{H}}(h, \mu, R) = \frac{\xi(\mu)}{2} R h^2 + \frac{\lambda(\mu)}{4} h^4 + \dots , \tag{2}$$

where R is the Ricci scalar that quantifies the curvature of spacetime [8], to ensure the renormalizability of the theory on a curved spacetime [9]. Therefore, assuming that vacuum decay was sufficiently suppressed during inflation constrains ξ accordingly.

1.2 Vacuum Decay and Bubble Nucleation

After the Higgs vacuum decays to its true minimum at a particular point in spacetime, the surrounding field follows in a runaway process that leads to the formation of a true-vacuum bubble, that expands with a velocity approaching the speed of light [7].

We cannot safely tell anything specific about the exotic physics inside the bubble, but it is assumed, within the SM context, that the violent decay process results in the interior collapsing into a singularity [10]. Given that the Higgs field is in the metastable EW state at the moment, no bubbles could have nucleated within our cosmological history. This argument enables us to constrain fundamental physics, by demanding that the potential should have been "metastable enough" in order to survive. Quantitatively, this means that the expectation value of the number of bubbles $\mathcal{N}$ has to obey

$$\langle \mathcal{N} \rangle \leq 1\,, \tag{3}$$

in order to be compatible with observations [7]. This quantity is given by the integral of the decay rate per spacetime volume Γ over the past lightcone

$$\langle \mathcal{N} \rangle = \int_{\text{past}} d^4x \sqrt{-g}\Gamma(x)\,, \tag{4}$$

where g is the determinant of the spacetime metric $g_{\mu\nu}$.

The early universe provides a fruitful context for these considerations, since various mechanisms could have enhanced the decay rate at such high scales [11–14], and the apparent curvature of spacetime means that the non-minimal term could have prominent effects on the stability of the EW vacuum. Therefore, we are motivated to study vacuum decay in the context of inflation, where spacetime was significantly curved. During inflation, the Universe undergoes a phase of exponential expansion due to the slow roll of a scalar field, i.e. the inflaton ϕ, in an approximately flat potential [15]. The duration of inflation is measured for convenience by e-foldings, defined as

$$N \equiv \ln\left(\frac{a_{\text{inf}}}{a}\right)\,, \tag{5}$$

with a corresponding to the scale factor and the index "inf" denoting the end of inflation. In order to comply with observations, a minimum number of approximately 60 e-foldings is required [15], meaning that

$$N_{\text{start}} \geq 60\,. \tag{6}$$

Expressing Eq. (4) in the inflationary context results in

$$\langle \mathcal{N} \rangle = \frac{4\pi}{3} \int_0^{N_{\text{start}}} dN \left(\frac{a_{\text{inf}}\left(\eta_0 - \eta\left(N\right)\right)}{e^N}\right)^3 \frac{\Gamma(N)}{H(N)}\,, \tag{7}$$

where we are integrating backwards in time from $N_{\text{inf}} = 0$, $\eta_0 - \eta(N)$ is the comoving radius of the spacetime volume in terms of the conformal time η between today and N e-folds before the end of inflation, and

$$H \equiv \frac{\dot{a}}{a} = \frac{da}{dt}\frac{1}{a} \tag{8}$$

is the Hubble rate that measures the acceleration of the expansion.

Hence, we identify two independent calculations that coalesce into a comprehensive computation of lower bounds on the Higgs curvature coupling,

$$\xi \geq \xi_{\langle\mathcal{N}\rangle=1}\,, \tag{9}$$

by imposing the requirement given by Eq. (4) on the integral in Eq. (7). One stream involves the calculation of the relevant cosmological quantities with respect to the lightcone volume, i.e. the scale factor, conformal time, and Hubble rate. The dynamic evolution of these quantities is calculated beyond the simplifying slow-roll approximation and depends on the choice of the inflationary model, the definition of the end of inflation, and its total duration [2]. In our case, we chose to consider *Starobinsky Inflation* [16], due to its minimality and favourable compliance with observational data [17], with the potential

$$V_{\text{I}}(\phi) = \frac{3M^2 M_P^4}{4}\left(1 - e^{-\sqrt{\frac{2}{3}}\frac{\phi}{M_P}}\right)^2\,, \tag{10}$$

where $M = 1.1 \times 10^{-5}$ is a dimensionless constant fixed by the CMB anisotropies and $M_P = 2.435 \times 10^{18}$ GeV is the reduced Planck mass. At the same time, it is necessary to compute the time-dependent decay rate Γ in de Sitter (dS) space according to the shape of the effective Higgs potential during inflation. We utilise the Hawking-Moss (HM) [18] instant on solution for the estimation of Γ,

$$\Gamma \approx \left(\frac{R}{12}\right)^2 e^{-\frac{384\pi^2 \Delta V_{\text{H}}}{R^2}}\,, \tag{11}$$

where ΔV_{H} is the height of the potential barrier between the two vacuum, since this is the dominant process in the high Hubble scale regime during the inflationary epoch. For commentary on the comparison between the HM, the Coleman-de Lucia, and the stochastic formalism approaches, we refer the reader to [19].

2 The Effective Higgs Potential in Curved Spacetime

2.1 Loop Corrections and Renomalisation Group Improvement

Incorporating loop corrections to the tree-level potential (2) leads to

$$V_{\mathrm{H}}(h, \mu, R) = \frac{\xi}{2} R h^2 + \frac{\lambda}{4} h^4 + \frac{\alpha}{144} R^2 + \Delta V_{\mathrm{loops}}\,, \tag{12}$$

where the Higgs mass term has been neglected (because it is negligible compared to the high Hubble scales during inflation), a subdominant gravitational term is radiatively generated, and a loop term from the entire SM particle spectrum is included, consisting of 3-loop flat space corrections and 1-loop dS corrections

$$\Delta V_{\mathrm{loops}} = \frac{1}{64\pi^2} \sum_{i=1}^{31} \left\{ n_i \mathcal{M}_i^4 \left[\log\left(\frac{|\mathcal{M}_i^2|}{\mu^2} \right) - d_i \right] + \frac{n_i' R^2}{144} \log\left(\frac{|\mathcal{M}_i^2|}{\mu^2} \right) \right\}, \tag{13}$$

where $\mathcal{M}_i(h, \mu, R)$ is the effective mass, and $n_i,\ d_i,\ n_i'$ are fixed parameters [20].

We wish to simplify (12) and eliminate its scale dependence since it is not physical, but only an arbitrary dimensionful constant [5]. Whether the perturbative expansion (13) converges safely is sensitive to the chosen value of μ, but no singular option guarantees this for all field values [19]. The usual approximation

$$\mu = h \tag{14}$$

is inadequate in this scenario, since it neglects the dominant contribution from curvature [7]. To address this, the function

$$\mu^2 = a h^2 + b R\,, \tag{15}$$

was proposed in [21], with a, b being fitted parameters. Employing this scale choice means that direct loop corrections do not entirely cancel out and must be incorporated into the effective potential [19]. In our case, we utilise the most optimal and field-dependent scale choice via the procedure of Renormalization Group (RG) improvement, where $\mu = \mu_*(h, R)$ is the solution of the equation

$$\Delta V_{\mathrm{loops}}(h, \mu_*, R) = 0\,, \tag{16}$$

resulting in the RG improved effective Higgs potential [2]

$$V_{\mathrm{H}}^{\mathrm{RGI}}(h, R) = \frac{\xi(\mu_*(h, R))}{2} R h^2 + \frac{\lambda(\mu_*(h, R))}{4} h^4 + \frac{\alpha(\mu_*(h, R))}{144} R^2\,. \tag{17}$$

2.2 Embedding the Effective Potential in $R + R^2$ Gravity

In the case of Starobinky inflation, which arises naturally by including a quadratic curvature term in the gravitational action [16], embedding the entire SM in $R + R^2$ gravity makes the computation of the RG improved effective potential more demanding. The action of the Higgs field in the Jordan frame, where gravity is modified by the addition of the geometric R^2-term to the Einstein-Hilbert action of General Relativity (GR), is given by

$$S = \int d^4x\sqrt{-g_J}\left[\frac{M_P^2}{2}\left(1-\frac{\xi h^2}{M_P^2}\right)R_J + \frac{1}{12M^2}R_J^2 + \frac{1}{2}g_J^{\mu\nu}\partial_\mu h\partial_\nu h - \frac{\lambda}{4}h^4\right], \tag{18}$$

where J denotes the value of each quantity in this frame. After a Weyl transformation to the Einstein frame

$$g_{J\,\mu\nu} = e^{-\sqrt{\frac{2}{3}}\frac{\phi}{M_P}}g_{\mu\nu}\,, \tag{19}$$

where gravity behaves according to GR, an additional scalar field ϕ in the matter sector carries the extra degree of freedom and plays the role of the inflation.

In order to be able to make appropriate comparisons and draw meaningful conclusions, we have to express the action (18) in a diagonal form. Firstly, we perform the field redefinition

$$\tilde{h} = e^{-\frac{1}{2}\sqrt{\frac{2}{3}}\frac{\phi}{M_P}}h\,, \tag{20}$$

that enables the potential's RGI in the dS limit [1] according to the prescription described in Sect. 2.1. Two further redefinitions, one regarding the inflaton

$$\phi = \tilde{\phi} - M_P\sqrt{\frac{3}{2}}\ln\left[1-\left(\xi-\frac{1}{6}\right)\left(\frac{\tilde{h}}{M_P}\right)^2\right], \tag{21}$$

and the other for the Higgs field

$$\tilde{h} \approx \rho\left[1+\left(\xi-\frac{1}{6}\right)^2\left(\frac{\rho}{M_P}\right)^2 + \mathcal{O}(\rho^4)\right]. \tag{22}$$

result in the approximately canonical Lagrangian given by

$$\mathcal{L} \approx \frac{M_P^2}{2}R + \frac{1}{2}\partial_\mu\tilde{\phi}\partial^\mu\tilde{\phi} + \frac{1}{2}\partial_\mu\rho\partial^\mu\rho - \tilde{U}(\tilde{\phi},\rho)\,. \tag{23}$$

The total potential $\tilde{U}$ includes the Starobinsky potential (10)

$$\tilde{U}(\tilde{\phi}, \rho) = V_{\rm I}(\tilde{\phi}) + m_{\rm eff}^2(\tilde{\phi}, \mu_*)\frac{\rho^2}{2} + \lambda_{\rm eff}(\tilde{\phi}, \mu_*)\frac{\rho^4}{4} + \frac{\alpha(\mu_*)}{144}R^2(\tilde{\phi}) + \mathcal{O}\left(\frac{\rho^6}{M_P^2}\right), \tag{24}$$

and the other terms resemble the effective Higgs potential (17), with the addition of a Planck suppressed sextic term and the $\tilde{\phi}$-dependent contributions to the effective quadratic and quartic couplings given by

$$m_{\rm eff}^2 = \xi R + 3M^2 M_P^2\left(\xi - \frac{1}{6}\right)\left(1 - e^{-\sqrt{\frac{2}{3}}\frac{\tilde{\phi}}{M_P}}\right)e^{-\sqrt{\frac{2}{3}}\frac{\tilde{\phi}}{M_P}} + \frac{(\xi - \frac{1}{6})}{M_P^2}\partial_\mu\tilde{\phi}\partial^\mu\tilde{\phi}, \tag{25}$$

$$\lambda_{\rm eff} = \lambda + \frac{3M^2(\xi - \frac{1}{6})^2}{e^{2\sqrt{\frac{2}{3}}\frac{\tilde{\phi}}{M_P}}} + \frac{4\left[\xi R + \Delta m_1^2\right](\xi - \frac{1}{6})^2}{M_P^2} + \frac{4(\xi - \frac{1}{6})^3}{M_P^4}\partial_\mu\tilde{\phi}\partial^\mu\tilde{\phi}, \tag{26}$$

respectively, where Δm_1^2 corresponds to the middle term of $m_{\rm eff}^2$ in Eq. (25).

The HM formalism is valid for $0 < \xi_{\rm EW} < 1/6$, meaning that the inflaton-dependent terms in Eq. (25) are negative and thus act against the stabilising ξR term. Therefore, the lower constraints on ξ are stronger in this case when compared to the field theory case of [2], since the Higgs curvature term needs to counter these additional terms in order to ensure the survivability of the metastable EW vacuum. The ξ-bounds derived with this prescription for varying top quark mass and with different definitions for the end of inflation are shown in Fig. 3. In the $R + R^2$ context, we have obtained $\xi_{\rm EW} \geq 0.13$, in contrast to the field theory bound of $\xi_{\rm EW} \geq 0.06$ for input values of $m_t = 172.76$ GeV, $m_h = 125.10$ GeV, and $\ddot{a} = 0$ defining the endpoint of inflation.

In general, bubbles nucleate towards the end of inflation, but the additional terms in Eq. (25) postpone bubble production further. As spacetime departs from dS in this regime, these results loose validity because the decay rate Γ and the effective RGI potential were calculated in the dS limit. Hence, if we adopt a more conservative assumption for the inflationary finale, we can quote ξ-bounds that are weaker but more reliable. For example, if we perform the same calculation assuming inflation lasts until $\frac{\ddot{a}}{a} = \frac{H^2}{2}$, the constraint on the non-minimal coupling is $\xi_{\rm EW} \geq 0.1$.

3 Conclusions

With this overview of our most recent investigation [1], we showcased a fruitful method for constraining fundamental physics via the synergy between SM physics and early universe cosmology. More specifically, we studied the experimentally suggested feature of Higgs vacuum metastability in the context of the observationally

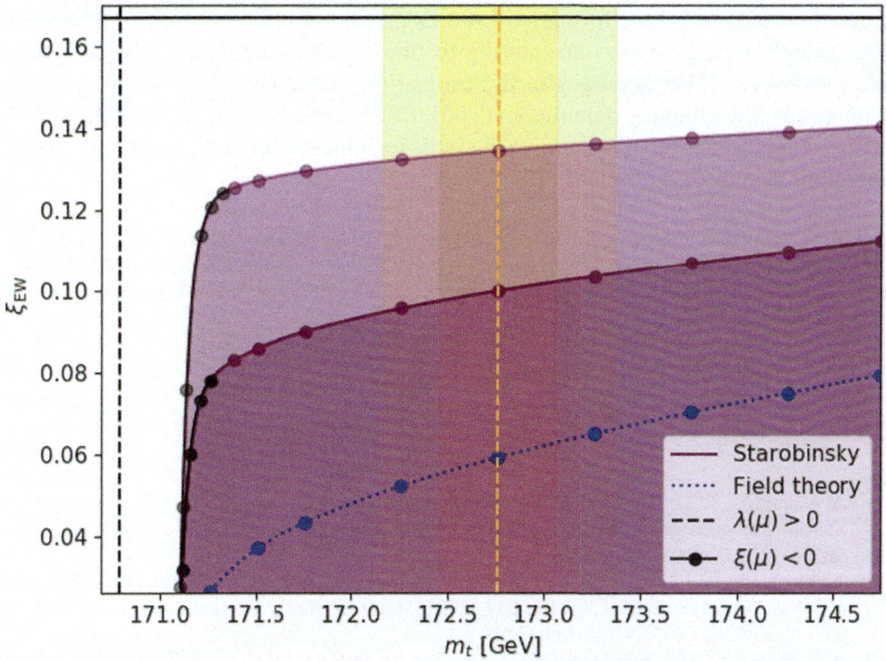

Fig. 3 Lower constraints on the Higgs curvature coupling at the electroweak scale $\xi_{\rm EW}$ for varying top quark mass m_t (vertical dashed orange line: central value, shades: $\pm\sigma_{m_t}, \pm 2\sigma_{m_t}$). The parameter space below each curve is excluded, with the end of inflation set at $\dot{H}/H^2 = -1/4$ and $\dot{H}/H^2 = -1$ for the dark and light purple curves, respectively. The black parts highlight the value below which the ξ turns negative as it runs. The field theory bounds [2] shown by the dotted blue curve are included for comparison. The vertical dashed black line is at the threshold of m_t below which λ remains positive, and the horizontal black line shows the conformal point $\xi = 1/6$

supported model of Starobinsky inflation. The novelty of our approach lies in the fact that we went beyond the dS description of the inflationary spacetime when calculating the relevant cosmological quantities, and we took into account the time dependence of the Hubble rate instead of setting it as a constant free parameter. In addition, we extended the range of similar studies [22, 23], by improving also the calculation of the effective potential to 3-loops, with curvature corrections in dS to 1-loop, and a robust RGI prescription.

By employing the analytical and numerical techniques outlined in [1], we derived more robust ξ-bounds than the dS studies [2, 7]

$$\xi_{\rm EW} \geq 0.1 > 0.06\,, \tag{27}$$

through the computation of the most accurate RG improved effective Higgs potential at the time, which consists of additional destabilising terms. The evaluation of $\Gamma_{\rm HM}$ and the effective potential in the dS regime begin to break down when reaching the time of prominent bubble formation towards the end of inflation. Hence, it becomes increasingly crucial to start accounting for the reheating dynamics [24, 25], if we are to achieve higher accuracy in the estimation of the Higgs curvature coupling. Conversely, the effective potential matches the field theory scenario at earlier times, where slow-roll applies, resulting in the same insights and indications against eternal inflation, akin to those found in [2].

Acknowledgements The author acknowledges the supervision of A. Rajantie and T. Markkanen, along with the support of an STFC PhD studentship, for the study [1] that was presented in HEP2023 - 40th Conference on Recent Developments in High Energy Physics and Cosmology at the University of Ioannina.

References

1. A. Mantziris, T. Markkanen, A. Rajantie, The effective Higgs potential and vacuum decay in Starobinsky inflation. JCAP **10**, 073 (2022). https://doi.org/10.1088/1475-7516/2022/10/073. [arXiv:2207.00696 [astro-ph.CO]]
2. A. Mantziris, T. Markkanen, A. Rajantie, Vacuum decay constraints on the Higgs curvature coupling from inflation. JCAP **03**, 077 (2021). https://doi.org/10.1088/1475-7516/2021/03/077. [arXiv:2011.03763 [astro-ph.CO]]
3. P.A. Zyla et al. [Particle Data Group], Review of particle physics. PTEP **2020**(8), 083C01 (2020). https://doi.org/10.1093/ptep/ptaa104
4. A. Rajantie, Higgs cosmolog. Phil. Trans. Roy. Soc. Lond. A **376**(2114) (2018) 20170128 https://doi.org/10.1098/rsta.2017.0128
5. T. Lancaster, S.J. Blundell, Quantum Field Theory for the Gifted Amateur. (Oxford University Press, 2014). ISBN 978-0-19-969933-9
6. V. De Luca, A. Kehagias, A. Riotto, On the cosmological stability of the Higgs instability. JCAP **09**, 055 (2022). https://doi.org/10.1088/1475-7516/2022/09/055 [arXiv:2205.10240 [hep-ph]]
7. T. Markkanen, A. Rajantie, S. Stopyra, Cosmological aspects of higgs vacuum metastability. Front. Astron. Space Sci. **5**, 40 (2018). https://doi.org/10.3389/fspas.2018.00040 [arXiv:1809.06923 [astro-ph.CO]]
8. T. Markkanen, S. Nurmi, A. Rajantie, Do metric fluctuations affect the Higgs dynamics during inflation? JCAP **12**, 026 (2017). https://doi.org/10.1088/1475-7516/2017/12/026 [arXiv:1707.00866 [hep-ph]]
9. C.G. Callan Jr., S.R. Coleman, R. Jackiw, A New improved energy - momentum tensor. Ann. Phys. **59**, 42–73 (1970). https://doi.org/10.1016/0003-4916(70)90394-5
10. N. Tetradis, Exact solutions for Vacuum Decay in Unbounded Potentials. [arXiv:2302.12132 [hep-ph]]
11. A. Eichhorn, M. Fairbairn, T. Markkanen, A. Rajantie, Proceedings, Higgs cosmology: Newport Pagnell, Buckinghamshire, UK, March 27-28 (2017)
12. J.S. Cruz, S. Brandt, M. Urban, Quantum and gradient corrections to false vacuum decay on a de Sitter background. Phys. Rev. D **106**(6), 065001 (2022). https://doi.org/10.1103/PhysRevD.106.065001 [arXiv:2205.10136 [gr-qc]]
13. S. Vicentini, New bounds on vacuum decay in de Sitter space. [arXiv:2205.11036 [gr-qc]]
14. O. Lebedev, The Higgs portal to cosmology. Prog. Part. Nucl. Phys. **120**, 103881 (2021). https://doi.org/10.1016/j.ppnp.2021.103881. [arXiv:2104.03342 [hep-ph]]
15. A.R. Liddle, D.H. Lyth, *Cosmological Inflation and Large Scale Structure* (Cambridge University Press, 2000). ISBN
16. A.A. Starobinsky, Spectrum of relict gravitational radiation and the early state of the universe. JETP Lett. **30**, 682–685 (1979)

17. Y. Akrami et al. [Planck], Planck 2018 results. X. Constraints on inflation. Astron. Astrophys. **641**, A10 (2020) https://doi.org/10.1051/0004-6361/201833887 [arXiv:1807.06211 [astro-ph.CO]]
18. S.W. Hawking, I.G. Moss, Supercooled phase transitions in the very early universe. Phys. Lett. B **110**, 35–38 (1982). https://doi.org/10.1016/0370-2693(82)90946-7
19. A. Mantziris, Constraining the Higgs curvature coupling from vacuum decay during inflation. Ph.D. Thesis, https://doi.org/10.25560/99632
20. T. Markkanen, S. Nurmi, A. Rajantie, S. Stopyra, The 1-loop effective potential for the Standard Model in curved spacetime. JHEP **06**, 040 (2018). https://doi.org/10.1007/JHEP06(2018)040. [arXiv:1804.02020 [hep-ph]]
21. M. Herranen, T. Markkanen, S. Nurmi, A. Rajantie, Spacetime curvature and the Higgs stability during inflation. Phys. Rev. Lett. **113**(21), 211102 (2014). https://doi.org/10.1103/PhysRevLett.113.211102[arXiv:1407.3141 [hep-ph]]
22. J. Fumagalli, S. Renaux-Petel, J.W. Ronayne, Higgs vacuum (in)stability during inflation: the dangerous relevance of de Sitter departure and Planck-suppressed operators. JHEP **02**, 142 (2020). https://doi.org/10.1007/JHEP02(2020)142. [arXiv:1910.13430 [hep-ph]]
23. Q. Li, T. Moroi, K. Nakayama, W. Yin, Instability of the electroweak vacuum in Starobinsky inflation. JHEP **09**, 102 (2022). https://doi.org/10.1007/JHEP09(2022)102. [arXiv:2206.05926 [hep-ph]]
24. J. Kost, C.S. Shin, T. Terada, Massless preheating and electroweak vacuum metastability. Phys. Rev. D **105**(4), 043508 (2022). https://doi.org/10.1103/PhysRevD.105.043508 [arXiv:2105.06939 [hep-ph]]
25. D.G. Figueroa, A. Florio, T. Opferkuch, B.A. Stefanek, Dynamics of Non-minimally Coupled Scalar Fields in the Jordan Frame. [arXiv:2112.08388 [astro-ph.CO]]

Search for New Particles at CERN on the Zooniverse Citizen-Science Platform

Stylianos Angelidakis

Abstract The REINFORCE EU project engages and supports citizens to cooperate with researchers and actively contribute to the development of new knowledge for the needs of science and society. REINFORCE offers four "discovery demonstrators" in different areas of physics. The infrastructure of all demonstrators is based on Zooniverse, the most popular citizen-science platform. The demonstrator titled *Search for new particles at CERN* introduces citizens to research performed at the LHC of CERN, adopting a three-stage workflow. The first two stages use simulated data to train citizens, but also to allow for a quantitative assessment of their performance and comparison with machine-based algorithms. The third stage uses real data from the ATLAS Open-Data subset, providing two research paths: (a) study of Higgs boson decays to two photons, one of which could be converted to an electron-positron pair due to interactions with detector material and (b) search for yet undiscovered long-lived particles, predicted by certain beyond-the-standard-model theories. Since the launch of the project in October 2021, the demonstrator has attracted about 4000 volunteers.

1 Introduction

REINFORCE [1] (Research Infrastructures FOR Citizens in Europe) is a Research & Innovation Project funded by the European Union's Horizon 2020 SwafS (Science with and for Society) framework programme. The project aims to engage citizens in research conducted in Large Research Infrastructures using a participatory design methodology that takes into account the special characteristics of different target groups. By leveraging open data and expert support, citizens undertake their own inquiries, fostering collaboration between citizens and experts within the REINFORCE framework.

The REINFORCE project utilises four Large Scale Citizen-Science demonstrators developed by the consortium partners on the Zooniverse platform [2] to bridge

S. Angelidakis (✉)
National and Kapodistrian University of Athens, Athens, Greece
e-mail: stylianos.angelidakis@cern.ch

J. Strologas and L. Perivolaropoulos (eds.), *HEP2023*, Springer Proceedings in Physics 333 https://doi.org/10.1007/978-3-031-76856-9_18

the gap between frontier science and society. Each demonstrator engages citizens in research experiences by providing tasks which are essential for the needs of scientific research. Furthermore, efforts have been made to extend participation to sense-disabled and senior citizens, enhancing inclusivity in scientific inference. The REINFORCE demonstrators are titled: (a) *GWitch Hunters* [3], (b) *Deep Sea Explorers* [4], (c) *New Particle Search at CERN* [5], and (d) *Cosmic Muon Images* [6].

2 New Particle Search at CERN

New Particle Search at CERN introduces non-expert citizens to research performed at the Large Hadron Collider (LHC) of CERN. The workflow, described in the sections below, is designed around two topics of research interest: (a) the search for the hypothetical Long-Lived Particles (LLPs) [7], and (b) the study of Higgs boson [8] decays to photons using simulated and real data from proton-proton (*pp*) collisions at the LHC, recorded with the ATLAS detector [9]. Both topics are the focus of many research projects of the ATLAS collaboration.

By exploiting the infrastructures of the Zooniverse platform, as well as the interactive capabilities of HYPATIA [10] (HYbrid Pupil's Analysis Tool for Interactions in ATLAS), citizens are invited to perform visual data analysis following a three-stage workflow, using just their mouse pointer. In each stage, they are guided throughout a specific task by recorded tutorials and help sections including several examples. Furthermore, they are encouraged to discuss their questions and findings on the talk forums of the platform thereby establishing a community of actively engaged citizens.

3 Stage 1—Displaced Vertex Identification

In Stage 1 of the demonstrator, citizens inspect stationary images derived from a simulated sample of LLPs produced in *pp* collisions at the LHC. Each image corresponds to a single *pp* collision event recorded by the ATLAS detector. Due to their long life-time, LLPs travel a significant distance in the detector before decaying into stable particles. Therefore, their characteristic signature is a Displaced Vertex (DV) from which product particles originate. This DV is distant from the *pp* interaction point, unlike the majority of known unstable particles produced at the LHC, which can be associated to a Primary Vertex (PV) close to the interaction point.

Figure 1 shows one of the stationary images presented to the citizens. Reconstructed tracks of charged particles from a *pp* collision event are displayed on two perpendicular views (transverse and longitudinal) of the ATLAS inner detector. Each track is drawn on both views with the same unique color so that tracks from the two views can be matched to obtain an understanding of the 3D trajectory of the particle. Citizens are invited to identify, using the mouse pointer, the coordinates of any DV

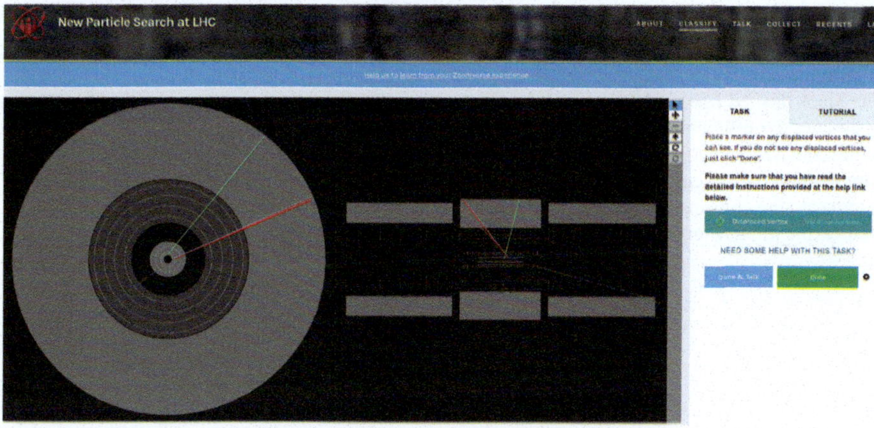

Fig. 1 A *pp* collision event containing a DV, as displayed on Zooniverse. Two views of the ATLAS inner detector are shown, transverse (left) and longitudinal (right)

in the event, i.e. the location where the simulated LLP decayed, in both views of the inner detector (two clicks per DV). Preselection of the displayed tracks has already been applied following the typical strategy of LLP searches, which includes removal of tracks associated to a PV, with few additional requirements facilitating detection by human eye. DV candidates can be identified as the common origin of any two or more tracks.

Citizens' selections are assessed by Zooniverse in real-time, based on the true location of each DV uploaded to the platform, enabling immediate feedback after processing each event. The goal of Stage 1 is to compare the accuracy and efficiency of human participants with that of an automated algorithm performing the same task. Therefore, simulated events are utilised, as they provide the true position of each DV. The same events are analysed by both the algorithm and citizens for an accurate and impartial comparison.

In the official duration of the project (from its launch on Zooniverse on October 19, 2021 to the project end date of November 30, 2022), 1962 registered users participated in Stage 1 of the demonstrator providing 173017 classified events (today Stage 1 engages over 2300 registered users). Their selections have been analysed in terms of both identification efficiency (fraction of DVs correctly identified over the total number of DVs encountered) and purity (fraction of DVs correctly identified over total number of DVs reported by the citizen).

An important finding from the analysis of the results is that combining citizens' responses for each given event results in significantly higher identification efficiency than that of individual citizens. As depicted in Fig. 2, when considering events classified by at least 10 highly active citizens (who have performed over 200 classification with over 50% efficiency), the mean of the reported coordinates (i.e. the "consensus") yields an identification efficiency of 93%, which is on par with the efficiency

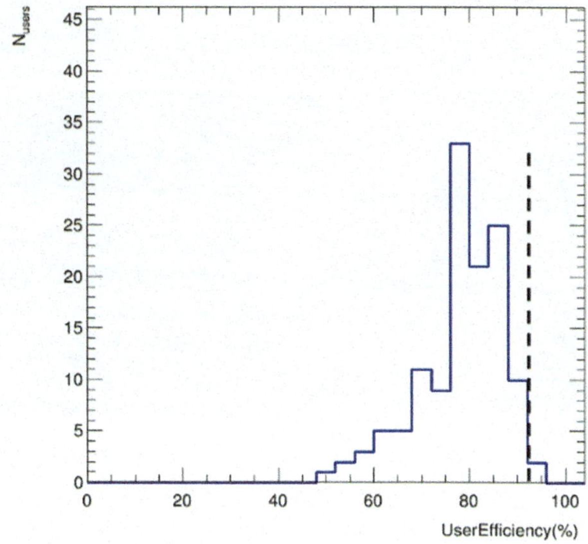

Fig. 2 Distribution of the DV identification efficiency of individual active citizens. The black dashed line designates the efficiency achieved when using the combination of citizens' responses (the consensus) as an identification criterion

of 94% achieved by the automated algorithm. In contrast, the average efficiency of individual citizens is approximately 79%. Such combination of independent citizen selections could be further exploited in citizen-science.

4 Stage 2—Particle Identification

In Stage 2, citizens engage with the interactive event-display environment of HYPATIA to identify particles interacting within the ATLAS detector. In each event (an example is depicted in Fig. 3), participants select particle tracks in the ATLAS inner detector or clusters of energy deposited in the electromagnetic calorimeter, classifying them as muons, electrons, and converted or non-converted photons. To achieve this, they inspect various properties of the tracks or clusters, including momentum, charge, and direction, all conveniently displayed by HYPATIA. Simulated events sourced from the ATLAS Open Data set [11] are utilized in this stage since simulation provides precise knowledge of the particles produced, enabling us to determine the accuracy of participants' selections. Participant selections, automatically recorded by HYPATIA, undergo thorough analysis and comparison with those of a Machine-Learning (ML) algorithm trained on the same data, ensuring an objective evaluation. Notably, Stage 2 has been thoughtfully adapted to accommodate participants with visual impairments, employing SonoUno's capabilities to sonify the entire task [12].

In the active duration of the project, citizens classified 36690 particles with an overall classification efficiency of 80.4%. Each identification was assessed based on the true content of the simulated event and compared to the output of the ML

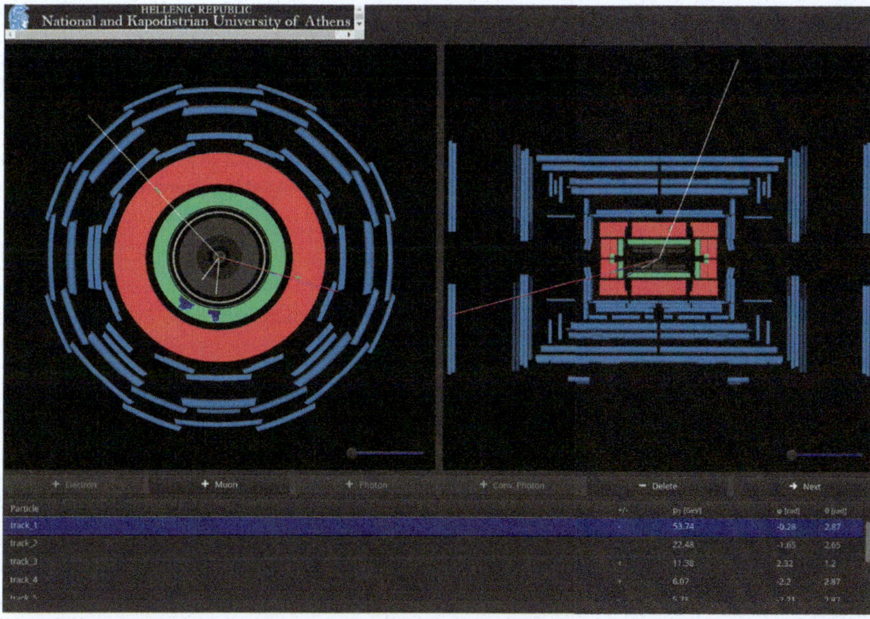

Fig. 3 HYPATIA display of a simulated event containing two electrons and two muons, in Stage 2. Citizens use the mouse pointer to select tracks (straight-lines) and clusters in the electromagnetic calorimeter (blue regions on the green layer), and classify them as electrons, muons, photons or converted photons

Table 1 Comparison of citizens' performance to that of the ML algorithm in terms of particle identification efficiency and purity

Particle type	User efficiency (%)	ML efficiency (%)	User purity (%)	ML purity (%)
Electron	90	97	88	78
Photon	96	97	80	78
Conv. photon	62	95	35	81

algorithm. Table 1 summarizes the results in terms of the efficiency and purity for each type of particle, both for citizens and the ML algorithm. The comparison shows that the identification ability of citizens does not surpass that of a dedicated ML algorithm. It is noteworthy though that for electrons and photons, their efficiency is not significantly far behind. Moreover, in terms of purity percentages, citizens seem to perform better in cases of electrons and photons. However, it's important to underscore that the primary aim of this stage is not to enhance the ML algorithm, but rather to sufficiently train citizens for the more complex tasks of the third stage, which involves further interaction with the event displays and the calculation of kinematic qualities.

5 Stage 3—Introduction to Real Data Analysis

In Stage 3, citizens delve into the analysis of real data collected with the ATLAS detector during the LHC's second operational phase (Run-2), drawn from the 13 TeV ATLAS Open Data set. The stage unfolds in two distinct sub-stages: Stage 3a, titled *Study of Higgs Bosons*, where participants explore decays of the SM Higgs boson into photons; and Stage 3b, titled *Discovery of Long-Lived Particles*, wherein they scrutinize Displaced Vertices (DVs) previously identified by ATLAS and look for consistency with the decay signatures of Long-Lived Particles (LLPs). Both sub-stages necessitate active engagement with the event display of `HYPATIA`. This interaction allows citizens not only to select particle traces but also to compute kinematic variables and probe properties of the sought-after LLPs. To facilitate these tasks, `HYPATIA` has been tailored to furnish the requisite information and functionalities for each sub-stage.

In Stage 3a (Fig. 4), citizens examine a collection of Higgs boson candidates, searching for signatures of their decay into two photons. They also consider scenarios where photons convert into electron pairs before reaching the electromagnetic calorimeter by interacting with the detector. By selecting any pair of clusters, HYPATIA calculates their invariant mass in real-time to help citizens determine if they are

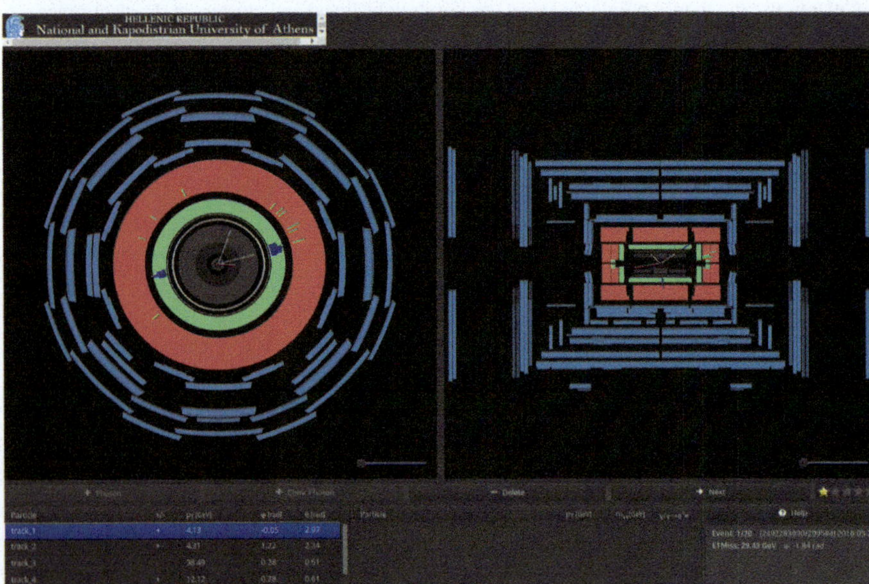

Fig. 4 HYPATIA display of a real collision event containing an electron and a photon candidate, in Stage 3a. Citizens use the mouse pointer to select pairs of photons or converted photons combatible with the products of a Higgs boson decay, and use the star-rating system to evaluate the event according to the provided guidelines

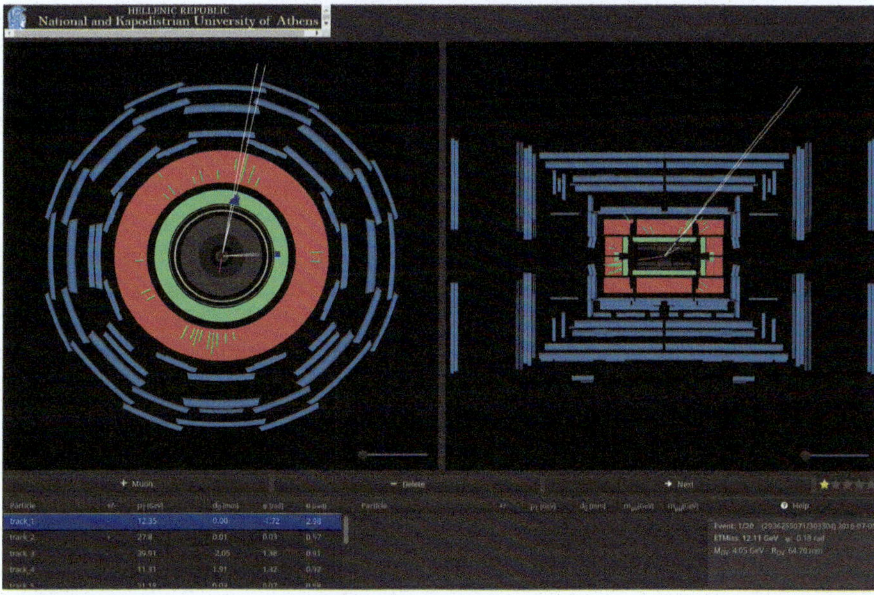

Fig. 5 HYPATIA display of a real collision event, in Stage 3b. Citizens use the mouse pointer to identify the characteristic DV from the decay of a LLP into a muon and hadronic jet, and use the star-rating system to evaluate the event according to the provided guidelines

compatible to the products a Higgs boson decay. HYPATIA also features a star-rating system allowing citizens to rate events with 1 to 5 stars, based on any additional occurrences, such as the presence of extra particles beyond those anticipated from the Higgs boson decay.

In Stage 3b (Fig. 5), citizens search for LLPs decaying to a muon and a hadronic jet by identifying their characteristic DV and tagging the track of the produced muon. Following the DV identification, they study kinematic properties of the decay products provided by HYPATIA, namely the invariant mass and the position of the DV, as well as the momentum of the muon, and rate the event using the star-rating system according to the likelihood that the DV corresponds to a LLP decay rather than arising from background processes.

In both parts of Stage 3, citizens have been performing equally well, having identified according to the given instructions 21369 candidate photon pairs with invariant mass close to the mass of the Higgs boson (Stage 3a) and 46502 DVs (Stage 3b) in the official duration of the project. In Stage 3a, citizens have rated a few hundred events with 5 stars, as likely to contain complex Higgs boson production mechanisms, whereas in stage 3b, multiple users reported certain events as candidates for LLP decays. These events were examined by the expert scientists and discussed with the citizens on the Zooniverse talk forum.

6 Conclusion

The REINFORCE demonstrator *New Particle Search at CERN* has successfully attracted thousands of citizens (2684 registered to date) interested in further understanding and contributing to frontier scientific research, and has managed to bring them together as an active community of citizen-scientists (about 3000 messages have been exchanged between either citizens or citizens and scientists on the Zooniverse talk forum). The data gathered from all three stages of the demonstrator have been analysed to probe the extent at which citizens could contribute to science and highlight possible directions in order to maximize the benefit of such contribution to science and to society. Finally, the incorporation of sonification ensures that citizens with visual impairments can equally participate in this scientific endeavor.

Acknowledgements The REINFORCE project has received funding from the European Union's Horizon 2020 project call H2020-SwafS-2018-2020 funded project Grant Agreement no. 872859.

References

1. REINFORCE project webpage: https://www.reinforceeu.eu
2. Zooniverse platform webpage: https://www.zooniverse.org
3. GWitch Hunters webpage: https://reinforceeu.eu/demonstrators/gwitchhunters
4. Deep Sea Explorers webpage: https://reinforceeu.eu/demonstrators/deep-sea-explorers
5. New Particle Search At CERN webpage: https://reinforceeu.eu/index.php/demonstrators/new-particle-search-cern
6. Cosmic Muon Images webpage: https://reinforceeu.eu/demonstrators/cosmic-muon-images
7. ATLAS Collaboration, "Search for long-lived massive particles in events with a displaced vertex and a muon with large impact parameter in *pp* collisions at $\sqrt{s} = 13$ TeV", Phys. Rev. D **102** 032006 (2020)
8. ATLAS Collaboration, Observation of a new particle in the search for the Standard Model Higgs boson with the ATLAS detector at the LHC. Phys. Lett. B **716**, 1–29 (2012)
9. ATLAS Collaboration, The ATLAS Experiment at the CERN Large Hadron Collider. JINST **3**, S08003 (2008)
10. C. Kourkoumelis, S. Vourakis, HYPATIA - An online tool for ATLAS event visualization. Phys. Educ. **49**, 21–32 (2014)
11. ATLAS Open Data webpage: http://opendata.atlas.cern
12. J. Casado et al., SonoUno: a user-centred approach to sonification. Proc. Int. Astron.Union **15**(S367), 120–123 (2019)

Multijet Cross Sections and Ratios in pp Collisions at $\sqrt{s} = 13$ TeV with CMS

Argyro Ziaka

Abstract A measurement of inclusive differential multijet cross sections and ratios is presented. The analysis is based on data from proton-proton (pp) collisions collected by the CMS [1] experiment at a center-of-mass energy of $\sqrt{s} = 13$ TeV, corresponding to an integrated luminosity of 59.82 fb^{-1} collected during the LHC Run2. The ratios from the corresponding differential multijet cross sections with different multiplicities R_{mn} are also calculated, where m, n stand for the multiplicity of the numerator and denominator. Jets are reconstructed using the anti-k_{T} clustering algorithm with a jet cone size of $R = 0.8$. The analysis considers events with jets with transverse momentum $p_{\text{T}} > 150\text{GeV}$ and absolute rapidity within $|y| < 2.5$. The inclusive differential multijet cross sections and their ratios R_{32}, R_{43}, and R_{42} are measured as a function of the average transverse momentum of the two leading jets. Experimental data are compared to predictions from simulations using various Monte Carlo event generators.

1 Introduction

1.1 Motivation and Observables

The pp collisions at the LHC are viewed as interactions among their constituent partons, (anti-)quarks and gluons. These interactions can be either hard or soft collisions, corresponding to high and low momentum transfer respectively. The probability for the final partons to radiate additional gluons, forming multijets in the final state after hadronisation is high. The multijet event cross section $\sigma_{\text{i-jet}}$ that describes a pp collision is proportional to the strong coupling constant as $\alpha_{\text{S}}^{\text{i}}$, where i is an integer representing the jet multiplicity. In the context of perturbative quantum chromodynamics (pQCD), the scattering cross section can be expressed as a sum of α_{S} terms

On behalf of the CMS Collaboration.

A. Ziaka (✉)
University of Ioannina, Ioannina, Greece
e-mail: argyro.ziaka@cern.ch

J. Strologas and L. Perivolaropoulos (eds.), *HEP2023*, Springer Proceedings in Physics 333 https://doi.org/10.1007/978-3-031-76856-9_19

of increasing power, convoluted with parton distribution functions (PDFs) inside the proton. The cases that are usually studied include the production of at least two jets and thus, terms of the form α_S^1 are not common. The lowest order α_S^2 amounts to the production of two-parton final states, whereas the higher order terms α_S^3 and α_S^4 represent the production of three and four partons in the final states [2]. Figure 1 shows the running of α_S, i.e. $\alpha_S(Q)$, derived from the latest jet-based CMS α_S running test [3], where a 2-loop solution for the Renormalisation Group Equations was used.

The current content is part of a Master Thesis at the University of Ioannina, which is archived at the Repository of UOI "Olympias" [4]. The main goal of the analysis is the measurement of the differential 2, 3, 4 and 5-jet event cross sections:

$$\frac{d\sigma}{dH_{T,2}/2} = \frac{1}{\mathcal{L}} \frac{N_{ev}}{\Delta(H_{T,2}/2)} \tag{1}$$

where $\mathcal{L}$ is the effective integrated luminosity, N_{ev} is the number of events counted in a bin and $\Delta(H_{T,2}/2)$ is the bin width. The observables are measured as a function of the average transverse momentum of the leading and sub-leading jets, $H_{T,2}/2 = \frac{1}{2}(p_{T,1} + p_{T,2})$, where $p_{T,1}$ and $p_{T,2}$ denote the transverse momenta of the two leading jets. The measurements are normalised by the bin width $(dH_{T,2}/2)$ and hence, are reported in units of (pb/GeV). Since the inclusive differential multijet cross sections are measured for multiplicity of 2, 3, 4 and 5, the number of events N_{ev} will stand for the counts of 2, 3, 4 and 5-jet events in each bin.

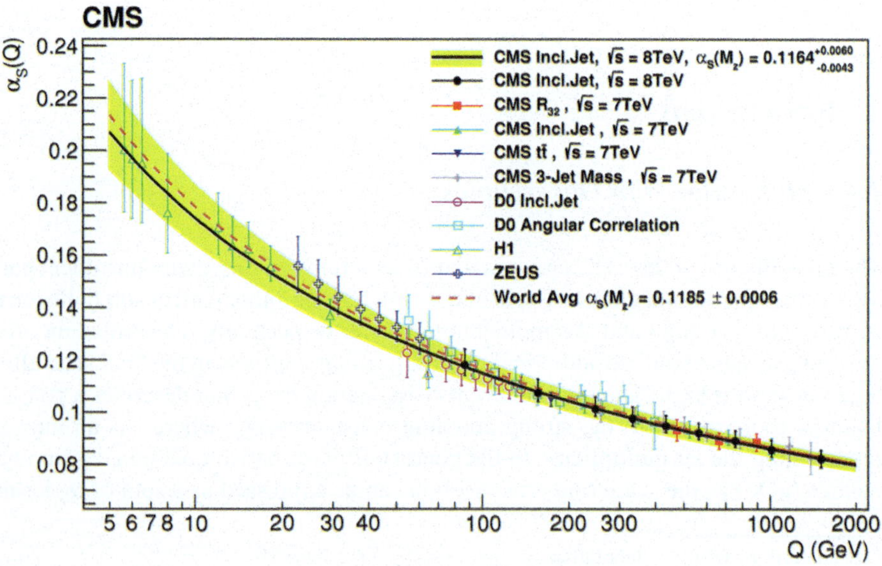

Fig. 1 Running of the strong coupling constant $\alpha_S(Q)$ as determined in [3]

More often than not, instead of studying inclusive jet cross sections, it is useful to calculate their ratios, where many theoretical and experimental uncertainties due to luminosity, scale and PDF dependence etc., may cancel between numerator and denominator. Ultimately, these ratios can be used to determine the strong coupling constant α_S. Thus, ratios of the form R_{mn} are studied, which stand for the ratio of the differential cross section of selected inclusive m-jet events to that of inclusive n-jet events in each bin, with m>n:

$$R_{\mathrm{mn}} = \frac{\left[\frac{d\sigma}{d(H_{\mathrm{T},2}/2)}\right]_{(\mathrm{m-jet})}}{\left[\frac{d\sigma}{d(H_{\mathrm{T},2}/2)}\right]_{(\mathrm{n-jet})}} = \frac{\sigma_{\mathrm{m-jet}}}{\sigma_{\mathrm{n-jet}}} \tag{2}$$

These ratios are proportional to $\alpha_S^{\mathrm{m-n}}$. More specifically, the observables R_{32}, R_{43} and R_{42} are calculated, using the distributions that correspond to the inclusive 2, 3 and 4-jet cross sections. Hence, the first two of these ratios are proportional to α_S, whereas the last ratio is proportional to α_S^2.

Previous analyses of the ratio R_{32} as a function of the average transverse momentum of the two leading p_{T} jets in the event was performed at $\sqrt{s} = 7$ TeV [5] by the CMS Collaboration and lead to an extraction of $\alpha_S(M_Z) = 0.1148 \pm 0.0014$ (exp.) ± 0.0018 (PDF) ± 0.0050 (theory), where the dominant uncertainty stems from the estimation of higher-order corrections to the next-to-leading order (NLO) prediction. Another measurement of the ratio R_{32} was performed by the CMS Collaboration at $\sqrt{s} = 8$ TeV [2], again, as a function of the average transverse momentum of the two

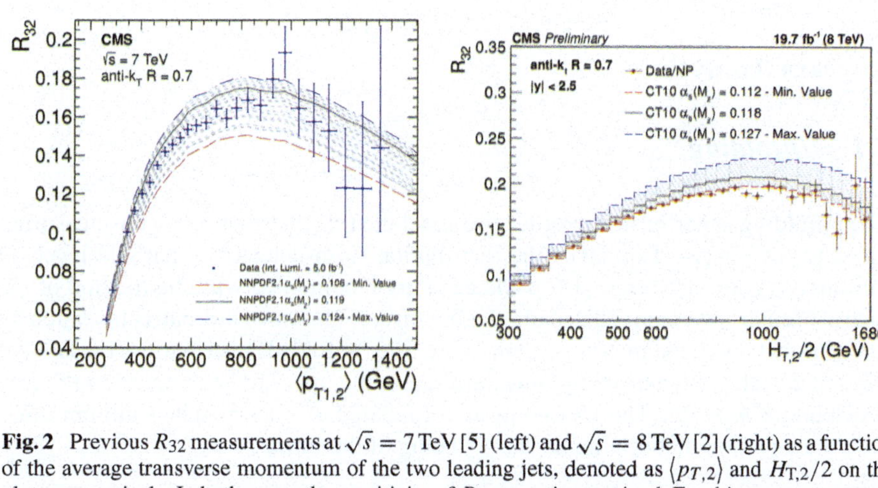

Fig. 2 Previous R_{32} measurements at $\sqrt{s} = 7$ TeV [5] (left) and $\sqrt{s} = 8$ TeV [2] (right) as a function of the average transverse momentum of the two leading jets, denoted as $\langle p_{T,2} \rangle$ and $H_{\mathrm{T},2}/2$ on the plots, respectively. In both cases, the sensitivity of R_{32} to α_S is examined. For this purpose, a range of α_S values was used for a given PDF set; the red line represents the minimum, the blue line the maximum, the continuous black line the central value of this range, whereas the black dotted lines stand for the intermediate values. The points represent the data

leading jets in the event, giving $\alpha_S(M_Z) = 0.1150 \pm 0.0010$ (exp) ± 0.0013 (PDF) ± 0.0015 (NP) $^{+0.0050}_{-0.0000}$ (scale). These are depicted in Fig. 2.

1.2 Event Selection

The measurements are based on the data collected by the CMS detector at the CERN Large Hadron Collider during 2018, for pp collisions with center of mass energy of 13 TeV, corresponding to a luminosity of 59.82 fb^{-1}. Furthermore, the Monte Carlo (MC) technique was used for the simulation; the PYTHIA 8 [6] tune CP5 [7] event generator is an official CMS MC sample, whereas the simulated events from the other two PYTHIA 8 tunes, CUETP8M1 [8] and CUETP8M2T4 [9], as well as HERWIG++ [10] were analysed using the RIVET (Robust Independent Validation of Experiment and Theory) [11] toolkit.

Events were selected requiring all jets to have $p_T > 150$ GeV and $|y| < 2.5$. The type of jets used in the analysis was PFCHS jets, meaning that they were reconstructed with the Particle Flow algorithm, which combines the information from the individual subdetectors, along with the charged-hadron substraction method (CHS) so that contributions from pile-up are mitigated [12]. Moreover, the anti-k_T clustering algorithm [13] was used in the current analysis, that repeatedly recombines objects based on their momentum space distance, with a distance parameter, that defines the jet radius, set at $R = 0.8$. In general, the distance parameter is given as: $\Delta R = \sqrt{(\Delta\phi)^2 + (\Delta\eta)^2}$.

2 Data Analysis

2.1 Unfolding

The unfolding is performed in order to extract the true distribution by unsmeasuring the detector effects. This takes the distribution from detector to particle level. In this analysis the TUNFOLD [14] package is used, combined with the method of the matrix pseudo-inversion. This method relies on the solution of the matrix inversion by using more bins in the measured than in the true (unfolded) distribution, i.e. $m > n$. More precisely, twice as many bins are used in the measured compared to the true distribution: $m = 2n$. The Unfolding is formulated as a least square minimisation with the equation that minimises the χ^2 being of the form:

$$\chi^2 = (\mathbf{Ax} + \mathbf{b} - \mathbf{y})^{\mathbf{T}}(\mathbf{V})^{-\mathbf{1}}(\mathbf{Ax} + \mathbf{y} - \mathbf{b}) \quad (3)$$

where $\mathbf{A}$ is the m $\times$ n response matrix that can be interpreted as a density function, $\mathbf{x}$ is the unknown true distribution that needs to be determined, the vector $\mathbf{b}$ describes the expected number of background events, **y** is the measured distribution and **V** is the covariance matrix that describes the statistical covariance among the bins of the measurement [15].

For the easier treatment of correlations and migrations, the unfolding was performed on the exclusive 2, 3, 4 and inclusive 5-jet cross sections, where each event is filled in only one distribution. TUnfoldBinning has been used and therefore, the x axis has switched from the $H_{T,2}/2$ bins to global bin IDs, i.e. to the bins of each distribution will be assigned an integer, starting from 1, and all spectra will be depicted consecutively on the same axis. The enumeration begins from the exclusive 2-jet cross section and continues with the bins of the next distribution. As a result, the x axis no longer has a physical binning. Consequently, the events will be dealt with a 2D unfolding, where the treatment of correlations is simpler. The reason is that pinpointing migrations from one bin to another, but also among distributions, is much easier. On the contrary, by filling the inclusive jet distributions with different multiplicity, one would be faced with multiple migrations. The result is shown in Fig. 3 along with the corresponding Response Matrix. The Response Matrix (RM) for the inclusive multijet cross section distributions was built using PYTHIA 8 with the CP5 tune. The RM was manufactured on an event-by-event basis by matching the jets between particle level (GEN jets) and detector level (PF jets) in the simulated events.

The correlations are represented by the dimensionless version of the covariance, included in the correlation matrix. The Correlation Matrix was produced (Fig. 4) for the exclusive 2, 3, 4 and inclusive 5-jet cross sections, but also for the R_{mn} ratios. In the former case, the Correlation Matrix contains the main diagonal but also off-diagonal elements, and portrays non-negligible correlations among the $H_{T,2}/2$ bins but also small correlations and anti-correlations among distributions with different

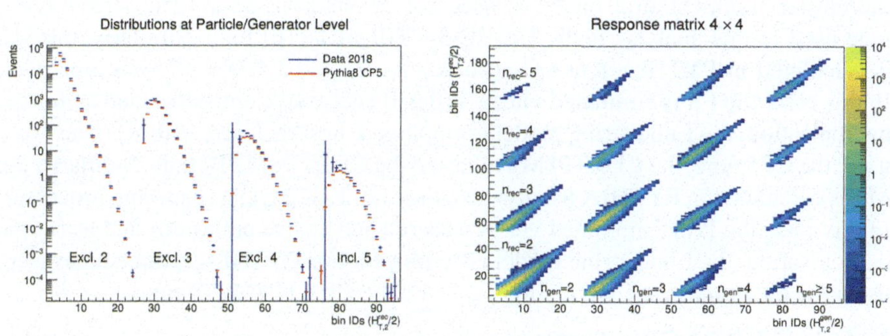

Fig. 3 Exclusive 2, 3, 4 and inclusive 5-jet cross sections of the 2018 Data and PYTHIA 8 at particle and generated level respectively (left) and the corresponding Response Matrix (right) built using PYTHIA 8 simulated events. The horizontal axis corresponds to generator-level jet p_T and the vertical axis to reconstructed-level jet p_T. The RM contains the transverse momentum p_T on the horizontal and vertical axes of each cell. The color band covers a range between 10^{-4} and 10^4 and indicates the migrations from a particle (generator)-level bin to the corresponding detector (reconstructed)-level bin to occur

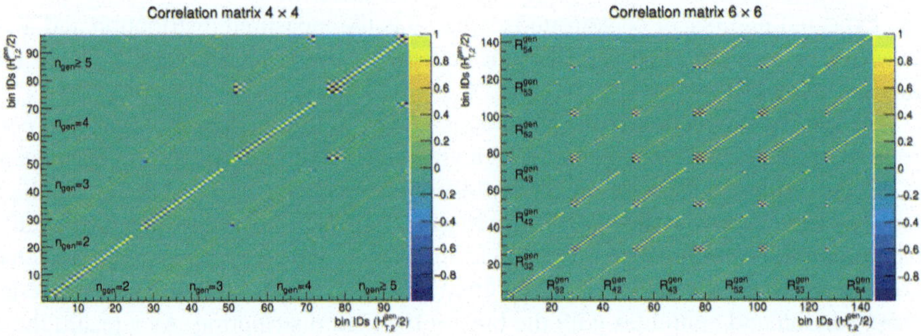

Fig. 4 Correlation matrices after the unfolding for the exclusive 2, 3, 4 and inclusive 5-jet distributions (left) and the R_{mn} observables (right) at particle level. The value 1 corresponds to fully correlated bins, while the negative values describe anti-correlated bins

multiplicity. In the latter case, the Correlation Matrix includes the R_{mn} ratios at particle level on both the x and y axis; one can deduce that there are strong correlations and anti-correlations among the $H_{\mathrm{T,2}}/2$ bins, but also among the R_{mn} ratios.

2.2 *Particle Level Measurement*

The experimental data are compared to predictions from HERWIG++ and PYTHIA 8 MC event generators. The HERWIG++ event generator computes the matrix element (ME) at leading order (LO) accuracy for $2 \rightarrow 2$ scattering processes, with the parton shower (PS) being simulated through angular-ordered successive emissions, while the cluster fragmentation model is used for the hadronisation. The HERWIG++ tune used in this analysis is the UE-EE-5-CTEQ6L1 [16] tune. Similarly, the ME is calculated in PYTHIA 8 at LO accuracy for $2 \rightarrow 2$ QCD scattering processes. In this case, the PS is simulated via p_{T}-ordered successive emissions and as for the hadronisation, the Lund string model is employed. For PYTHIA 8, three tunes were used: the CP5 tune, the CUETP8M1 tune and the CUETP8M2T4 tune. Similarly, the ME is calculated in PYTHIA 8 at LO accuracy for $2 \rightarrow 2$ QCD scattering processes. In this case, the PS is simulated via p_{T}-ordered successive emissions and as for the hadronisation, the Lund string model is employed. For PYTHIA 8, three tunes were used: the CP5 tune, the CUETP8M1 tune and the CUETP8M2T4 tune.

Inclusive Multijet Cross Sections. In Fig. 5 the inclusive 2 (top left), 3 (top right), 4 (bottom left) and 5- (bottom right) jet cross sections are depicted. The particle level data for the cross section measurements are compared to the LO MC event generators PYTHIA 8 tune CP5, PYTHIA 8 tune CUETP8M1, PYTHIA 8 tune CUETP8M2T4 and HERWIG++ tune UE-EE-5-CTEQ6L1. On the bottom panel of each plot, the ratios between data and Monte Carlo predictions are shown, with the statistical uncertainties drawn with black error bars.

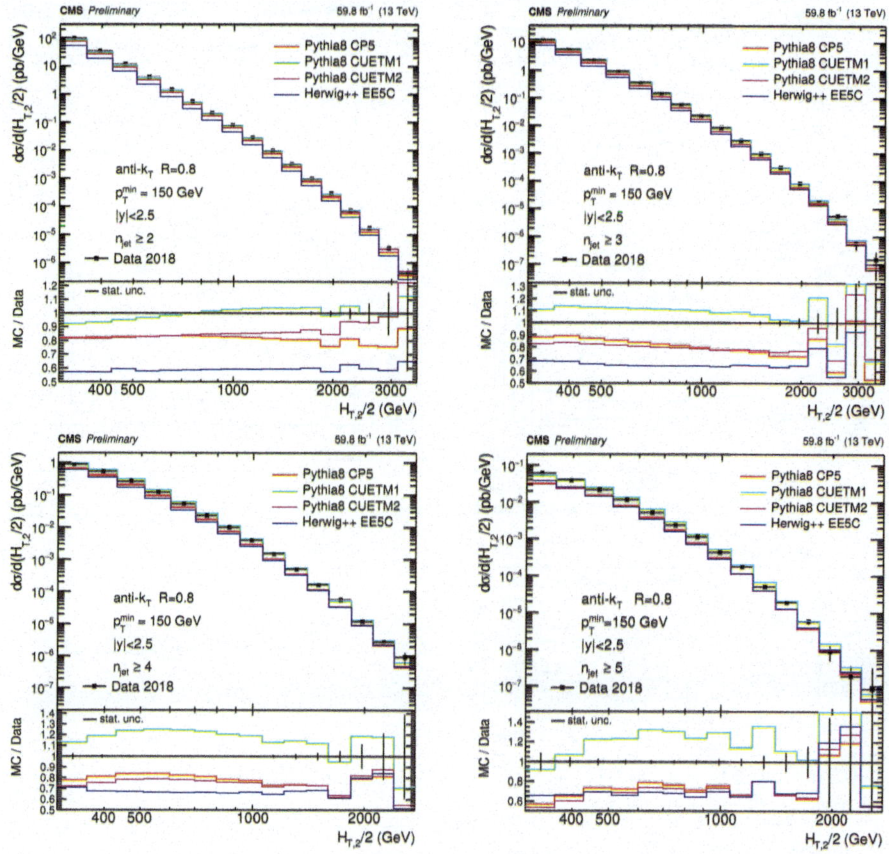

Fig. 5 Comparison of the inclusive differential 2 (top left), 3 (top right) 4 (bottom left) and 5-jet (bottom right) cross section between Data and LO Monte Carlo predictions at particle level obtained from HERWIG++ and PYTHIA 8. The data are drawn with black markers and the MC predictions with solid colored lines

The distributions exhibit the expected declining behavior. For the n$\geq$ 2 jets case, the predictions from (LO) PYTHIA 8 CP5 and CUETP8M2T4 provide a satisfying description of the measurement, while the predictions from (LO) HERWIG++ underestimate it by $\sim$ 35% in all cases. However, when moving to higher multiplicity, the agreement between the experimental data and the MC predictions becomes worse, thus, the LO MC event generators no longer describe the measurement with great accuracy.

R_{mn} Ratios. In Fig. 6 the $R_{\rm mn}$ ratios are portrayed, that include the R_{32} (top left), the R_{43} (top right) and the R_{42} (bottom) observables. The particle level data for the $R_{\rm mn}$ observables are compared to the LO MC event generators PYTHIA 8 tune CP5, PYTHIA 8 tune CUETP8M1, PYTHIA 8 tune CUETP8M2T4 and HERWIG++ tune

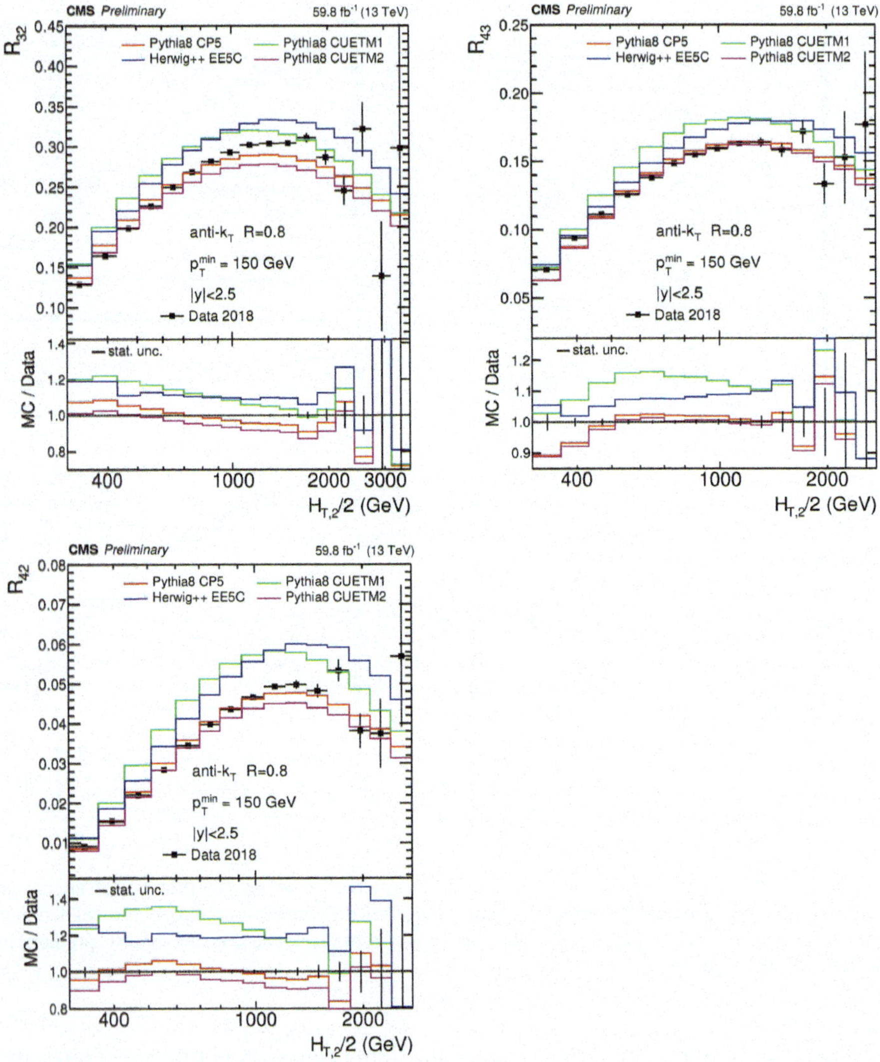

Fig. 6 Comparison of the R_{32} (top left), R_{43} (top right) and R_{42} (bottom) observable between Data and LO Monte Carlo predictions at particle level

UE-EE-5-CTEQ6L1. On the bottom panel of each plot, the ratios between data and MC predictions are shown, with the statistical uncertainties drawn with black errors.

From 300 GeV up to around 1000 GeV, the R_{mn} distributions increase, since the probability for the production of an additional jet becomes higher, due to the fact that the phase space for its production broadens. The distributions reach a plateau until around 1600 GeV and then they start to decrease.

The predictions from (LO) PYTHIA 8 CP5 and CUETP8M2 are in nice agreement with the data $\sim 5-10\%$. The corresponding predictions from HERWIG++ deviate from the data by $\sim 20\%$ for R_{32} and R_{42} and $\sim$10% for R_{43}. It is also evident that some MC event generators overestimate and others underestimate the measurement, while in some regions, the prediction from the simulation lies outside the statistical error bars of the data. The statistical uncertainties that originate from the MC are negligible compared to the respective uncertainties from the data and for this reason they are not used in the comparison. However, they are intended to be included in the total systematic uncertainties at later stage of the analysis. The quality of the agreement lies on the fact that at this stage, only LO MC event generators have been used. For next-to-leading (NLO) and next-to-next-to-leading order (NNLO) event generators, the agreement with the data is expected to be better.

3 Summary

The measurement of the normalised inclusive differential 2, 3, 4 and 5-jet cross sections as well as the extraction of the R_{32}, R_{43} and R_{42} observables are presented. These measurements were performed as a function of the average transverse momentum of the leading and sub-leading jets, $H_{T,2}/2$, and were based on data from pp collisions collected by the CMS experiment at a center-of-mass energy of 13 TeVand corresponding to an integrated luminosity of 59.82 fb^{-1}. The experimental data are compared to predictions from Monte Carlo event generators; more specifically, PYTHIA 8 with tunes CP5, CUETP8M1 and CUETP8M2T4 and HERWIG++ with tune UE-EE-5-CTEQ6L1 were used.

The inclusive differential multijet cross sections are best described by PYTHIA8 CP5 and CUETP8M2T4, however as the multiplicity increases, all MC event generators fail to provide an accurate description. As for the R_{mn} observables, they are in good agreement with PYTHIA 8 with tune CUETP8M2T4 and CP5. In all cases, HERWIG++ and PYTHIA 8 provide the worst description of the R_{mn} observables.

Acknowledgements The research project was supported by the Hellenic Foundation for Research and Innovation (H.F.R.I.) under the "1st Call for H.F.R.I. Research Projects to support Faculty Members & Researchers and the Procurement of High-cost research equipment grant" (Project Number: 2288).

H.F.R.I.
Hellenic Foundation for Research & Innovation

References

1. C.M.S. Collaboration, The CMS experiment at the CERN LHC. JINST **3**, S08004 (2008). https://doi.org/10.1088/1748-0221/3/08/S08004
2. CMS Collaboration, Determination of the strong coupling constant from the measurement of inclusive multijet event cross sections in pp collisions at $\sqrt{s} = 8$ TeV. CMS-PAS-SMP-16-008 (2016). https://inspirehep.net/literature/1514769
3. V. Khachatryan et al. [CMS Collaboration]: measurement and QCD analysis of double-differential inclusive jet cross sections in pp collisions at $\sqrt{s} = 8$ TeV and cross section ratios to 2.76 and 7 TeV. JHEP **03**, 156 (2017). https://doi.org/10.1007/JHEP03(2017)156
4. A. Ziaka, Multijet cross section measurements up to the multiplicity of five jets with data from the CMS experiment at the LHC in proton - proton collisions at the center of mass energy of 13 TeV, MSc Thesis, University of Ioannina, Greece (2023). https://olympias.lib.uoi.gr/jspui/handle/123456789/32984
5. S. Chatrchyan et al., [CMS Collaboration]: Measurement of the Ratio of the Inclusive 3-Jet Cross Section to the Inclusive 2-Jet Cross Section in pp Collisions at $\sqrt{s} = 7$ TeV and First Determination of the Strong Coupling Constant in the TeV Range. Eur. Phys. J. C. **73**, 2604 (2013). https://doi.org/10.1140/epjc/s10052-013-2604-6
6. T. Sjöstrand, S. Ask, J.R. Christiansen, R. Corke, N. Desai, P. Ilten, S. Mrenna, S.Prestel, C.O. Rasmussen, P.Z. Skands, An introduction to PYTHIA 8.2. Comput. Phys. Commun. **191**, 159–177 (2015). https://doi.org/10.1016/j.cpc.2015.01.024
7. A.M. Sirunyan, A. Tumasyan, W. Adam et al., Extraction and validation of a new set of CMS PYTHIA8 tunes from underlying-event measurements. Eur. Phys. J. C **80**, 4 (2020). https://doi.org/10.1140/epjc/s10052-019-7499-4
8. V. Khachatryan, A.M. Sirunyan, A. Tumasyan et al., Event generator tunes obtained from underlying event and multiparton scattering measurements. Eur. Phys. J. C **76**, 155 (2016). https://doi.org/10.1140/epjc/s10052-016-3988-x
9. CMS Collaboration, Investigations of the impact of the parton shower tuning in Pythia 8 in the modelling of $t\bar{t}$ at $\sqrt{s} = 8$ and 13 TeV. Technical report, CERN, Geneva (2016). https://cds.cern.ch/record/2235192
10. M. Bahr et al., Herwig++ physics and manual. Eur. Phys. J. C **58**, 639–707 (2008). https://doi.org/10.1140/epjc/s10052-008-0798-9
11. C. Bierlich et al., Robust independent validation of experiment and theory: rivet version 3. SciPost Phys. **8**, 026 (2020). https://doi.org/10.21468/SciPostPhys.8.2.026
12. K. Rabbertz, *Jet Physics at the LHC: The Strong Force beyond the TeV Scale*. (Springer International Publishing, 2017)
13. M. Cacciari et al., The anti-k_t jet clustering algorithm. JHEP **04**, 063 (2008). https://doi.org/10.1088/1126-6708/2008/04/063
14. S. Schmitt, TUnfold, an algorithm for correcting migration effects in high energy physics. J. Instrum **7**, T10003 (2012). https://doi.org/10.1088/1748-0221/7/10/T10003
15. S. Schmitt, Data Unfolding Methods in High Energy Physics. EPJ Web of Conf. **137**, 11008 (2017). https://doi.org/10.1051/epjconf/201713711008
16. Michael H. Seymour, Andrzej Siodmok, Constraining MPI models using σ_{eff} and recent Tevatron and LHC Underlying Event data. JHEP **10**, 113 (2013). https://doi.org/10.1007/JHEP10(2013)113

Multi-differential Measurement of the Dijet Cross Section in Proton-Proton Collisions at $\sqrt{s} = 13$ TeV with CMS

Polidamas Georgios Kosmoglou Kioseoglou

Abstract A measurement of the dijet production cross section is reported based on an integrated luminosity of 36.3fb^{-1} of proton–proton collision data collected in 2016 at $\sqrt{s} = 13$ TeV by the CMS detector at the CERN LHC. Jets are reconstructed with the anti-k_T algorithm using distance parameters of $R = 0.4$ and $R = 0.8$ and differential cross sections are measured as a function of the kinematic properties of the two jets with largest transverse momenta p_T. Double-differential (2D) measurements are presented as a function of the largest absolute rapidity $|y|_{max}$ of the two jets and the dijet invariant mass $m_{1,2}$. Triple-differential (3D) measurements are presented as a function of the dijet rapidity separation y^*, the total boost y_b of the dijet system, and either $m_{1,2}$ or the average dijet transverse momentum $\langle p_T \rangle_{1,2}$ as the third variable. The measured cross sections are unfolded in order to correct for detector effects and are compared with fixed-order calculations derived at next-to-next-to-leading order in perturbative quantum chromodynamics. The impact of the 2D and 3D measurements on the determination of the parton distribution functions is investigated. The inclusion of the 2D cross sections leads to a more precise determination of the strong coupling constant (α_S), which is measured to be $\alpha_S(m_Z) = 0.1179 \pm 0.0019$.

Keywords SM · Proton · Jets · QCD · pQCD · PDFs

1 Dijet Measurements

Jets are among the most frequently encountered objects in high energy proton-proton (pp) collisions. They provide an understanding of the underlying physics that describes their production and the interactions between the particles that comprise

---

Polidamas Georgios Kosmoglou Kioseoglou on behalf of the CMS Collaboration.

P. G. K. Kioseoglou (✉)
University of Ioannina, Ioannina, Greece
e-mail: polidamas.kosmoglou@cern.ch

J. Strologas and L. Perivolaropoulos (eds.), *HEP2023*, Springer Proceedings in Physics 333 https://doi.org/10.1007/978-3-031-76856-9_20

them. Precise dijet measurements can be exploited to further constrain existing parton distribution functions (PDFs) and study α_S.

The inclusive dijet production cross section is measured in proton-proton collisions at the CERN LHC with data collected by the CMS detector [1] in 2016 at $\sqrt{s} = 13\,\text{TeV}$, corresponding to an integrated luminosity of $36.3\,fb^{-1}$. The double-differential (2D) measurement is performed as a function of the largest absolute rapidity $|y|_{max}$ of the two jets with the largest p_T in the event and their invariant mass $m_{1,2}$. The triple-differential (3D) measurement is then taken as a function of the rapidity separation $y^* = |y_1 - y_2|/2$, the total boost of the dijet system $y_b = |y_1 + y_2|/2$, and either the invariant mass $m_{1,2}$ or the average transverse momentum $\langle p_T \rangle_{1,2}$ of the two jets.

Jet reconstruction is performed with the anti-k_T clustering algorithm [2]. Two different values of the jet distance parameter R are used, 0.4 and 0.8, amounting to six dijet measurements (two 2D measurements with respect to $m_{1,2}$ and four 3D measurements with respect to either $m_{1,2}$ or $\langle p_T \rangle_{1,2}$) in total.

1.1 Data and Simulation Samples

An extensive set of jet-related triggers are used to efficiently collect and later recombine the recorded data. Single-jet (dijet) triggers are used for the 2D (3D) measurements each requiring at least one jet (two jets) with a p_T (average p_T) above a certain threshold (40, 60, 80, 140, 200, 260, 320, 400, 450, 500 GeV). Triggers are prescaled by a factor to keep under control the data collection rate in different phase space regions, resulting in lower effective integrated luminosities. Simulated samples are generated with the Pythia8 [3] Monte Carlo (MC) event generator at LO in perturbative quantum chromodynamics (pQCD). The CUETP8M1 [4] tune is used to emulate the hadronization and the multi parton interactions. The CMS detector is simulated using the GEANT4 package [5].

1.2 Event Selection

In the offline analysis only a subset of the overall dataset is used. Specifically an event is accepted if the dijet system respects: $p_{T,1} \geq 100\,\text{GeV}$, $p_{T,2} \geq 50\,\text{GeV}$ and $|y_1| < 2.5\,(3.0)$, $|y_2| < 2.5\,(3.0)$ for the 2D (for the 3D) measurement. Events that pass the selection phase are then corrected for a series of known effects. Jet four momentum is calibrated. The data are further corrected to treat detector inefficiencies that were observed in certain phase space regions. Since the MC samples are simulated through more ideal detector conditions the pile up profile information is reweighted to match the one observed in data. Lastly, simulated samples are smeared according to the detector resolution and are cleaned from unphysical events.

1.3 Particle-Level Distributions

To obtain the "truth level" cross sections the measured distributions have to be unfolded. Key element of this operation is the construction of a response or probability matrix which holds the probability of an event generated in a given bin i, to end up in bin j once recorded from the detector. It is build from the MC sample accessing the information both at generator and reconstructed level to calculate $m_{1,2}$. The TUnfold [6] package is used and no explicit regularization is applied. The problem is treated with the pseudo-matrix inversion approach, choosing a larger number of bins for the reconstructed distributions than for the particle-level distributions. Because the resolution is larger at outer rapidity regions a coarser particle-level binning is chosen for the two outermost $|y|_{max}$ regions for the 2D measurements, and the corresponding nine outermost $(y^*,\ y_b)$ regions for the 3D measurements. In the end, it is said that distributions are unsmeared, i.e., event migrations are restored. With that respect migrations are treated equally among all observable bins resorting to multi-dimensional unfolding.

1.4 Experimental Uncertainties

Retrieving event migrations is no trivial task. After unfolding bins are correlated due to such migrations and the statistical uncertainties need to be treated during error propagation. For this reason, the initial covariance matrix obtained from data, in it's trivial form, is propagated through the procedure to account for these non-negligible correlations. Similarly, systematic effects are also propagated for both data and simulation. For data, variations are created based on nominal distributions while for simulation, variations are taken on the response matrix. Unfolding is performed and the difference between the resulted up/down variations, for a given source, to the nominal one is attributed to an asymmetric systematic uncertainty for the respective source. The final experimental uncertainty is then taken as the statistical plus all systematic uncertainties in quadrature.

2 Fixed Order Theory Predictions

Fixed order (FO) predictions are obtained with the NNLOJET program [7] (revision 5918) as FASTNLO [8, 9] (version 2.3) interpolation grids within the APPLFAST interface [10] (version 0.046). The matrix element (ME) is evaluated up to next-to-next-to-leading order (NNLO) accuracy in pQCD with the renormalization and factorization scales chosen to be equal to the dijet invariant mass $\mu_R = \mu_F = m_{1,2}$. Figure 1, left shows the prediction as obtained for the 2D measurement for $1 <$

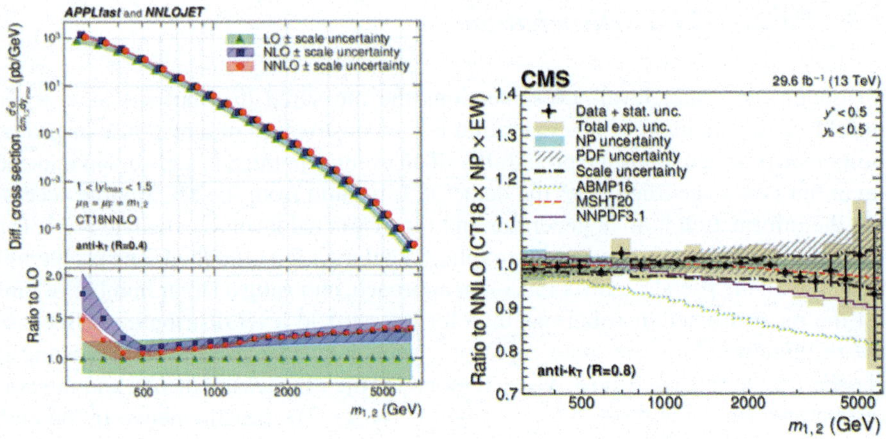

Fig. 1 FO predictions at NNLO for the 2D measurement as a function of $m_{1,2}$ for $1 < |y|_{max} < 1.5$ [10] (left). Ratio of data over theory for the 3D measurement as a function of $m_{1,2}$ for $y^* < 0.5,\ y_b < 0.5$ [15] (right)

$|y|_{max} < 1.5$ at each order in pQCD, up to NNLO. The bands display the scale uncertainty where a good perturbative convergence is observed as points from a higher order calculation mostly lie inside the band from the previous, lower order, calculation.

The produced ME is combined with the CT18 at NNLO PDF set [11] to obtain the parton level cross sections. Nonperturbative (NP) effects, as is the hadronization and multiparton interactions, need to be taken into account to correctly compare data to theory. An extensive set of MC configurations have been used for this purpose, to evaluate the correction factors later applied to theory. Additionally, the theory is corrected for extra electroweak (EW) radiation (virtual exchange of soft or collinear $W^{\pm}$ or Z bosons) [12], an effect mostly dominant in the region above 1 TeV. It is noted that the current predictions are based on the leading-color (N_C) and leading-flavor-number (N_F) approximations [13, 14]. Subleading-color contributions have not yet been tried for this set of observables so their impact still remains unknown. Thus these measurements provide an opportunity to test such calculations.

Three sources of theoretical uncertainties are investigated, namely the scale, PDF and NP, with the total uncertainty taken as the sum in quadrature of the effects of these sources. A scale uncertainty is introduced to account for the missing higher orders in pQCD and is estimated from a six scale variation of μ_R, μ_F around their nominal values as (1/2, 1/2), (1/2, 1), (1, 1/2), (2, 1), (1, 2), and (2, 2). The PDF uncertainty is taken as 68% confidence intervals following the prescriptions given in the respective reference [11]. Lastly, the NP uncertainty is calculated by taking the envelope of the nonperturbative contribution as estimated by different MC event generators.

2.1 *Data to Theory Comparison*

The experimental data are compared to the NNLO predictions, and the agreement is found to be good. The fractional difference remains below 10% in most phase space regions, within respective uncertainties, for all six measurement configurations. Best agreement is generally displayed by the larger jet cone size ($R = 0.8$). FO predictions are additionally combined with other PDF sets, namely ABMP16 [16], MSHT20 [17] and NNPDF3.1 [18], and compared to the CT18 baseline. MSHT20 seems to display the best agreement to the CT18 PDF set. Again, one such case of a data to theory comparison is provided in Fig. 1, right. The points represent the data to theory ratio around of which the statistical and the total experimental uncertainties are displayed. Around unity different bands illustrate individual theory uncertainties while continuous lines represent ratio to theories, between a given PDF set and the CT18.

3 Quantum Chromodynamic Analysis

Aiming to further constrain the proton PDFs, a choice had to be made for picking the right measurements to undertake the QCD analysis. A profile was created, to exploit the best agreement displayed by the NNLO pQCD predictions and to reach the highest energy scales possible. Two measurements succeed in meeting these requirements, a double-differential and a triple-differential, given as a function of $m_{1,2}$ and using the larger cone size for the jets $R = 0.8$. The strategy followed is the same as in past HERAPDF analyses [19] where HERA deep inelastic scattering (DIS) data are used and are complemented with the CMS dijet cross sections.

Both PDF-only (with $\alpha_S(m_Z) = 0.118$ fixed) and PDF+α_S fits are performed with the XFITTER program (version 2.0.1) [20] interfaced to FASTNLO. The PDFs are parame-trized to the proton momentum fraction x as:

$$xf(x, \mu_{F,0}^2) = A_f x^{B_f}(1-x)^{C_f}(1 + D_f x + E_f x^2) \tag{1}$$

where f corresponds to the individual distributions that express the proton structure, namely the gluon distribution $g(x)$, the up and down valence quark distributions $u_v(x)$, $d_v(x)$ and the up- and down-type sea antiquark distributions $\overline{U}(x)$, $\overline{D}(x)$. PDFs are determined at a starting scale of $\mu_{F,0}^2 = 1.9\,\text{GeV}^2$ and subsequently evolved at the scale of the top quark mass m_t.

Due to the overall rules and constrains applied, the fit initially starts with ten A, B and C parameters which are evaluated during the procedure. Additional D and E parameters are then added based on a χ^2 minimization scan. The final parametrization is shown in Table 1. Note that, amazingly, the fit seems to prefer the exact same parametrization with either the inclusion of the 2D or 3D cross sections.

The fits have been performed in two different scenarios, once by exploiting all available phase space and once by restricting it. The former scenario showed that including the forward rapidity regions resulted in cases that the partial χ^2 divided

Table 1 Final PDF parametrization obtained after the χ^2 minimization scan for the fits of HERA DIS data together with the CMS 2D or 3D dijet cross sections [15]

PDF	Fitted data sets HERA DIS + CMS dijets
$xg(x, \mu_{F,0}^2)$	$A_g x^{B_g}(1-x)^{C_g}$
$xu_v(x, \mu_{F,0}^2)$	$A_{u_v} x^{B_{u_v}}(1-x)^{C_{u_v}}(1 + D_{u_v}x + E_{u_v}x^2)$
$xd_v(x, \mu_{F,0}^2)$	$A_{d_v} x^{B_{d_v}}(1-x)^{C_{d_v}}$
$x\overline{U}(x, \mu_{F,0}^2)$	$A_{\overline{U}} x^{B_{\overline{U}}}(1-x)^{C_{\overline{U}}}(1 + D_{\overline{U}}x)$
$x\overline{D}(x, \mu_{F,0}^2)$	$A_{\overline{D}} x^{B_{\overline{D}}}(1-x)^{C_{\overline{D}}}$

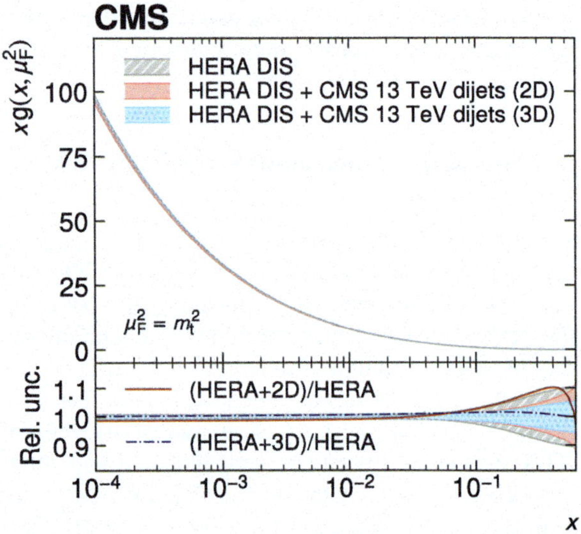

Fig. 2 The gluon PDF with the inclusion of 2D (3D) cross sections [15] in red (blue). The fit uncertainty is displayed around the central PDF value (top). The ratio between the inclusion of the CMS data and the HERA DIS only fits with the respective relative uncertainty (bottom)

by the number of data points was increased, reaching values up to ≈ 3. Therefore, in the second scenario, the phase space was restricted and some of the forward rapidity regions (one $|y|_{max}$ and six y_b, y^* bins) that displayed a poor fit quality were dropped. The resulted gluon PDF evolved at the scale of the m_t can be seen in Fig. 2 for the second scenario where only a subset of rapidity regions was included. There a comparison for the gluon PDF is performed between the plain HERA DIS result and the one obtained by including the 2D (3D) CMS dijet cross sections in red (blue).

Four sources of systematic uncertainties are taken into account as a total PDF uncertainty:

- the fit uncertainty, as estimated by the MC method, i.e., by performing a large number of alternative fits where the input data are randomly varied based on their statistical and systematic uncertainties, taking the standard deviation of the resulting PDFs as a final estimate,
- the scale uncertainty estimated by taking a six point variation envelope around the nominal value of μ_R and μ_F,

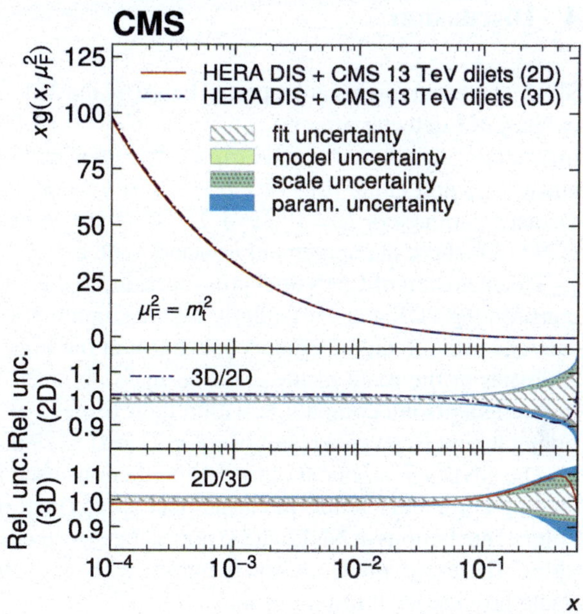

Fig. 3 The gluon PDF central value with the inclusion of 2D (3D) cross sections [15] in red (blue) (top). Individual uncertainty contributions are illustrated by bands with different styles indicating the relative uncertainty for each fit separately (bottom)

- the model uncertainty taken as the sum in quadrature of non-PDF parameter variations (e.g., $\mu_{F,0}$, the strangeness fraction f_s, the charm and bottom quark masses m_c and m_b),
- the parametrization uncertainty, by taking an envelope of all fits performed with one additional E or D parameter in addition to the final parametrization.

The effects of all sources are added in quadrature to the total PDF uncertainty with the exception of the effect of parametrization, which is added linearly. Figure 3 shows the individual uncertainty effects on the gluon PDF determination for the 2D (3D) measurement in red (blue).

For the determination of α_S, PDF-only fits are performed using the exact same parametrization as in the PDF+α_S fits where α_S is now taken as an additional parameter. The process is repeated two times, once with the inclusion of the 2D cross sections and once with the 3D cross sections. The resulted values for α_S at the scale of the Z boson mass m_Z are given below, respectively:

$$\alpha_S(m_Z) = 0.1179\ (15)_{fit}\ (8)_{scale}\ (8)_{model}\ (1)_{param.}\ |\ (19)_{total} \tag{2}$$

$$\alpha_S(m_Z) = 0.1181\ (13)_{fit}\ (9)_{scale}\ (6)_{model}\ (2)_{param.}\ |\ (22)_{total} \tag{3}$$

The two values are in good agreement taking uncertainties into account, with a more precise result given by the inclusion of the 2D cross sections. Comparing the values obtained with this study to the current α_S world average ($\alpha_S(m_Z) = 0.1179 \pm 0.0009$) they are also found to be in good agreement.

4 Discussion

Six dijet measurements are reported based on the 2016 dataset of 36.3 fb^{-1} recorded by the CMS detector after pp colissions at $\sqrt{s} = 13$ TeV [15]. The cross sections are measured both in a double and a triple differential manner as a function of $|y|_{max}$, $m_{1,2}$ and y^*, y_b and either $m_{1,2}$ or $\langle p_T \rangle_{1,2}$, while the jets are clustered for two different distance parameters $R = 0.4$ and 0.8. Data are compared to FO theory predictions at NNLO where the agreement is found to be in good.

The inclusion of CMS dijet cross sections to the HERA DIS data result in further constraining PDF sets. Particularly for the gluon PDF the inclusion of 2D or 3D cross sections leads to an uncertainty reduction in the high-x region ($x > 10^{-1}$). The α_S estimates at the scale of the Z boson mass obtained between the 2D and 3D dijet measurements are compatible within their respective uncertainties, with the former giving the more precise result of $\alpha_S(m_Z) = 0.1179 \pm 0.0019$ at NNLO.

The NNLO component of the FO predictions used in this analysis is based on the leading-color and leading-flavor-number approximations. The impact of subleading-color contributions at NNLO have not yet been tested for the $m_{1,2}$ and $\langle p_T \rangle_{1,2}$ observables. Therefore, the work that is unraveled here of this six-fold dijet measurement could provide ideal testing grounds for such studies.

Acknowledgements The research work was supported by the Hellenic Foundation for Research and Innovation (HFRI) under the 3rd Call for HFRI PhD Fellowships (Fellowship Number: 83154).

H.F.R.I.
Hellenic Foundation for Research & Innovation

References

1. CMS Collaboration, The CMS experiment at the CERN LHC. JINST **3**, S08004 (2008). https://doi.org/10.1088/1748-0221/3/08/S08004
2. M. Cacciari et al., The anti-kt jet clustering algorithm. JHEP **04**, 063 (2008). https://doi.org/10.1088/1126-6708/2008/04/063
3. T. Sjöstrand et al., A brief introduction to PYTHIA 8.1. Comput. Phys. Commun. **178**, 852 (2008). https://doi.org/10.1016/j.cpc.2008.01.036
4. CMS Collaboration, Event generator tunes obtained from underlying event and multiparton scattering measurements. Eur. Phys. J. C **76**, (2016). https://doi.org/10.1140/epjc/s10052-016-3988-x
5. GEANT4 Collaboration, Geant4-a simulation toolkit. Nucl. Instrum. Methods A **506**, 250 (2003). https://doi.org/10.1016/S0168-9002(03)01368-8
6. S. Schmitt, TUnfold, an algorithm for correcting migration effects in high energy physics. J. Instrum. **7**, T10003 (2012). https://doi.org/10.1088/1748-0221/7/10/T10003
7. T. Gehrmann et al., Jet cross sections and transverse momentume distributions with NNLOJET, in *PoS RADCOR2017*, vol. 290 (2018), p. 074. https://doi.org/10.22323/1.290.0074

8. T. Kluge et al., FastNLO: fast pQCD calculations for PDF fits, in *Proceedings, 14th International Workshop on Deep-Inelastic Scattering (DIS 2006)*, vol. 483 (2017). https://doi.org/10.1142/9789812706706_0110
9. D. Britzger et al., New features in version 2 of the fastNLO project, in *20th International Workshop on Deep-Inelastic Scattering and Related Subject*, vol. 217 (2012). http://www.arxiv.org/abs/1208.3641
10. D. Britzger et al., Calculations for deep inelastic scattering using fast interpolation grid techniques at NNLO in QCD and the extraction of α_S from HERA data. Eur. Phys. J. C **79**, 845 (2019). https://doi.org/10.1140/epjc/s10052-019-7351-x
11. T.-J. Hou et al., New CTEQ global analysis of quantum chromodynamics with high-precision data from the LHC. Phys. Rev. D **103**, 014013 (2021). https://doi.org/10.1103/PhysRevD.103.014013
12. S. Dittmaier et al., Weak radiative corrections to dijet production at hadron colliders. JHEP **11**, 095 (2012). https://doi.org/10.1007/JHEP11(2012)095
13. J. Currie et al., Precise predictions for dijet production at the LHC. Phys. Rev. Lett. **119**, 152001 (2017). https://doi.org/10.1103/PhysRevLett.119.152001
14. A. Gehrmann-De Ridder et al., Triple differential dijet cross section at the LHC. Phys. Rev. Lett. **123**, 102001 (2019). https://doi.org/10.1103/PhysRevLett.123.102001
15. CMS Collaboration, Multi-differential measurement of the dijet cross section in proton-proton collisions at $\sqrt{s} = 13$ TeV. CMS-PAS-SMP-21-008 (2022). https://cds.cern.ch/record/2843201
16. S. Alekhin et al., Parton distribution functions, α_S, and heavy-quark masses for LHC Run II. Phys. Rev. D **96**, 014011 (2017). https://doi.org/10.1103/PhysRevD.96.014011
17. S. Bailey et al., Parton distributions from LHC, HERA, Tevatron and fixed target data: MSHT20 PDFs. Eur. Phys. J. C **81**, 341 (2021). https://doi.org/10.1140/epjc/s10052-021-09057-0
18. NNPDF Collaboration, Parton distributions from high-precision collider data. Eur. Phys. J. C **77**, 663 (2017). https://doi.org/10.1140/epjc/s10052-017-5199-5
19. H1 and ZEUS Collaborations, Combination of measurements of inclusive deep inelastic $e \pm p$ scattering cross sections and QCD analysis of HERA data. Eur. Phys. J. C **75**, 580 (2015). https://doi.org/10.1140/epjc/s10052-015-3710-4
20. S. Alekhin et al., HERAFitter: open source QCD fit project. Eur. Phys. J. C **75**, 304 (2015). https://doi.org/10.1140/epjc/s10052-015-3480-z

ATLAS Highlights

Stylianos Angelidakis

Abstract This report provides a selection of recent results from the physics programme of the ATLAS collaboration, covering a wide spectrum of research areas. The analyses presented use proton-proton collision data from the LHC. With the increase of the integrated luminosity, improved inputs from theory and the exploitation of new techniques, ATLAS increases the precision of measurements and extends the search for new physics to remote regions of the phase space.

1 Introduction

The ATLAS collaboration [1] recently concluded the challenging Phase-I upgrade of the ATLAS detector, which took place during Long Shutdown 2 (LS2) of the LHC. Since July 2022, ATLAS has been recording and analyzing proton-proton (pp) collision data from LHC Run-3, at a center-of-mass energy of $\sqrt{s} = 13.6$ TeV, while engaging in multiple other fronts, including the completion of a large number of research projects analyzing the dataset collected during LHC Run-2, the validation of the New Small Wheel [2] (the new inner end-caps of the Muon Spectrometer installed during the Phase-I upgrade), and the preparations for the Phase-II upgrade and the High-Luminosity LHC (HL-LHC).

The following sections provide a summary of recent results published by the ATLAS collaboration. Preliminary results from the analysis of the first Run-3 data are included, while the majority of studies are based on the analysis of the large amount of data recorded during LHC Run-2. Among the Run-2 analyses presented below, most rely on the preliminary estimation of the integrated luminosity for the recorded pp collisions, which is 139.0 ± 2.4 fb^{-1} [3]. The latest results, however, benefit from

Stylianos Angelidakis on behalf of the ATLAS collaboration.

S. Angelidakis (✉)
National and Kapodistrian University of Athens, Athens, Greece
e-mail: stylianos.angelidakis@cern.ch

J. Strologas and L. Perivolaropoulos (eds.), *HEP2023*, Springer Proceedings in Physics 333 https://doi.org/10.1007/978-3-031-76856-9_21

the precise determination of the integrated luminosity, which was recently reported by ATLAS and is presented in Sect. 2.

2 Luminosity Determination in *pp* Collisions

The precise determination of the integrated luminosity is a key component of the ATLAS physics programme. ATLAS recently reported the measurement of the luminosity of the *pp* collision data sample delivered at $\sqrt{s} = 13$ TeV during Run-2 of the LHC [4]. The absolute luminosity scale is determined using van der Meer beam separation scans [5, 6] during dedicated running periods in each year, and extrapolated to the physics data-taking regime using complementary measurements from several luminosity-sensitive detectors. The total uncertainties in the integrated luminosities for each individual year of data-taking range from 0.9% to 1.1%, and are partially correlated between years. After standard data-quality selections, the full Run-2 *pp* data sample corresponds to an integrated luminosity of 140.1 ± 1.2 fb^{-1}, i.e. an uncertainty of 0.83%. This calibration is significantly more precise than those achieved by ATLAS for Run-1 (1.8% at $\sqrt{s} = 7$ TeV and 1.9% at $\sqrt{s} = 8$ TeV, and is the most precise luminosity measurement at any hadron collider to date, apart from some of the second-generation total cross-section experiments at the CERN ISR [7].

3 $H \rightarrow \gamma\gamma$ Fiducial Cross-Section

The Higgs boson production cross-section in the di-photon decay channel is measured using 31.4 fb^{-1} of *pp* collision data collected during LHC Run-3, at $\sqrt{s} = 13.6$ TeV [8]. Signal event candidates are characterized from the two leading in transverse energy, E_{T}, photons which are required to be isolated from other significant activity in the inner detector or the calorimeters, with invariant mass in the range 105 GeV $\leq m_{\gamma\gamma} \leq$ 160 GeV, and with the leading in E_{T} photon satisfying $E_{\mathrm{T}}/m_{\gamma\gamma} > 0.35$ while the sub-leading photon satisfies $E_{\mathrm{T}}/m_{\gamma\gamma} > 0.25$. In order to reduce model dependence, the measurement is restricted to a particle-level phase space that matches closely the reconstruction-level photon kinematic selection, and it is corrected for detector effects. The cross-section, extracted by an unbinned maximum likelihood fit to the $m_{\gamma\gamma}$ spectrum, is measured to be $\sigma_{\mathrm{fid}}(pp \rightarrow H \rightarrow \gamma\gamma) = 76^{+14}_{-13}$ fb $= 76 \pm 11(\mathrm{stat})\,^{+9}_{-7}(\mathrm{syst})$ fb, in agreement with the Standard Model (SM) prediction of 67.5 ± 3.4 fb. Assuming the SM acceptance and di-photon branching fraction, the fiducial measurement is extrapolated to the full phase space yielding a total cross-section of $\sigma(pp \rightarrow H) = 67^{+13}_{-12}$ pb at 13.6 TeV, to be compared with the SM prediction of 59.8 ± 2.6 pb.

4 Higgs Boson Mass

The mass of the Higgs boson is measured in the $H \rightarrow ZZ^* \rightarrow 4\ell$ decay channel [9] using pp collision data from the LHC at $\sqrt{s} = 13$ TeV recorded by the ATLAS detector between 2015 and 2018, corresponding to an integrated luminosity of 139 fb^{-1}. The measured value of the Higgs boson mass is $124.99 \pm 0.18(\text{stat.}) \pm 0.04(\text{syst.})$ GeV and is based on improved momentum-scale calibration for muons relative to previous publications. The measurement also employs an analytic model that takes into account the invariant-mass resolution of the four-lepton system on a per-event basis and the output of a deep neural network discriminating signal from background events. This measurement is combined with the corresponding measurement using 7 and 8 TeV pp collision data, resulting in a Higgs boson mass measurement of $124.94 \pm 0.17(\text{stat.}) \pm 0.03(\text{syst.})$ GeV.

5 Higgs Boson Width

ATLAS reports the search for the off-shell production of the Higgs boson using 139 fb^{-1} of pp collision data at $\sqrt{s} = 13$ TeV [10]. The observable signature is a pair of on-shell Z bosons, which result from the production and subsequent decay of a virtual Higgs boson and the interference of that process with the continuum ZZ production. The two observable final states are $ZZ \rightarrow 4\ell$ and $ZZ \rightarrow 2\ell 2\nu$ with $\ell = e$ or μ. In the $ZZ \rightarrow 2\ell 2\nu$ analyses, the final observable is the reconstructed transverse mass, m_{T}^{ZZ}, of the ZZ system, whereas in the case of $ZZ \rightarrow 4\ell$, a dense Neural Network is used to enhance analysis sensitivity. From the combined statistical interpretation, the background-only hypothesis is rejected with an observed (expected) significance of 3.3 (2.2) standard deviations, representing experimental evidence for off-shell Higgs boson production. Assuming that no new particles contribute to the production of the virtual Higgs boson, its total width can be deduced from the measurement of its off-shell production cross-section. The measured total width of the Higgs boson is $4.5^{+3.3}_{-2.5}$ MeV, compatible with the SM prediction of $4.1^{+3.2}_{-3.5}$ MeV and the observed (expected) upper limit on the total width is 10.5 (10.9) MeV at 95% Confidence Level (CL).

6 Higgs Boson Self-coupling

A major priority of the LHC physics program is the measurement of the Higgs boson self-interaction, which is directly accessible via the production of Higgs boson pairs (double-Higgs production). In this study of ATLAS [11], constraints on the Higgs boson self-coupling are set by combining double-Higgs boson analyses in the $b\bar{b}b\bar{b}$, $b\bar{b}\tau^+\tau^-$ and $b\bar{b}\gamma\gamma$ decay channels with single-Higgs boson analyses targeting the $\gamma\gamma$, ZZ^*, WW^*, $\tau^+\tau^-$ and $b\bar{b}$ decay channels. The data used in these analyses were

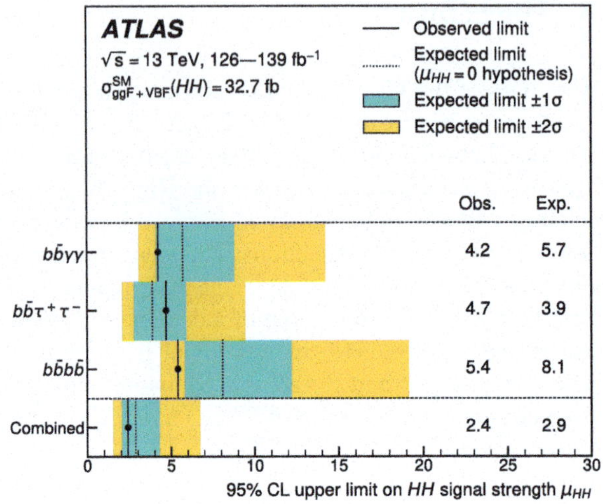

Fig. 1 Observed and expected 95% CL upper limits on the signal strength for double-Higgs production from the $b\bar{b}b\bar{b}$, $b\bar{b}\tau^+\tau^-$ and $b\bar{b}\gamma\gamma$ decay channels, and their statistical combination [11]. The value $m_H = 125.09$ GeV is assumed to derive the predicted SM cross-section

recorded by ATLAS during LHC Run-2 and correspond to an integrated luminosity of 126–139fb^{-1}. The combination of the double-Higgs analyses sets an upper limit of $\mu_{HH} < 2.4$ at 95% CL on the double-Higgs production cross-section normalised to the SM prediction (Fig. 1). Combining the single-Higgs and double-Higgs analyses, with the assumption that new physics affects only the Higgs boson self-coupling, values outside the interval $-0.4 < \kappa_\lambda = \lambda_{HHH}/\lambda_{HHH}^{SM} < 6.3$, where λ_{HHH} is the trilinear self-coupling of the Higgs boson and λ_{HHH}^{SM} is the value predicted by the SM for a Higgs boson mass of $m_H = 125.09$ GeV [12], are excluded at 95% CL. Finally, results with fewer assumptions are extracted from the combined single-Higgs and double-Higgs analyses by introducing additional coupling modifiers to account for Higgs boson interactions with the other SM particles. In this relaxed scenario, the constraint becomes $-1.4 < \kappa_\lambda < 6.1$ at 95% CL.

7 Observation of Four-Top-Quark Production in the Multilepton Final State

In this study [13], ATLAS reports the observation of four-top-quark ($tt\bar{t}\bar{t}$) production in pp collisions at the LHC. The analysis is performed using an integrated luminosity of 140.1 fb^{-1}, at $\sqrt{s} = 13$ TeV. Events containing two leptons with the same electric charge or at least three leptons (electrons or muons) are selected. Event kinematics are used to separate signal from background through a multivariate discriminant, and dedicated control regions are used to constrain the dominant backgrounds. The observed (expected) significance of the measured $tt\bar{t}\bar{t}$ signal with respect to the standard model (SM) background-only hypothesis is 6.1 (4.3) standard deviations. The $tt\bar{t}\bar{t}$ production cross section is measured to be $22.5^{+6.6}_{-5.5}$ fb $= 22.5^{+4.7}_{-4.3}$(stat) $^{+4.6}_{-3.4}$(syst) fb, consistent with the SM prediction of 12.0 ± 2.4 fb within 1.8 standard deviations. Data are also used to set limits on the three-top-quark production cross section, being an irreducible background not measured previously, and to constrain the top-Higgs Yukawa coupling and effective field theory operator coefficients that affect $tt\bar{t}\bar{t}$ production.

8 Measurement of α_s

The coupling constant of the strong interaction, α_s, is one of the fundamental parameters of the SM and is the least precisely known among the fundamental couplings in nature. Improving the precision of the measurement of α_s is important in order to reduce the associated theoretical uncertainty which enters into all cross-section calculations for processes at the LHC, and affects several key observables at e^+e^- colliders. In this study [14], α_s is determined from the transverse-momentum distribution of Z bosons, decaying to leptons (electrons or muons), which are produced in pp collision at $\sqrt{s} = 8$ TeV at the LHC and recorded by the ATLAS experiment. The Z-boson cross sections are measured in the full phase space of the decay leptons using 15 million electron and muon pairs, in a dataset corresponding to an integrated luminosity of 20.2 fb^{-1}. The analysis is based on predictions evaluated at third order in perturbative QCD, supplemented by the resummation of logarithmically-enhanced contributions in the small transverse-momentum region of the lepton pairs. The determined value of the strong coupling at the reference scale corresponding to the Z-boson mass is $\alpha_s(m_Z) = 0.11828^{+0.00084}_{-0.00088}$. This is the most precise experimental determination of $\alpha_s(m_Z)$ to date, as illustrated in Fig. 2.

9 Measurement of the W Boson Mass

The W boson mass is one of the most interesting fundamental parameters of the Standard Model of particle physics, as it allows for model-independent probes for the discovery of new physics. In this work [15], pp collision data recorded by the ATLAS detector in 2011, at $\sqrt{s} = 7$ TeV, that were used for the first W boson mass measurement at the LHC, have been reanalyzed with an advanced fitting technique based on a profile likelihood approach. This allows for a reduction of several systematic uncertainties. Furthermore, advances in the modelling of the parton distribution functions (PDFs) of the proton in recent years have been taken into account and a more modern PDF set has been chosen as baseline. The updated measurement yields a value of $m_W = 80360 \pm 5(\text{stat.}) \pm 15(\text{syst.}) = 80360 \pm 16$ MeV. This result (Fig. 3), is compatible with the previously published value, with its uncertainty improved by 15%.

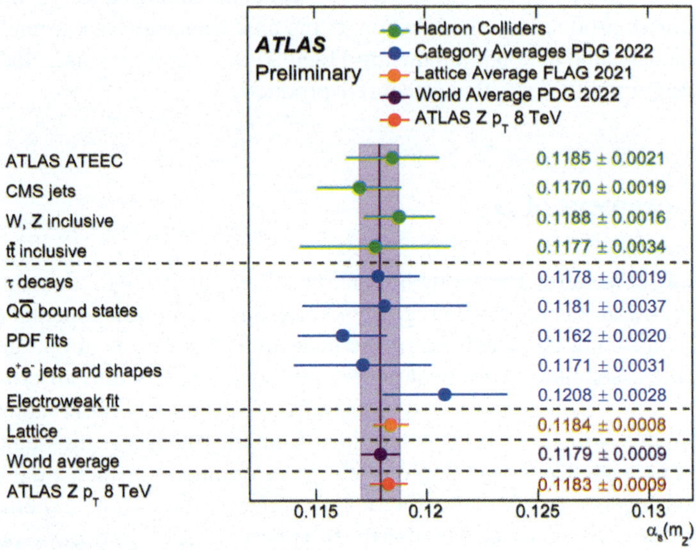

Fig. 2 Comparison of the determination of $\alpha_s(m_Z)$ from the Z-boson transverse-momentum distribution with other determinations [14]

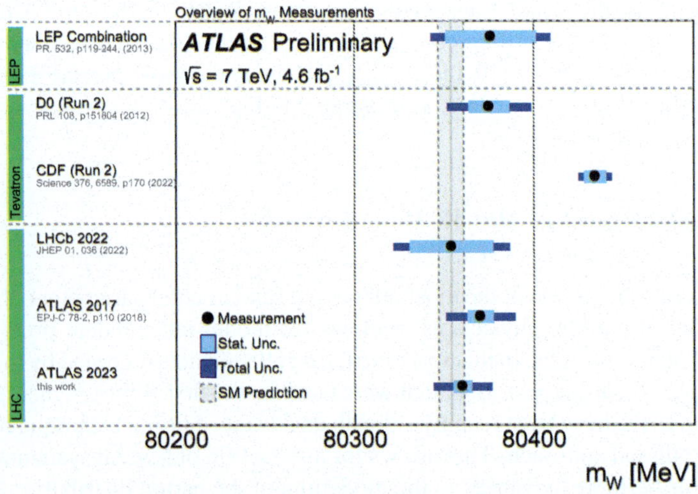

Fig. 3 Measured value of m_W compared to SM prediction from the global electroweak fit, and to the values measured at LEP, Tevatron and the LHC [15]

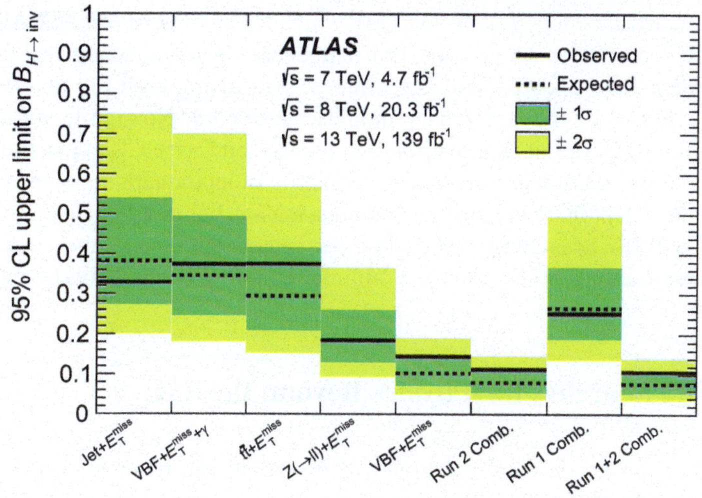

Fig. 4 The observed and expected upper limits on $B_{H\to\text{inv}}$ at 95% CL set by analyses targetting different final states, along with their statistical combination [16]

10 Searches for Decays of the Higgs Boson to Invisible Particles

Extensions of the SM predict the production of dark matter particles at the LHC. Light dark matter particles, for example, could be produced from decays of the Higgs boson that would be invisible to the detector. ATLAS recently published the combination of searches for "$H\to$ invisible" decays where multiple production modes of the SM Higgs boson are considered [16]. These searches are performed using 139 fb^{-1} of pp collisions at $\sqrt{s} = 13$ TeV at the LHC. In combination with results at $\sqrt{s} = 7$ TeV and 8 TeV, as shown in Fig. 4, an upper limit on the $H\to$ invisible branching ratio ($B_{H\to\text{inv}}$) of 0.107 (0.077) at the 95% CL is observed (expected). The results are also interpreted in the context of models where the 125 GeV Higgs boson acts as a portal to dark matter, and limits are set on the scattering cross-section of weakly interacting massive particles and nucleons.

11 Search for Pair Production of Third-Generation Leptoquarks Decaying to a b-Quark and a τ-Lepton

ATLAS has published searches for leptoquarks (LQs) coupling to the first, second and third generations of quarks and leptons [17]. In each generation, LQs are classified into up-type and down-type LQs with different electric charges. For instance, in the third generation they are split into up-type leptopquarks (LQu), which decay to $b\tau$

or $t\nu$, and down-type leptopquarks (LQd), which decay to $t\tau$ or $b\nu$. ATLAS recently updated the search for an up-type LQ pair decaying to $b\tau$ using the full Run-2 data sample and an updated analysis strategy, prioritising high LQ masses that are not yet excluded for the benchmark models considered. No significant deviations from the Standard Model expectation are observed and 95% CL upper limits on the production cross-section are derived as a function of leptoquark mass and branching ratio into the τ-lepton. For scalar leptoquarks, masses below 1490 GeV are excluded assuming a 100% branching ratio, while for vector leptoquarks the corresponding limit is 1690 GeV (1960 GeV) in the minimal-coupling (Yang-Mills) scenario.

12 More Searches for Physics Beyond the SM

Discovering physics beyond the Standard Model is a primary goal of research at the LHC. In addition to the studies presented above, many searches by the ATLAS collaboration have recently reported results from the analysis of pp collisions at $\sqrt{s} = 13$ TeV probing a considerable fraction of the parameter space for a wide variety of benchmark models. A sample these results are summarized in the Figs. 5 and 6.

Figure 5a summarizes the exclusion limits on the mass of particles predicted by SUSY from searches that focus on the pair production of gluinos (supersymmetric partner of the gluon). Different decay modes of gluinos to the lightest supersymmetfic

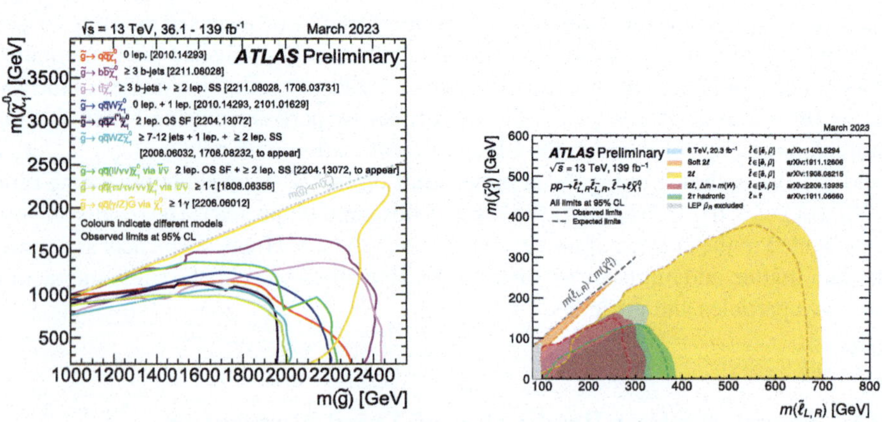

(a) (b)

Fig. 5 Exclusion limits at 95% CL based on 13 TeV data: **a** in the (gluino, lightest neutralino) mass plane for different simplified models featuring the decay of the gluino to the lightest supersymmetric particle (lightest neutralino or gravitino), **b** in the (slepton, lightest neutralino) mass plane for different analyses probing the direct production of sleptons with decays to lepton and the lightest neutralino [18]

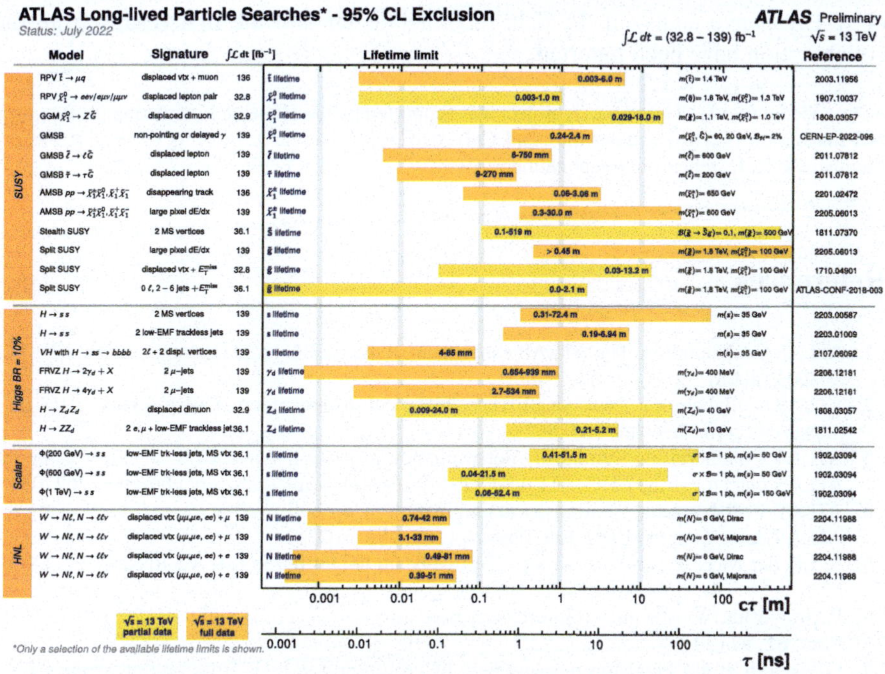

Fig. 6 Ranges of new particle lifetimes excluded at the 95% CL [19]

particle (LSP), which is assumed to be the lightest neutralino or gravitino, are probed assuming that they proceed with a 100% branching ratio. Similarly, Fig. 5b presents exclusion limits from SUSY analyses probing the electroweak production of sleptons with decays to lepton, neutralino. For a massless neutralino, masses up to 700 GeV are excluded assuming three generations of mass-degenerate sleptons. Finally, a representative set of most sensitive recent results from searches of Long-Lived Particles are summarized in Fig. 6.

13 Conclusion

The ATLAS collaboration presently concludes the analysis of the large amount of data recorded during LHC Run-2. With the exploitation of new analysis techniques and improved theoretical inputs, SM measurements have been significantly improved in precision, whereas extended regions of the phase-space have been excluded for direct evidence of new physics. Signs of new physics have yet to be observed at the LHC, however, the ongoing LHC Run-3 is expected to at least double the size of the available data sample, which will further extend the precision reach and give access to additional rare SM processes. Already the performance and analysis groups of ATLAS have analyzed the data recorded so far and the first results from the collaboration have been reported.

Acknowledgements The research project was supported by the Hellenic Foundation for Research and Innovation (H.F.R.I.) under the "2nd Call for H.F.R.I. Research Projects to support Faculty Members & Researchers" (Project Number: 04612).

References

1. ATLAS Collaboration, The ATLAS experiment at the CERN large Hadron collider. JINST **3**, S08003 (2008)
2. ATLAS Collaboration, New small wheel technical design report. CERN-LHCC-2013-006. https://cds.cern.ch/record/1552862
3. ATLAS Collaboration, Luminosity determination in *pp* collisions at $\sqrt{s} = 13$TeV using the ATLAS detector at the LHC. ATLAS-CONF-2019-021. https://cds.cern.ch/record/2677054
4. ATLAS Collaboration, Luminosity determination in *pp* collisions at $\sqrt{s} = 13$TeV using the ATLAS detector at the LHC. Eur. Phys. J. C **83**, 982 (2023)
5. S. van der Meer, Calibration of the effective beam height in the ISR. CERN-ISR-PO-68-31. https://cds.cern.ch/record/296752
6. P. Grafström, W. Kozanecki, Luminosity determination at proton colliders. Prog. Part. Nucl. Phys. **81**, 97 (2015)
7. U. Amaldi et al., Precision measurement of proton-proton total cross section at the CERN intersecting storage rings. Nucl. Phys. B. **145**, 367 (1978)
8. ATLAS Collaboration, Measurement of the $H\gamma\gamma$ fiducial cross-section in *pp* collisions at $\sqrt{s} = 13.6$TeV with the ATLAS detector. ATLAS-CONF-2023-003. http://cds.cern.ch/record/2853060
9. ATLAS Collaboration, Measurement of the Higgs boson mass in the decay channel using 139 $\text{fb}^{-}1$ of $\sqrt{s} = 13$TeV *pp* collisions recorded by the ATLAS detector at the LHC. Phys. Lett. B **843**, 137880 (2023)
10. ATLAS Collaboration, Evidence of off-shell Higgs boson production from ZZ leptonic decay channels and constraints on its total width with the ATLAS detector. Phys. Lett. B. arXiv: 2304.01532 [hep-ex]
11. ATLAS Collaboration, Constraints on the Higgs boson self-coupling from single- and double-Higgs production with the ATLAS detector using pp collisions at $\sqrt{s} = 13$TeV. Phys. Lett. B **843**, 137745 (2023)
12. ATLAS Collaboration, CMS Collaboration, "Combined Measurement of the Higgs Boson Mass in *pp* Collisions at $\sqrt{s} = 7$ and 8 TeV with the ATLAS and CMS Experiments", Phys. Rev. Lett. **114**, 191803 (2015)
13. ATLAS Collaboration, Observation of four-top-quark production in the multilepton final state with the ATLAS detector. Eur. Phys. J. C **83**, 496 (2023)
14. ATLAS Collaboration, A precise determination of the strong-coupling constant from the recoil of Z bosons with the ATLAS experiment at $\sqrt{s} = 8$TeV. ATLAS-CONF-2023-015. http://cds.cern.ch/record/2855244
15. ATLAS Collaboration, Improved W boson mass measurement using $\sqrt{s} = 7$TeV Proton-Proton Collisions with the ATLAS Detector. ATLAS-CONF-2023-004. http://cds.cern.ch/record/2853290
16. ATLAS Collaboration, Combination of searches for invisible decays of the Higgs boson using $139\text{fb}^{-}1$ of proton-proton collision data at $\sqrt{s} = 13$TeV collected with the ATLAS experiment. Phys. Lett. B **842**, 137963 (2023)

17. ATLAS Collaboration, Search for pair production of third-generation leptoquarks decaying into a bottom quark and a τ-lepton with the ATLAS detector. Eur. Phys. J. C. arXiv: 2303.01294 [hep-ex]
18. ATLAS Collaboration, SUSY March 2023 Summary Plot Update. ATL-PHYS-PUB-2023-005. https://atlas.web.cern.ch/Atlas/GROUPS/PHYSICS/PUBNOTES/ATL-PHYS-PUB-2023-005
19. ATLAS Collaboration, Summary plots for heavy particle searches and long-lived particle searches - March 2023. ATL-PHYS-PUB-2023-008. https://atlas.web.cern.ch/Atlas/GROUPS/PHYSICS/PUBNOTES/ATL-PHYS-PUB-2022-034

Cosmic Inflation in Metric-Affine Gravity

Ioannis D. Gialamas and Kyriakos Tamvakis

Abstract In the context of metric-affine gravity theories, where the metric and connection are independent, we examine actions involving quadratic terms in the Ricci scalar curvature and the Holst invariant. These actions are non-minimally coupled to a scalar field. We explore the behavior of the corresponding effective metric theory, which includes an extra dynamic pseudoscalar degree of freedom. Detailed analysis of inflationary predictions reveals compliance with recent observations across various parameters, potentially allowing a higher tensor-to-scalar ratio. The spectral index's direction of change varies based on parameter positioning.

Keywords Metric-affine quadratic gravity · Holst invariant · Inflation

1 Introduction

The present framework of modern cosmology consists of classical General Relativity (GR) as a theory of gravitation and Quantum Field Theory (QFT) as the theory of matter. A common working assumption is that the quantum aspects of gravitation can be ignored for energies below the Planck energy of 10^{19} GeV and, therefore, gravity can be treated classically. In contrast, the full quantum character of particle interactions is considered within QFT. The quantum interactions of the matter fields coupled to the classical gravitational field introduce modifications to the standard GR action with cosmological implications. Such are non-minimal couplings of the inflaton field or higher power terms of the Ricci curvature in models of cosmological inflation. The metric-affine formulation of gravity, where the metric and the

---

I. D. Gialamas
Laboratory of High Energy and Computational Physics, National Institute of Chemical Physics and Biophysics, Tallinn, Estonia

K. Tamvakis (✉)
Physics Department, University of Ioannina, Ioannina, Greece
e-mail: tamvakis@uoi.gr

J. Strologas and L. Perivolaropoulos (eds.), *HEP2023*, Springer Proceedings in Physics 333 https://doi.org/10.1007/978-3-031-76856-9_22

connection are independent variables, although equivalent to the standard (metric) GR in the case of the Einstein–Hilbert action, leads to different predictions when the above corrections are included.

2 Metric Versus Metric-Affine Formulation of Gravity

The General Relativity Principle states that all laws of physics should be invariant under general coordinate transformations. To implement such a principle we need to introduce a metric $g_{\mu\nu}$, which has to transform as

$$g'_{\alpha\beta}(x') = \left(\frac{\partial x^\mu}{\partial x'^\alpha}\right)\left(\frac{\partial x^\nu}{\partial x'^\beta}\right) g_{\mu\nu}(x)\,, \tag{1}$$

as well as a Connection $\Gamma_{\mu\ \nu}^{\ \rho}$ in order to define covariant derivatives of tensors. In the standard metric formulation of gravity the connection is not an independent quantity but it is given by the Levi–Civita relation as

$$\Gamma_{\mu\ \nu}^{\ \rho}\big|_{LC} = \frac{1}{2} g^{\rho\sigma}\left(\partial_\mu g_{\rho\nu} + \partial_\nu g_{\mu\sigma} - \partial_\sigma g_{\mu\nu}\right)\,. \tag{2}$$

In contrast, in the so-called metric-affine theories of gravity the connection is an independent variable not related to the metric through (2). Note that $D_\mu g_{\nu\rho}\big|_{LC} = 0$ (metricity), while $D_\mu g_{\nu\rho} \neq 0$ in general for a metric-affine theory.

2.1 *The Distortion Tensor*

The difference between the independent connection of a metric-affine theory and the corresponding Levi–Civita one is a tensor called the distortion tensor

$$C_{\mu\ \nu}^{\ \rho} = \Gamma_{\mu\ \nu}^{\ \rho} - \Gamma_{\mu\ \nu}^{\ \rho}\big|_{LC}\,. \tag{3}$$

The distortion tensor vanishes for metric theories. The Torsion is given by $T_{\mu\ \nu}^{\ \rho} = C_{\mu\ \nu}^{\ \rho} - C_{\nu\ \mu}^{\ \rho} = 2C_{[\mu\ \nu]}^{\ \ \rho} = \Gamma_{\mu\ \nu}^{\ \rho} - \Gamma_{\nu\ \mu}^{\ \rho}$, while the curvature tensor of a metric-affine theory is defined as

$$\mathcal{R}_{\mu\nu\ \sigma}^{\ \ \ \rho} = \partial_\mu \Gamma_{\nu\ \sigma}^{\ \rho} - \partial_\nu \Gamma_{\mu\ \sigma}^{\ \rho} + \Gamma_{\mu\ \lambda}^{\ \rho}\Gamma_{\nu\ \sigma}^{\ \lambda} - \Gamma_{\nu\ \lambda}^{\ \rho}\Gamma_{\mu\ \sigma}^{\ \lambda}\,. \tag{4}$$

The following two scalars, linear in the Riemann tensor, can be defined as

$$\mathcal{R} = \mathcal{R}_{\mu\nu}{}^{\rho}{}_{\sigma}\delta^{\mu}_{\rho}g^{\nu\sigma} = \mathcal{R}_{\mu\nu}{}^{\mu\nu}\,, \tag{5}$$

$$\tilde{\mathcal{R}} = (-g)^{-1/2}\epsilon^{\mu\nu}{}_{\rho}{}^{\sigma}\mathcal{R}_{\mu\nu}{}^{\rho}{}_{\sigma} = (-g)^{-1/2}\epsilon^{\mu\nu\rho\sigma}\mathcal{R}_{\mu\nu\rho\sigma}\,. \tag{6}$$

The first corresponds to the usual Ricci scalar, while $\tilde{\mathcal{R}}$ is the so-called Holst invariant, which vanishes identically in a metric theory due to the symmetry in the lower indices of the Levi–Civita connection. The Holst invariant commonly appears in Loop Quantum Gravity [1] and has been studied extensively in inflationary cosmology [2–10].

The following expressions of the curvature scalars can be written in terms of the distortion

$$\mathcal{R} = R + D_{\mu}C_{\nu}{}^{\mu\nu} - D_{\nu}C_{\mu}{}^{\mu\nu} + C_{\mu}{}^{\mu}{}_{\lambda}C_{\nu}{}^{\lambda\nu} - C_{\nu}{}^{\mu}{}_{\lambda}C_{\mu}{}^{\lambda\nu}\,, \tag{7}$$

$$\tilde{\mathcal{R}} = 2(-g)^{-1/2}\epsilon^{\mu\nu\rho\sigma}\left(D_{\mu}C_{\nu\rho\sigma} + C_{\mu\rho\lambda}C_{\nu}{}^{\lambda}{}_{\sigma}\right)\,, \tag{8}$$

where $R = R[g]$ is the standard metric Ricci scalar and the covariant derivatives are taken with respect to $\Gamma_{\mu}{}^{\rho}{}_{\nu}\big|_{LC}$.

The metric-affine version of the Einstein–Hilbert action is

$$\begin{aligned}\mathcal{S}_{EH} &= \frac{1}{2}\int d^4x\sqrt{-g}\mathcal{R}\\ &= \frac{1}{2}\int d^4x\sqrt{-g}\left\{R + D_{\mu}C_{\nu}{}^{\mu\nu} - D_{\nu}C_{\mu}{}^{\mu\nu} + C_{\mu}{}^{\mu}{}_{\lambda}C_{\nu}{}^{\lambda\nu} - C_{\nu}{}^{\mu}{}_{\lambda}C_{\mu}{}^{\lambda\nu}\right\}.\end{aligned} \tag{9}$$

Variation with respect to the distortion gives

$$\frac{\delta\mathcal{S}}{\delta C} = 0 \implies \delta^{\alpha}_{\beta}C_{\nu\gamma}{}^{\nu} + \delta^{\alpha}_{\gamma}C_{\nu}{}^{\nu}{}_{\beta} - C_{\beta\gamma}{}^{\alpha} - C_{\gamma}{}^{\alpha}{}_{\beta} = 0\,. \tag{10}$$

This equation possesses the general solution $C_{\mu\nu\rho} = U_{\mu}g_{\nu\rho}$, in terms of the arbitrary vector U_{μ}. Upon inserting this solution into the action, the terms that rely on C disappear. Consequently, $\mathcal{S}_{EH}$ is equivalent to the conventional metric GR. Nevertheless, this equivalence does not extend to quadratic actions.

3 Non-minimal Coupling to Scalars

One can derive the metric-equivalent of any metric-affine theory based on an action, where gravity couples to a scalar field,

$$\mathcal{S} = \int d^4x\,\sqrt{-g}\left\{\frac{1}{2}\Omega^2(\phi)\mathcal{R} + \mathcal{L}(\phi, g_{\mu\nu}, \partial_{\mu}\phi)\right\}. \tag{11}$$

Note that any $F(\mathcal{R})$ theory can also be set in this form, i.e.

$$\mathcal{S}=\int d^4x\sqrt{-g}\left\{\frac{1}{2}F'(\chi)\mathcal{R}-V(\chi)\right\}\,,\quad \textit{where}\quad V(\chi)=\frac{1}{2}\left(\chi F'(\chi)-F(\chi)\right)\,, \tag{12}$$

in terms of the auxiliary scalar χ.

Substituting the expression of $\mathcal{R}$ in terms of the distortion, we obtain

$$\mathcal{S}=\int d^4x\,\sqrt{-g}\,\left\{\frac{1}{2}\Omega^2(\phi)R(g)+\frac{1}{2}\Omega^2(\phi)\left(D_\mu C_\nu{}^{\mu\nu}-D_\nu C_\mu{}^{\mu\nu}\right.\right.$$
$$\left.\left.+C_\mu{}^\mu{}_\lambda C_\nu{}^{\lambda\nu}-C_\nu{}^\mu{}_\lambda C_\mu{}^{\lambda\nu}\right)+\mathcal{L}(\phi,g_{\mu\nu},\partial_\mu\phi)\right\}\,. \tag{13}$$

Variation with respect to the distortion gives us the equation

$$\Omega^2\left(\delta^\alpha_\beta C_{\nu\gamma}{}^\nu+\delta^\alpha_\gamma C_\nu{}^\nu{}_\beta-C_{\beta\gamma}{}^\alpha-C_\gamma{}^\alpha{}_\beta\right)=\delta^\alpha_\gamma\partial_\beta\Omega^2-\delta^\alpha_\beta\partial_\gamma\Omega^2\,, \tag{14}$$

with a solution (up to a term $U_\mu\, g_{\nu\rho}$ of an arbitrary vector U_μ)

$$C_\mu{}^\rho{}_\nu=\frac{1}{2}\left(g^\rho_\mu\partial_\nu\ln\Omega^2+g^\rho_\nu\partial_\mu\ln\Omega^2-g_{\mu\nu}\partial^\rho\ln\Omega^2\right)\,. \tag{15}$$

Substituting the distortion C back into the action we obtain

$$\mathcal{S}=\int d^4x\,\sqrt{-g}\left\{\frac{1}{2}\Omega^2(\phi)R(g)\,+\,\frac{3}{4}\frac{(\nabla\Omega^2)^2}{\Omega^2}\,+\,\mathcal{L}(\phi,g_{\mu\nu},\partial_\mu\phi)\right\}\,. \tag{16}$$

This is a metric theory and the appearing connection is the Levi–Civita one. Note that the extra term has the form of the extra kinetic term that appears when we Weyl-rescale the metric theory to the Einstein frame, albeit with an opposite sign. The inequivalence of the two formulations rests on this term. Only in the case of the Einstein–Hilbert action this term vanishes and the two formulations are equivalent.

4 Quadratic Metric-Affine Theories

Consider the following metric-affine generalization of the Starobinsky model [11]

$$\mathcal{S}=\int d^4x\sqrt{-g}\left\{\frac{1}{2}\alpha\mathcal{R}+\frac{1}{2}\beta\tilde{\mathcal{R}}+\frac{1}{4}\gamma\mathcal{R}^2+\frac{1}{4}\delta\tilde{\mathcal{R}}^2\right\}\,, \tag{17}$$

where $\mathcal{R}$ is the Ricci scalar curvature and $\tilde{\mathcal{R}}$ is the Holst invariant. This is a general quadratic action of these scalars. Nevertheless, Palatini inflationary models (i.e. metric-affine models with zero torsion) of a scalar field in the presence of an R^2

term have been studied with predictions compatible with existing cosmological data [12–50].

In what follows we shall use Planck-mass units taking $\alpha = 1$. An equivalent way to express the action is in terms of the auxiliary scalars χ and ζ as

$$\mathcal{S} = \int d^4x\sqrt{-g}\left\{\frac{1}{2}(1+\gamma\chi)\mathcal{R} + \frac{1}{2}(\beta+\delta\zeta)\tilde{\mathcal{R}} - \frac{1}{4}\left(\gamma\chi^2+\delta\zeta^2\right)\right\}. \tag{18}$$

Next, we may use the expressions of $\mathcal{R}$ and $\tilde{\mathcal{R}}$ in terms of the distortion C, given in (7), (8), and obtain the equivalent metric theory. Variation with respect to $C_{\alpha}^{\ \beta\gamma}$ gives[1]

$$\frac{\Omega^2}{2}\left(\delta^{\alpha}_{\beta}C_{\nu\gamma}^{\ \ \nu} + \delta^{\alpha}_{\gamma}C^{\nu}_{\ \nu\ \beta} - C_{\beta\gamma}^{\ \ \alpha} - C_{\gamma\ \beta}^{\ \alpha}\right) - \frac{\overline{\Omega}^2}{\sqrt{-g}}\left(\epsilon^{\mu\alpha\sigma}{}_{\beta}C_{\mu\gamma\sigma} + \epsilon^{\mu\alpha\sigma}{}_{\gamma}C_{\mu\sigma\beta}\right) = J_{\beta\ \gamma}^{\ \alpha}, \tag{19}$$

where

$$J_{\beta\ \gamma}^{\ \alpha} = \frac{1}{2}\delta^{\alpha}_{\gamma}\partial_{\beta}\Omega^2 - \frac{1}{2}\delta^{\alpha}_{\beta}\partial_{\gamma}\Omega^2 + \frac{\epsilon^{\mu\alpha}{}_{\beta\gamma}}{\sqrt{-g}}\partial_{\mu}\overline{\Omega}^2, \tag{20}$$

and $\Omega^2 \equiv 1+\gamma\chi$, $\ \overline{\Omega}^2 = \beta+\delta\zeta$. Note that in the previous equation the RHS is antisymmetric in the lower indices, i.e. $J_{\beta\ \gamma}^{\ \alpha} = -J_{\gamma\ \beta}^{\ \alpha}$. Then, we obtain the following solution

$$C_{\mu\nu\rho} = \frac{g_{\mu\nu}}{2\Delta}\left(\Omega^2\partial_{\rho}\Omega^2 + 4\overline{\Omega}^2\partial_{\rho}\overline{\Omega}^2\right) - \frac{g_{\mu\rho}}{2\Delta}\left(\Omega^2\partial_{\nu}\Omega^2 + 4\overline{\Omega}^2\partial_{\nu}\overline{\Omega}^2\right)$$
$$+\frac{\epsilon_{\mu\nu\rho\sigma}}{\Delta\sqrt{-g}}\left(\Omega^2\partial^{\sigma}\overline{\Omega}^2 - \overline{\Omega}^2\partial^{\sigma}\Omega^2\right), \tag{21}$$

where $\Delta \equiv \Omega^4 + 4\overline{\Omega}^4$. Substituting C back into the action, we obtain the corresponding metric action

$$\mathcal{S} = \int d^4x\sqrt{-g}\left\{\frac{1}{2}\Omega^2 R(g) + \frac{3}{4}\frac{(\nabla\Omega^2)^2}{\Omega^2} - \frac{3}{\Omega^2\Delta}\left(\Omega^2\nabla\overline{\Omega}^2 - \overline{\Omega}^2\nabla\Omega^2\right)^2\right.$$
$$\left. -\frac{1}{4\gamma}(\Omega^2-1)^2 - \frac{1}{4\delta}(\overline{\Omega}^2-\beta)^2\right\}. \tag{22}$$

[1] See [51] for a general solution of this type of equations and [8, 9] for more details on the solution adopted here.

4.1 The Einstein Frame

The Weyl rescaling $g_{\mu\nu} = \Omega^{-2}\bar{g}_{\mu\nu}$ takes us to the Einstein frame. Note that the Ricci scalar is rescaled as $R(g) = \Omega^2\bar{R} - 6\Omega^3\Box\Omega^{-1}$. The action is

$$\mathcal{S} = \int \mathrm{d}^4x\sqrt{-\bar{g}}\left\{\frac{1}{2}\bar{R}(\bar{g}) - \frac{3}{\Omega^4\Delta}\left(\Omega^2\bar{\nabla}\overline{\Omega}^2 - \overline{\Omega}^2\bar{\nabla}\Omega^2\right)^2\right.$$
$$\left.-\frac{1}{\Omega^4}\left(\frac{1}{\gamma}(\Omega^2-1)^2 + \frac{1}{\delta}(\overline{\Omega}^2-\beta)^2\right)\right\}. \tag{23}$$

Introducing the field $\sigma \equiv \overline{\Omega}^2/(2\Omega^2)$, the scalar part of the Lagrangian becomes

$$\mathcal{L} = -\frac{12(\bar{\nabla}\sigma)^2}{(1+16\sigma^2)} - \frac{1}{4}\left(\frac{1}{\gamma}(\Omega^{-2}-1)^2 + \frac{1}{\delta}(2\sigma - \beta\Omega^{-2})^2\right). \tag{24}$$

Variation with respect to the non-dynamical Ω^2 gives

$$\frac{\delta\mathcal{L}}{\delta\Omega^2} = 0 \Longrightarrow \Omega^{-2} = \frac{\delta + 2\beta\gamma\sigma}{\delta + \beta^2\gamma} \Longrightarrow \mathcal{L} = -\frac{12(\bar{\nabla}\sigma)^2}{(1+16\sigma^2)} - \frac{1}{4}\frac{(2\sigma-\beta)^2}{(\delta+\beta^2\gamma)}. \tag{25}$$

The theory can be expressed in terms of a canonical scalar s defined by $\sigma = \frac{1}{4}\sinh(\sqrt{2/3}\,s)$ as

$$\mathcal{L} = -\frac{1}{2}(\nabla s)^2 - \frac{1}{16}\frac{\left(\sinh(\sqrt{2/3}\,s) - 2\beta\right)^2}{(\delta+\beta^2\gamma)}. \tag{26}$$

At least one of γ and δ has to be included in order to generate the additional pseudoscalar degree of freedom represented by σ. The inflationary behaviour of this model has been studied by Pardisi and Salvio [6]. Note that the parameters γ and δ, associated with $\mathcal{R}^2$ and $\tilde{\mathcal{R}}^2$, can only have a secondary role in a possible inflationary behaviour, which would be controlled by β. See also Table 1 for more details.

5 Coupling to a Fundamental Scalar

We consider a scalar ϕ coupled to quadratic metric-affine gravity non-minimally. The action is

$$\mathcal{S} = \int \mathrm{d}^4x\sqrt{-g}\left\{\frac{1}{2}f(\phi)\mathcal{R} + \frac{1}{2}h(\phi)\tilde{\mathcal{R}} + \frac{\gamma}{4}\mathcal{R}^2 + \frac{\delta}{4}\tilde{\mathcal{R}}^2 + \mathcal{L}_\phi\right\}, \tag{27}$$

Table 1 List of different combinations of the curvature scalars in metric-affine gravity (left column) and the equivalent metric theory (right column). Note that the dynamical scalar field σ is not present in all cases, and the viability of inflation varies

Metric-affine gravity	Metric gravity
$\mathcal{R}$	GR
$\mathcal{R}+\tilde{\mathcal{R}}$	GR
$\mathcal{R}+\mathcal{R}^2$	GR
$\mathcal{R}+\tilde{\mathcal{R}}^2$	σ, No inflation
$\mathcal{R}+\tilde{\mathcal{R}}+\mathcal{R}^2$	σ, Inflation possible
$\mathcal{R}+\tilde{\mathcal{R}}+\tilde{\mathcal{R}}^2$	σ, Inflation possible
$\mathcal{R}+\tilde{\mathcal{R}}^2+\mathcal{R}^2$	σ, No inflation
$\mathcal{R}+\tilde{\mathcal{R}}+\mathcal{R}^2+\tilde{\mathcal{R}}^2$	σ, Inflation possible

with

$$\mathcal{L}_\phi = -\frac{1}{2}g^{\mu\nu}\partial_\mu\phi\partial_\nu\phi - V(\phi)\,. \tag{28}$$

Introducing the auxiliaries χ and ζ, we arrive at

$$\mathcal{S} = \int \mathrm{d}^4x\sqrt{-g}\left\{\frac{1}{2}\Omega^2\mathcal{R} + \frac{1}{2}\overline{\Omega}^2\tilde{\mathcal{R}} - \frac{1}{4}\left(\frac{1}{\gamma}(\Omega^2 - f(\phi))^2 + \frac{1}{\delta}(\overline{\Omega}^2 - h(\phi))^2\right) + \mathcal{L}_\phi\right\}, \tag{29}$$

with $\Omega^2 = \gamma\chi + f(\phi)$ and $\overline{\Omega}^2 = \delta\zeta + h(\phi)$. Rewriting the action in terms of the distortion and solving for it we arrive at the action of the corresponding metric theory in the Jordan frame. The Weyl rescaling $g_{\mu\nu} \to \Omega^{-2}g_{\mu\nu}$ takes the action into the Einstein frame and after introducing the auxiliary field $\sigma = \overline{\Omega}^2/(2\Omega^2)$, we get the action in the form

$$\mathcal{S} = \int \mathrm{d}^4x\sqrt{-g}\,\left\{\frac{1}{2}R - \frac{12(\nabla\sigma)^2}{(1+16\sigma^2)} - \frac{1}{2}\frac{(\nabla\phi)^2}{\Omega^2} - \frac{\sigma^2}{\delta} - \frac{1}{4\gamma\Omega^4}\left(f(\phi) - \Omega^2\right)^2 \right.$$
$$\left. - \frac{h(\phi)}{4\delta\Omega^4}\left(h(\phi) - 4\sigma\Omega^2\right) - \frac{V(\phi)}{\Omega^4}\right\}. \tag{30}$$

Note that no kinetic term for Ω^2 appears. Solving for it we obtain

$$\frac{\delta\mathcal{S}}{\delta\Omega^2} = 0 \implies \Omega^2 = \frac{f(\phi)^2 + 4\gamma V(\phi) + \gamma h^2(\phi)/\delta}{f(\phi) + 2\gamma\sigma h(\phi)/\delta - \gamma(\nabla\phi)^2}\,. \tag{31}$$

Substituting Ω^2 into the action we get it in the form

$$\mathcal{S} = \int \mathrm{d}^4x\sqrt{-g}\left\{\frac{1}{2}R - \frac{1}{2}K_\phi(\phi,\sigma)(\nabla\phi)^2 + \frac{1}{4}L_\phi(\phi)(\nabla\phi)^4 - \frac{1}{2}K_\sigma(\sigma)(\nabla\sigma)^2 - U(\phi,\sigma)\right\} \tag{32}$$

where

$$\begin{aligned} K_\phi(\phi,\sigma) &= \frac{f(\phi)+2\gamma\sigma h(\phi)/\delta}{\gamma h(\phi)^2/\delta+f^2(\phi)+4\gamma V(\phi)}\,, \\ L_\phi(\phi) &= \frac{\gamma}{\gamma h(\phi)^2/\delta+f^2(\phi)+4\gamma V(\phi)}\,, \\ K_\sigma(\sigma) &= \frac{24}{1+16\sigma^2}\,, \\ V_{\rm eff}(\phi,\sigma) &= \frac{V(\phi)}{f^2(\phi)+4\gamma V(\phi)}+\frac{1}{\delta}\left(\frac{f^2(\phi)+4\gamma V(\phi)}{\gamma h^2(\phi)/\delta+f^2(\phi)+4\gamma V(\phi)}\right)(\sigma-\sigma_0(\phi))^2\,. \end{aligned} \tag{33}$$

where

$$\sigma_0(\phi) = \frac{h(\phi)f(\phi)/2}{f^2(\phi)+4\gamma V(\phi)}\,. \tag{34}$$

Note that the potential is positive-definite with a minimum line along $\sigma=\sigma_0(\phi)$.

Along the minimum line the potential is just $U_0(\phi)=\frac{V(\phi)}{f^2(\phi)+4\gamma V(\phi)}$. However the kinetic term of ϕ is modified by $-\frac{1}{2}K_\sigma(\sigma_0(\phi))(\nabla\sigma_0(\phi))^2$. The effective Lagrangian is

$$\mathcal{L}_{eff} = -\frac{1}{2}\overline{K}(\phi)(\nabla\phi)^2+\frac{1}{4}L(\phi)(\nabla\phi)^4-U_0(\phi)\,, \tag{35}$$

where $\overline{K}(\phi)$ is the modified kinetic function [8].

6 Inflation

In what follows we shall adopt the following leading terms of $f(\phi)$ and $h(\phi)$, namely

$$f(\phi)=1+\xi\phi^2,\quad h(\phi)=\bar{\xi}\phi+\bar{\xi}'\phi^3\,. \tag{36}$$

Note that $h(\phi)$ is chosen this way to counteract the parity-odd coupling $h(\phi)\tilde{\mathcal{R}}$. We also replace σ with the canonical field $\sigma_c=2\sqrt{6}\int\frac{d\sigma}{\sqrt{1+16\sigma^2}}$. In a FRW background the equations of motion read

$$\begin{aligned} &(K_\phi+3L_\phi\dot{\phi}^2)\ddot{\phi}+3H(K_\phi+L_\phi\dot{\phi}^2)\dot{\phi}+\dot{\phi}\dot{\sigma}_c\frac{\partial K_\phi}{\partial\sigma_c}+\left(\frac{1}{2}\frac{\partial K_\phi}{\partial\phi}+\frac{3}{4}\frac{\partial L_\phi}{\partial\phi}\dot{\phi}^2\right)\dot{\phi}^2+\frac{\partial V_{eff}}{\partial\phi}=0\,, \\ &\ddot{\sigma}_c+3H\dot{\sigma}_c-\frac{1}{2}\frac{\partial K_\phi}{\partial\sigma_c}\dot{\phi}^2+\frac{\partial V_{eff}}{\partial\sigma_c}=0\,,\quad H^2=\frac{1}{3}\rho\,,\quad \dot{H}=-\frac{1}{2}(\rho+p)\,, \end{aligned}$$

where ρ and p are the total energy density and pressure respectively. Starting with large initial values for the fields, the system, in a relatively short time, drops into the valley defined by the minimum line $\sigma=\sigma_0(\phi)$ and its further evolution is described by a single field effective Lagrangian. This can be easily seen from the left panel

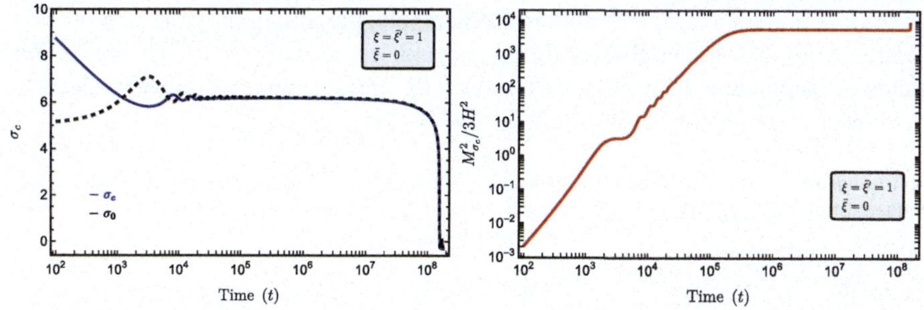

Fig. 1 **Left:** The evolution of the scalar field σ_c and the minimum direction $\sigma_0(\phi)$. **Right:** The evolution of the ratio $M^2_{\sigma_c}(\phi)/(3H^2)$. In both panels $\gamma = 10^6$, $\xi = \bar{\xi}' = 1$ and $\bar{\xi} = 0$

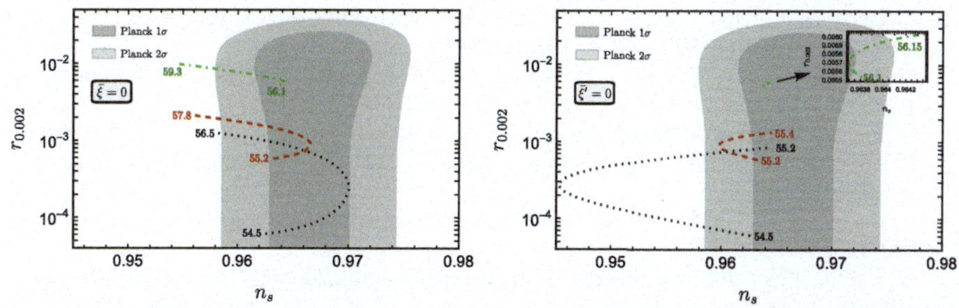

Fig. 2 Predictions of the model in the $n_s - r$ plane. **Left:** The case $\bar{\xi} = 0$, with running $\bar{\xi}'$. **Right:** The case $\bar{\xi}' = 0$, with running $\bar{\xi}$. In both panels, $\xi = 0.1$ (green), $\xi = 1$ (red) and $\xi = 10$ (black). The small numbers at the edges of the curves indicate the number of $e-$folds

of Fig. 1. In the right panel of Fig. 1 is displayed the relevant mass of the σ_c- field, $M^2_{\sigma_c}(\phi)$, in terms of the inflationary Hubble scale H. Following a numerical computation, it becomes evident that at times $t > 10^5$, the ratio $M^2_{\sigma_c}(\phi)/H^2 \sim 10^4$ (in the benchmark scenario). This finding provides additional support to the inference that the model is equivalent to a single-field inflationary model. In Fig. 2, we present the model's predictions using pivot scales of $0.05\,\text{Mpc}^{-1}$ for the spectral index n_s and $0.002\,\text{Mpc}^{-1}$ for the tensor-to-scalar ratio r. The regions filled with shading correspond to parameter ranges allowed at 68% and 95% confidence levels, as derived from [52, 53]. In the left (right) panel is displayed the case $\bar{\xi} = 0$ ($\bar{\xi}' = 0$).

7 Conclusions

Our investigation encompassed a broad quadratic metric-affine theory incorporating an additional dynamic degree of freedom, coupled non-minimally with a scalar field. Our exploration delved into the realm of two-field inflation, revealing its effective simplification to a single-field model. The model exhibited a potential very similar to the one in the Palatini-$\mathcal{R}^2$ models. Remarkably, the inflationary predictions originating from this model align with the latest observational constraints over a wide range of parameters. Moreover, this framework presents the possibility of elevating the tensor-to-scalar ratio.

Acknowledgements The work of IDG was supported by the Estonian Research Council grants SJD18, MOB3JD1202, RVTT3, RVTT7, and by the CoE program TK202 "Fundamental Universe".

References

1. C. Rovelli, Living Rev. Rel. **1**, 1 (1998). https://doi.org/10.12942/lrr-1998-1
2. M. Långvik, J.M. Ojanperä, S. Raatikainen, S. Rasanen, Phys. Rev. D **103**(8), 083514 (2021). https://doi.org/10.1103/PhysRevD.103.083514
3. M. Shaposhnikov, A. Shkerin, I. Timiryasov, S. Zell, JCAP **02**, 008 (2021). https://doi.org/10.1088/1475-7516/2021/10/E01. [Erratum: JCAP 10, E01 (2021)]
4. M. Piani, J. Rubio, JCAP **05**(05), 009 (2022). https://doi.org/10.1088/1475-7516/2022/05/009
5. C. Rigouzzo, S. Zell, Phys. Rev. D **106**(2), 024015 (2022). https://doi.org/10.1103/PhysRevD.106.024015
6. G. Pradisi, A. Salvio, Eur. Phys. J. C **82**(9), 840 (2022). https://doi.org/10.1140/epjc/s10052-022-10825-9
7. A. Salvio, Phys. Rev. D **106**(10), 103510 (2022). https://doi.org/10.1103/PhysRevD.106.103510
8. I.D. Gialamas, K. Tamvakis, JCAP **03**, 042 (2023). https://doi.org/10.1088/1475-7516/2023/03/042
9. I.D. Gialamas, H. Veermäe, Phys. Lett. B **844**, 138109 (2023). https://doi.org/10.1016/j.physletb.2023.138109
10. I.D. Gialamas, K. Tamvakis, Inflation in Weyl-invariant Einstein-Cartan gravity (2024). https://arxiv.org/abs/2410.16364
11. A.A. Starobinsky, Phys. Lett. B **91**, 99 (1980). https://doi.org/10.1016/0370-2693(80)90670-X
12. V.M. Enckell, K. Enqvist, S. Rasanen, L.P. Wahlman, JCAP **02**, 022 (2019). https://doi.org/10.1088/1475-7516/2019/02/022
13. I. Antoniadis, A. Karam, A. Lykkas, K. Tamvakis, JCAP **11**, 028 (2018). https://doi.org/10.1088/1475-7516/2018/11/028
14. F. Bombacigno, G. Montani, Eur. Phys. J. C **79**(5), 405 (2019). https://doi.org/10.1140/epjc/s10052-019-6918-x
15. I. Antoniadis, A. Karam, A. Lykkas, T. Pappas, K. Tamvakis, JCAP **03**, 005 (2019). https://doi.org/10.1088/1475-7516/2019/03/005
16. A. Edery, Y. Nakayama, Phys. Rev. D **99**(12), 124018 (2019). https://doi.org/10.1103/PhysRevD.99.124018
17. T. Tenkanen, Phys. Rev. D **101**(6), 063517 (2020). https://doi.org/10.1103/PhysRevD.101.063517
18. I.D. Gialamas, A. Lahanas, Phys. Rev. D **101**(8), 084007 (2020). https://doi.org/10.1103/PhysRevD.101.084007
19. A. Lloyd-Stubbs, J. McDonald, Phys. Rev. D **101**(12), 123515 (2020). https://doi.org/10.1103/PhysRevD.101.123515
20. I. Antoniadis, A. Lykkas, K. Tamvakis, JCAP **04**(04), 033 (2020). https://doi.org/10.1088/1475-7516/2020/04/033
21. D.M. Ghilencea, Eur. Phys. J. C **80**(12), 1147 (2020). https://doi.org/10.1140/epjc/s10052-020-08722-0

22. N. Das, S. Panda, JCAP **05**, 019 (2021). https://doi.org/10.1088/1475-7516/2021/05/019
23. Y. Tang, Y.L. Wu, Phys. Lett. B **809**, 135716 (2020). https://doi.org/10.1016/j.physletb.2020.135716
24. I.D. Gialamas, A. Karam, A. Racioppi, JCAP **11**, 014 (2020). https://doi.org/10.1088/1475-7516/2020/11/014
25. D. Iosifidis, A.C. Petkou, C.G. Tsagas, Gen. Rel. Grav. **51**(5), 66 (2019). https://doi.org/10.1007/s10714-019-2539-9
26. D. Iosifidis, L. Ravera, Class. Quant. Grav. **38**(11), 115003 (2021). https://doi.org/10.1088/1361-6382/abde1a
27. K. Dimopoulos, S. Sánchez López, Phys. Rev. D **103**(4), 043533 (2021). https://doi.org/10.1103/PhysRevD.103.043533
28. I.D. Gialamas, A. Karam, A. Lykkas, T.D. Pappas, Phys. Rev. D **102**(6), 063522 (2020). https://doi.org/10.1103/PhysRevD.102.063522
29. A. Karam, E. Tomberg, H. Veermäe, JCAP **06**, 023 (2021). https://doi.org/10.1088/1475-7516/2021/06/023
30. A. Lykkas, K. Tamvakis, JCAP **08**(043) (2021). https://doi.org/10.1088/1475-7516/2021/08/043
31. D.S.C. Gómez, Phys. Rev. D **104**(2), 024029 (2021). https://doi.org/10.1103/PhysRevD.104.024029
32. I.D. Gialamas, A. Karam, T.D. Pappas, V.C. Spanos, Phys. Rev. D **104**(2), 023521 (2021). https://doi.org/10.1103/PhysRevD.104.023521
33. I. Antoniadis, A. Guillen, K. Tamvakis, JHEP **08**, 018 (2021). https://doi.org/10.1007/JHEP05(2022)074. [Addendum: JHEP 05, 074 (2022)]
34. J. Annala, S. Rasanen, JCAP **09**, 032 (2021). https://doi.org/10.1088/1475-7516/2021/09/032
35. C. Dioguardi, A. Racioppi, E. Tomberg, JHEP **06**, 106 (2022). https://doi.org/10.1007/JHEP06(2022)106
36. I.D. Gialamas, A. Karam, T.D. Pappas, A. Racioppi, V.C. Spanos, J. Phys. Conf. Ser. **2105**(1), 012005 (2021). https://doi.org/10.1088/1742-6596/2105/1/012005
37. M. AlHallak, N. Chamoun, M.S. Eldaher, JCAP **10**, 001 (2022). https://doi.org/10.1088/1475-7516/2022/10/001
38. K. Dimopoulos, A. Karam, S. Sánchez López, E. Tomberg, Galaxies **10**(2), 57 (2022). https://doi.org/10.3390/galaxies10020057
39. C. Dioguardi, A. Racioppi, E. Tomberg (2022)
40. S. Panda, A.A. Tinwala, A. Vidyarthi, JCAP **01**, 029 (2023). https://doi.org/10.1088/1475-7516/2023/01/029
41. K. Dimopoulos, A. Karam, S. Sánchez López, E. Tomberg, JCAP **10**, 076 (2022). https://doi.org/10.1088/1475-7516/2022/10/076
42. R. Durrer, O. Sobol, S. Vilchinskii, Phys. Rev. D **106**(12), 123520 (2022). https://doi.org/10.1103/PhysRevD.106.123520
43. I. Antoniadis, A. Guillen, K. Tamvakis, JHEP **11**, 144 (2022). https://doi.org/10.1007/JHEP11(2022)144
44. A.B. Lahanas, Phys. Rev. D **106**(12), 123530 (2022). https://doi.org/10.1103/PhysRevD.106.123530
45. I.D. Gialamas, A. Karam, T.D. Pappas, Phys. Lett. B **840**, 137885 (2023). https://doi.org/10.1016/j.physletb.2023.137885
46. S. Panda, A. Rana, R. Thakur, Eur. Phys. J. C **83**(4), 297 (2023). https://doi.org/10.1140/epjc/s10052-023-11459-1
47. I.D. Gialamas, A. Karam, T.D. Pappas, E. Tomberg (2023). https://doi.org/10.1142/S0219887823300076
48. I.D. Gialamas, K. Tamvakis, Phys. Rev. D **107**(10), 104012 (2023). https://doi.org/10.1103/PhysRevD.107.104012
49. C. Dioguardi, A. Racioppi (2023)
50. I.D. Gialamas, T. Katsoulas, K. Tamvakis, Inflation and reheating in quadratic metric-affine gravity with derivative couplings. JCAP **06**, 005 (2024). https://doi.org/10.1088/1475-7516/2024/06/005. https://iopscience.iop.org/article/10.1088/1475-7516/2024/06/005

51. D. Iosifidis, Class. Quant. Grav. **39**(9), 095002 (2022). https://doi.org/10.1088/1361-6382/ac6058
52. Y. Akrami et al., Astron. Astrophys. **641**, A10 (2020). https://doi.org/10.1051/0004-6361/201833887
53. P.A.R. Ade et al., Phys. Rev. Lett. **127**(15), 151301 (2021). https://doi.org/10.1103/PhysRevLett.127.151301

Dark Matter Search with CAST and Beyond CAST

K. Zioutas, V. Anastassopoulos, A. Argiriou, G. Cantatore, S. Cetin, A. Gardikiotis, H. Haralambous, M. Karuza, A. Kryemadhi, M. Maroudas, A. Mastronikolis, C. Oikonomou, K. Ozbozduman, Y. K. Semertzidis, M. Tsagri, and I. Tsagris

Abstract The CERN Solar Axion Telescope (CAST) is searching axion particles since 1999. No axions have been detected yet by CAST. However, two publications in Nature show the CAST results which was initially operated as an axion helioscope and in the last few years was transformed into a Dark Matter (DM) axion haloscope. We recall that DM was discovered through unexpected cosmological observations by F. Zwicky in 1933. However, there are several observations in our solar neighborhood which cannot be explained conventionally. Examples include the physical mechanism behind the heating of the solar atmospheric layers, the triggering of solar flares, the quasi-ubiquitous 11-year solar cycle, but also the annual heating up of the

---

K. Zioutas · V. Anastassopoulos · A. Argiriou · A. Gardikiotis · M. Maroudas (✉) · M. Tsagri · I. Tsagris
Physics Department, University of Patras, Patras, Greece
e-mail: marios.maroudas@cern.ch

G. Cantatore
University and INFN Trieste, Trieste, Italy

S. Cetin
Istinye University, Istanbul, Türkiye

A. Gardikiotis · M. Maroudas
University of Hamburg, Hamburg, Germany

H. Haralambous · C. Oikonomou
Electrical Engineering Department, Frederick University, Nicosia, Cyprus

M. Karuza
University of Rijeka, Rijeka, Croatia

A. Kryemadhi
Messiah University, Mechanicsburg, PA, USA

A. Mastronikolis
Department of Physics and Astronomy, University of Manchester, Manchester, UK

K. Ozbozduman
Physics Department, Boğaziçi University, Istanbul, Türkiye

Y. K. Semertzidis
IBS/KAIST, Daejeon, Korea

J. Strologas and L. Perivolaropoulos (eds.), *HEP2023*, Springer Proceedings in Physics 333 https://doi.org/10.1007/978-3-031-76856-9_23

stratosphere in January, and the mysterious behavior of our ionosphere (at least since 1937) in December. Streaming DM with velocities of ~0.001c (c = speed of light) can provide a common solution through gravitational focusing by the solar-system bodies. This matches as the underlying physical process behind the solar cycle, which was actually initially suspected as a planetary-dependent signature. However, it is known since 1859, that planetary tidal forces are extremely small for such a planetary impact. Therefore, with the use of an overlooked "streaming invisible matter" an external impact can be possible reconciling all the investigated mysterious observations. Planetary relationships have been derived for both the dynamic Sun and Earth, reflecting multiple signatures for streaming DM. Therefore, anomalies within our solar system could be the manifestation of the dark Universe. Favorite particle candidates are anti-quark-nuggets (AQNs), magnetic monopoles, dark photons, or other proposals like the composite "pearls" and some stable baryons. The streaming DM scenario can enhance spatiotemporally the incoming DM flux favoring conditions for the direct DM detection of the not-so-invisible as originally anticipated dark Universe.

Keywords Axions · Dark sector · Planetary relationship · Invisible matter · Gravitational focusing · Solar system

1 Introduction

The proposition of "Dunkle Materie" (DM) by F. Zwicky in 1933 derived from unexpected cosmological observations. Nowadays, we have strong indications that our Universe is dominated by a mysterious DM, whose composition remains elusive. In between, its name became synonymous with something that does not emit, absorb, or reflect electromagnetic radiation, which makes it also difficult to detect. However, based on this work, this definition could be deceiving. Several counter-examples, presented here, could be originated by streaming DM, contradicting the aforementioned DM definition. The hypotheses of this work involve the combination of planetary (and solar) gravitational effects applied on non-relativistic "invisible massive particles" which can then be focused on the solar and planetary atmospheres (see Fig. 1). These particles could have a large cross-section, with baryonic matter and radiation while these DM constituents should interact also in the outer atmosphere before reaching the Earth's surface. This screening from the upper atmospheric region can be sizeable, thus suppressing the expected signals in the DM experiments underneath.

If an incoming stream lasts much longer than a corresponding planetary periodicity then over time, there would be a repetition of the alignment of a planet such as the Earth, with an incident invisible stream. This is not an unreasonable assumption for our solar system. Often, an observed periodicity of some solar system observable points either to a single planetary orbital periodicity or also to a synod of two or more planets, which results to a repeating signal enhancement. As an example, the 11-year

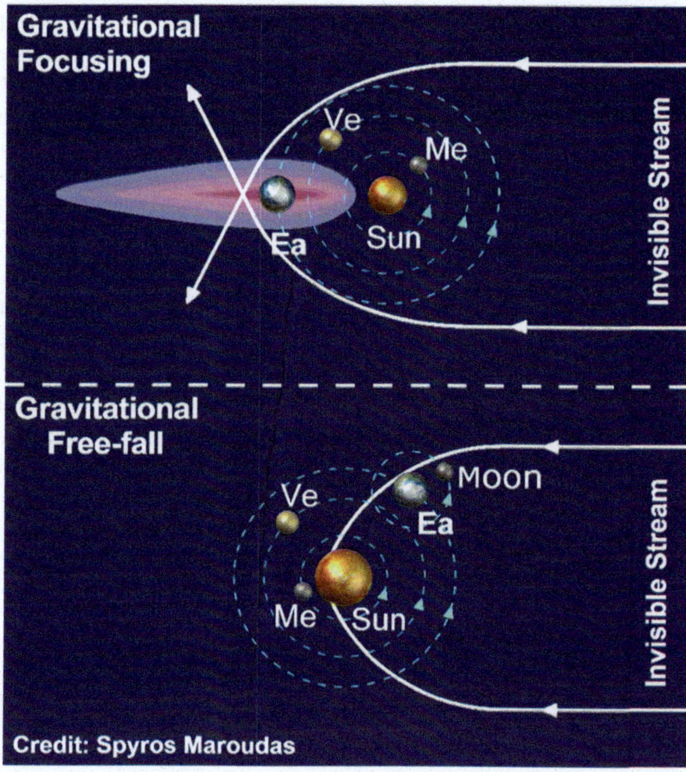

Fig. 1 Illustration of the gravitational focusing effects by the solar system on an incident DM stream. **Top**: gravitational focusing by the inner solar system. **Bottom**: Free-fall of incident low-speed streams towards the Sun can occasionally dominate planetary gravitational focusing since the flux enhancement increases with $(v_{\text{incident}}/v_{\text{escape}})^{-2}$, where v_{escape} is the escape velocity from the Sun and v_{incident} the initial particle velocity far away from the Sun. The flux arriving at the location of the Earth can also be gravitationally modulated by other intervening solar system bodies, like the Moon which can focus particles on the Earth with an incident velocity up to about 400 km/s [3, 8–10]

solar cycle seems to coincide with the synod of Jupiter-Earth-Venus; a coincidence which might not be a random one. This fact was actually suggestive for the scenario of streaming DM introduced in [1] and strengthened by several follow-up solar and terrestrial observations [2–7]. Following known physics, such an unexpected planetary dependence of any observation is deemed as a novel signature from the dark sector. Although a remote planetary force beyond the extremely feeble and smooth tidal force, does not exist, for the streaming DM scenario, the gravitational deflection of a stream is inversely proportional to the velocity squared [8]. This strongly benefits non-relativistic speeds as the ones which are assumed for the DM constituents which are on the level of $v \sim 0.001$ c.

Additionally, based on this approach, exo-solar planetary systems could also be explored since they also consist of orbiting gravitational lenses which, via gravitational lensing, can also focus dark sector constituents. As an example, the Moon can act as a gravitational lens focusing DM particles on the Earth's surface with velocities up to ~400 km/s which covers a large portion of the DM parameter space [3, 9, 10]. It is noted, that we refer to the applicable dark sector constituents as "invisible matter", in order to distinguish them from the common DM candidates like axions and Weakly Interacting Massive Particles (WIMPs).

To have efficiently large planetary gravitational lensing effects from the solar system bodies, invisible matter should consist, at least partly, in a form of streams. Recent cosmological simulations [11, 12] point to a very large number (up to $\sim 10^{14}$) of "fine-grained" DM streams within the Milky Way. Similarly, based on our working hypotheses, anomalous observations within our solar neighborhood also seem to point to the existence of streaming "invisible matter" (see also [1, 3]). Therefore, both independent findings of the cosmological "fine-grained" streams and the unusual observations within our solar neighborhood, converge towards some form of streaming DM, or invisible matter which can include also other candidates from the dark sector that have not been proposed yet.

2 Cast Searches for Solar and Dark Matter Axions

Following an idea from P. Sikivie in 1983, CAST is the so far constructed largest axion Helioscope, as can be seen from the exclusion limits given in Fig. 2 [38]. Therefore, along with the published CAST results [22] the derived constraints by CAST for solar axions dominate the direct axion detection sensitivity by the axion-photon coupling.

In 2019 CAST modified its design aiming to detect DM axions (see References [15, 22]). By installing microwave cavities inside the bore of its 8.8 T superconducting magnet, CAST was operated as a DM axion haloscope *à la* Sikivie, where DM axions are converted into photons inside the high-quality cavities with a resonant frequency that corresponds to the rest mass of the axion.

CAST-CAPP detector [39] has four rectangular cavities that can be tuned using a sophisticated locomotive mechanism consisting of two sapphire strips and piezoelectric motors in a wide tuning range of 660 MHz. CAST-CAPP makes use of two novel techniques for the detection of DM axions as well as transient events. The first one involves the usage of the phase-matching concept where the output power signals from the four identical cavities are combined coherently to increase the effective volume and therefore the detection sensitivity. The second one takes advantage of its state-of-the-art fast-tuning mechanism to cover the full frequency range within about 1 h. This novel detection strategy allows transient events such as axion streams and the cosmologically motivated axion mini-clusters to be probed, which, combined with the gravitational focusing effects of the solar system, can result in a temporally strongly enhanced flux density.

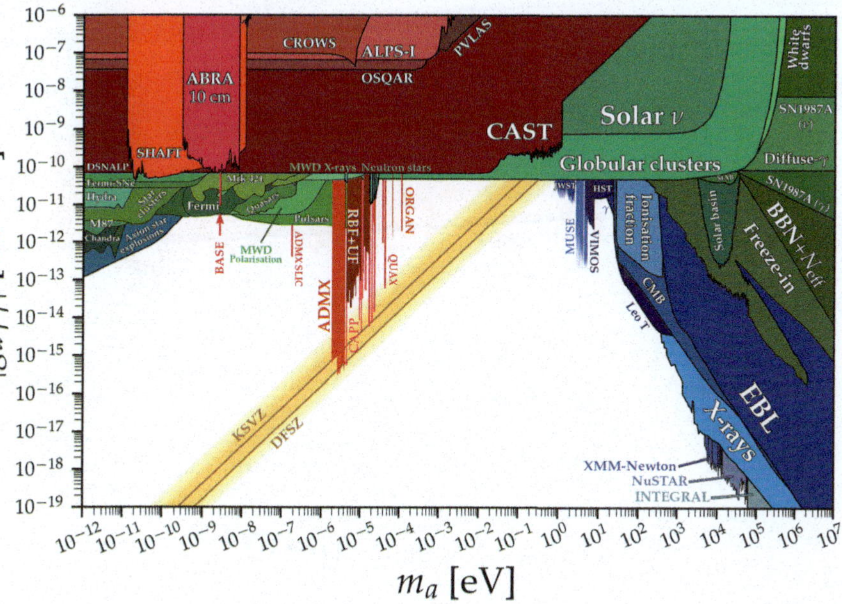

Fig. 2 Exclusion limits (at the 90% CL) for the axion electromagnetic coupling $g_{a\gamma\gamma}$ as a function of the mass of the axion m_a, obtained using various axion DM searches [38]. The contribution by CAST [22] in an axion rest mass range (even below ~100 meV/c^2) is dominating direct axion searches

CAST-CAPP acquired a total of about 172 h of data between 09/2019 and 06/2021 while scanning for axions between 4.77 and 5.43 GHz which correspond to axion masses between 19.74 and 22.47 μeV [39]. The exclusion limit on the axion-photon coupling as a function of the axion mass at a 90% confidence level is shown in Fig. 3. At the same time, an ongoing analysis is currently being performed, searching for possible streaming DM events as well as daily modulations which are expected for candidates like AQNs.

In addition to CAST-CAPP, CAST-RADES detector [40] was also installed inside one of the bores of CAST's superconducting magnet consisting of a single RF cavity comprised of alternating inductive irises searching for DM axions around 34.67 μeV.

Therefore, as seen from Figs. 2 and 3, CAST is prominently positioned in the axion landscape, while the cutting-edge technologies and novel detection strategies employed by its several detectors are of potential importance also for other DM axion searches and large-scale axion haloscopes in the future.

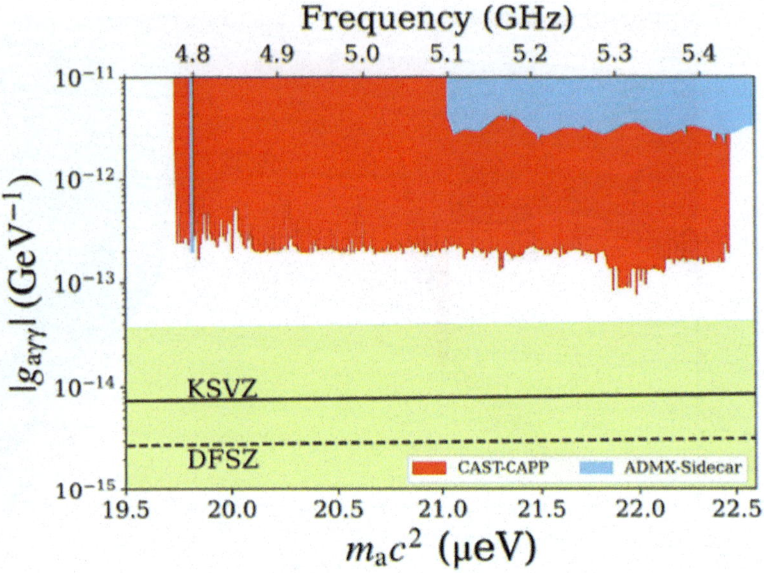

Fig. 3 Exclusion limit on the axion-photon coupling as a function of axion mass set by CAST-CAPP experiment (in red) at 90% confidence level. Other experimental axion exclusion results in the mass range between 19.5 and 22.5 μeV are also shown for comparison. The increased sensitivity around the 22 μeV region, is assigned to the usage of phase-matched cavities at this mass range [14, 39]

3 Some Striking Signatures from Space

Beyond CAST, the concept of this work aligns with the reasoning by Zwicky which pointed to the existence of DM on cosmological scales. In the last ~160 years, along with the quasi-ubiquitous 11-year solar cycle, multiple unexpected observations in the solar system can not be explained conventionally (see e.g. [3]). Based on the reasoning of this work, the dark Universe [1], could be behind these observations with its small-scale manifestation being overlooked. Based on observational evidence, we arrived to a class of "invisible" particle candidates from the dark sector, that should interact "strongly", i.e. with a large cross-section, with normal matter and are separate than the existing allowed parameter phase space for axions and WIMPs which have defied a direct observation since decades. Although axions or WIMPs do not fit in the approach of this work, they remain inspiring DM candidates.

The unexpected temperature of the solar corona along with its mysterious step-like temperature inversion, the unpredictable solar activity, the dynamic atmosphere of the Earth, and other similar unexpected observations could all be interpreted from invisible streams that can result in spatiotemporally strong flux enhancements due to gravitational lensing, by one or more solar system bodies, including the Sun [3]. It is mentioned, that the underlying dark constituents can have a diverse nature, and can even be theoretically not yet proposed, a fact that can make even more challenging their identification (see also below).

Some examples of energetic and conventionally unexplained observations include the flaring Sun along with its irradiance, and even more generally the dynamical behavior of the whole Sun [3, 13] as this is exhibited in the solar radio line (F10.7) at 10.7 cm wavelength which is widely accepted as a proxy of the solar behaviour. The most energetic planetary relationship is observed for the Sun's size variation during one solar cycle [13] by about 1 km, with the Sun being modulated in relative size by $\sim 10^{-6}$. As a comparison, we note that the energy that is required to lift 1 km of a thin photosphere layer ($\rho \approx 10^{-7}$ g/cm^3) by 1 km, is huge ($\sim 10^{30}$ ergs). Additionally, an unexpected planetary dependence was also found on the Sun's elemental composition [3, 14]. This observation has similarities with the solar EUV irradiation above ~20 eV photon energy which also showed a planetary relationship. It is noted that EUV irradiation is another manifestation of the mysterious heating of the solar corona [15]. Furthermore, the elemental magnetic bright points (MBPs) on the solar surface also demonstrate a planetary relationship [3]. In 1859 WOLF who was the first to observe sunspots (see [3] and Ref. [7] therein), was also the first suspecting a planetary influence behind the Sun's behavior [1]. However, a physical underlying process remained elusive within known physics.

Apart from the unexpected planetary relationships derived from the various solar observables, a number of terrestrial anomalous phenomena take place in the Earth's upper atmosphere, which are known since 1930s. As an example, the elusive mechanism behind the dynamical behavior of the ionosphere also demonstrates a planetary relationship [2]. More specifically, there is an annual anomaly, known since 1937 [16], where about 25% more atmospheric ionization is observed around December than six months later around June. We recall that the ionosphere is the most outer terrestrial region and is directly exposed to outer space. Subsequently, possible "invisible" dark sector constituents may first be manifested there, and most certainly if they interact via a large cross-section with normal matter (see e.g. Ref. [17]). Interestingly, this is feasible also for some DM constituents following recent publications like Ref. [17]. Therefore, the aforementioned requirement is not used to support the proposed scenario of this work. It is stressed that the underground direct DM experiments can have a very small sensitivity to such DM particles due to the unavoidable screening of "strongly" interacting dark constituents by the overhead Earth's layers including its atmosphere. Remarkably, the ionosphere's plasma state is observed by GPS satellites orbiting the Earth and continuously registering the ionospheric plasma column density for self-calibration purposes [18]. Thus, the observed striking planetary "dependency" of the ionosphere's plasma density, points to an exo-solar cause [2].

The not-so-randomly appearing behavior of the dynamic Earth including its atmosphere [14, 18], probably happens by accumulating energy deposition within long or short time intervals, similar to the dynamical solar behavior (see e.g., [2, 13]). Apparently, the speculated invisible stream(s) [12] or clusters (see e.g., [19, 20]) do not need to provide spatiotemporally the required entire energy. However, following pile-up effects, it can be the external trigger or catalyst. which matches the **tentatively** predicted streaming invisible matter hypothesis being thus of exo-solar origin [1].

Of note, for the second part of this work, only those observations are of potential interest which show, following known physics, a not extant planetary dependency. It is noted that more results are derived following this out-of-the-box approach, aiming to eventually expose the microscopic nature of the suspected stream(s) and/or clusters; therefore de-characterizing them as "invisible" [3]. In particular, while QCD-inspired axions seem not to fit in puzzling observations within the solar system, axion-like particles (ALPs) could be good candidates.

In addition, following the principle assumptions this work, it would be interesting to explore a possible similar behavior in exo-solar planetary systems [21]. With near-Earth exo-planetary systems, similar 'exo-planetary' relationships can be derived, or even also a cross-correlation with the dynamical behavior of our solar system could be made possible. The possible execution of such observations could expand our horizon towards the dark Universe, confirming the working hypotheses behind such a scenario.

In summary, we have pinpointed a simple feature being a common signature behind unexplained observations within our solar neighborhood. Notably, even the commonly assumed DM particles with a velocity of ~0.001c, while being in the form of streams, can be efficiently gravitationally focused or deflected within our solar system [1, 2, 12, 19, 20].

4 Dark Matter Direct Detection

The CAST experiment has been inspiring for this novel approach referring to observationally driven novel DM searches. So far, studying a number of diverse peaking planetary relationships of several observables [3, 13, 14] could point to optimum spatiotemporal conditions to enhance the direct DM detection sensitivity, not only in deep underground experiments but also in the upper atmosphere. Earth-related time windows of opportunity can be annually fixed; similarly, other planetary peaking relationships taking place in different time periods are also of potential interest.

A regular flux enhancement of streaming DM due to gravity effects by the solar system bodies can reach several orders of magnitude, with flux amplification factors of up to about 10^5–10^8 being realistic [8, 11, 23]. For example, in January in the northern hemisphere, the continuously monitored [24, 25] annually peaking stratospheric temperature (at a height of around 40 km) shows a striking planetary dependency. In the future, probably more planetary relationships will be revealed, which can be utilized in unconventional DM detection schemes (see e.g. [7, 23–25]).

Interestingly, National Aeronautics and Space Administration (NASA) has constructed scientific balloons (see [26–29]) that can be used for measurements in the upper stratosphere with a payload of up to a few tons and a working time of several months. This is inspiring for the reasoning of this work, as statistically significant planetary relationships have already been detected for the upper stratosphere's temperature [7] as well as the ionosphere's degree of ionization [2]. More specifically, for the case of the stratosphere [7], a significant peaking planetary behavior has been observed using the heliocentric longitudinal positions of Mercury and Venus. Then, when the orbital positions of Venus and Mercury are combined, a clear peaking relationship for the stratospheric temperature variation in early January in the northern hemisphere [7], might pave the way to perform even direct DM searches in the upper atmosphere [7, 24] contrary to the currently preferred underground searches. Of note, DM constituents interacting with large cross section could be detected preferentially in the upper atmosphere, given the performance of new balloons [26–29]. Future investigations could choose additional sites in the atmosphere instead of underground, aiming to directly search for DM up there. Thus, planetary lensing combined eventually with Earth's gravitational self-focusing due to its inner mass distribution has the potential to temporally strongly enhance the local DM flux by up to several orders of magnitude, thus suggesting new perspectives for DM research.

5 Summary

Based on observational data, we conclude that *a planetary relationship can be the novel signature pointing on its own at exo-solar impact for a certain observable.* Until now, the proposed viable common explanation for a number of observations with diverse behavior together with the conventionally unexpected planetary dependency is based on the gravitational focusing of streaming "invisible" matter from one solar system body to another, including the Sun and the Moon. We **tentatively** assume constituents in form of streams from the dark Universe, which could interact with a large cross-section with ordinary matter. At the moment, we can only speculate about the possible particle candidates which are already suggestive for even more searches of otherwise unexpected planetary relationships.

Several obvious implications exist for running as well as future DM experiments. As a result, we propose that all experiments searching for direct DM signatures should perform a statistical re-analysis following the reasoning underlying this approach or modify their data acquisition procedure accordingly for future measurements (see Ref. [3]). If a planetary dependency is found also in direct DM searches, this will strengthen the concept of "invisible streams" in our vicinity.

With this work, we are aiming to highlight such new signatures in DM searches, which are probably inherent also to other observations. Concluding, a diverse number of observations depict planetary relationships that may eventually enable us to pinpoint the otherwise "invisible" components of the dark Universe which have so far not been detected in direct searches. Some examples of observationally-fitting existing "invisible candidates" include:

(a) Axion Quark Nuggets (AQNs), which have been proposed by Zhitnitsky [30] (see also [31–34]). These peculiar composite macroscopic particles inspire several investigations ranging from the mysterious solar corona heating to the direct detection of relativistic (~0.6 c) DM axions by axion haloscopes such as CAST.

(b) Magnetic monopoles, whose interaction with the ubiquitous magnetic fields make different energy deposition scenarios of potential interest.
(c) Dark photons, which can resonantly convert to real photons if the local plasma density fits in the rest mass of the hidden photon. In contrast with axions or axion-like particles, the kinetic mixing between real photons with the hidden sector photons does not require the presence of an external magnetic field for the conversion.
(d) Pearls [35, 36]. A quantitative investigation which is already initiated for AQNs from CAST, would clarify whether these composite particles also fit to at least a number of the performed observations.
(e) Other composites with long lifetimes [37], or other dark sector constituents yet to be proposed.

6 Conclusion

An expected signal amplification by several orders of magnitude derived from the gravitational focusing effects of the solar system bodies including the self-focusing effects by the Earth [9, 10, 23] could initiate a long-awaited breakthrough for direct DM detection. The interaction strength of some components of "invisible streaming matter" with baryonic matter could be significant [17], thus opening new windows of opportunity in the field of DM.

At the moment, the most inspiring particle candidates fitting several observations are AQNs, magnetic monopoles, and dark photons. However, more candidates (see above) should be investigated if they suite the reasoning of this work. Therefore, along with Zwicky's interpretation of large-scale observations that have uncovered the invisible matter in the cosmos, also insisting anomalies/mysteries within the solar system might be the overlooked manifestation of the dark Universe. Therefore, they deserve further attention by DM axion haloscopes or other detection ideas being in the preparation phase. If more DM constituents exist in the form of streams, the concept presented here could open the horizon in the search for DM.

Acknowledgements Y.K.S acknowledges the support from IBS-R017-D1 of the Republic of Korea.

References

1. K. Zioutas, M. Tsagri, Y.K. Semertzidis, T. Papaevangelou, D.H.H. Hoffmann, V. Anastassopoulos, Mod. Phys. Lett. A **29**, 1440008 (2014)
2. S. Bertolucci, K. Zioutas, S. Hofmann, M. Maroudas, Phys. Dark Univ. **17**, 13 (2017)
3. K. Zioutas, V. Anastassopoulos, A. Argiriou, G. Cantatore, S.A. Cetin, A. Gardikiotis, D.H.H. Hoffmann, S. Hofmann, M. Karuza, A. Kryemadhi, M. Maroudas et al., Phys. Sci. Forum **2**, 10 (2021)

4. K. Zioutas, E. Valachovic, Biophys. Rev. Lett. **13**, 75 (2018)
5. K. Zioutas, E. Valachovic, M. Maroudas, Biophys. Rev. Lett. **14**, 11 (2019)
6. K. Zioutas, M. Maroudas, S. Hofmann, A. Kryemadhi, E. Matteson, Biophys. Rev. Lett. **15**, 275 (2020)
7. K. Zioutas, A. Argiriou, H. Fischer, S. Hofmann, M. Maroudas, A. Pappa, Y.K. Semertzidis, Phys. Dark Univ. **28**, 100497 (2020)
8. D.H.H. Hoffmann, J. Jacoby, K. Zioutas, Astropart. Phys. **20**, 73 (2003)
9. Y. Sofue, Galaxies **8**(2), 42 (2020)
10. G. Prezeau, ApJ **814**, 122 (2015)
11. A. Kryemadhi, M. Maroudas, A. Mastronikolis, K. Zioutas, Phys. Rev. **D108**, 123043 (2023). https://doi.org/10.1103/PhysRevD.108.123043. arXiv:2210.07367
12. M. Vogelsberger, S.D.M. White, Mon. Not. R. Astron. Soc. **413**, 1419 (2011)
13. K. Zioutas, M. Maroudas, A. Kosovichev, Symmetry **14**, 325 (2022)
14. M. Maroudas, *Signals for Invisible Matter from Solar-Terrestrial Observations*. Ph.D. thesis (University of Patras, Greece, 2022)
15. K. Murawski, Z.E. Musielak, S. Poedts, A.K. Srivastava, L. Kadowaki, Accepted Astrophys. Space Sci. (2022). https://doi.org/10.48550/arXiv.2211.12289
16. E.V. Appleton, Proc. Roy. Soc. London A **162**, 451 (1937)
17. T. Emken, R. Essig, C. Kouvaris, M. Sholapurkar, J. Cosmol. Astropart. Phys. **9**, 70 (2019)
18. K. Zioutas, V. Anastassopoulos, A. Argiriou, G. Cantatore, S. Cetin, H. Fischer, A. Gardikiotis, H. Haralambous, D.H.H. Hoffmann, S. Hofmann, M. Karuza, A. Kryemadhi, M. Maroudas, A. Mastronikolis, C. Oikonomou, K. Ozbozduman, Y.K. Semertzidis, Proceedings 15th International Conference on Meteorology, Climatology and Atmospheric Physics—COMECAP (Ioannina, Greece, p. 728, 2021). https://www.conferre.tv/comecap2020
19. I.I. Tkachev, Phys. Lett. **261**, 289 (1991)
20. E.W. Kolb, I.I. Tkachev, Phys. Rev. Lett. **71**, 3051 (1993)
21. M. Perryman, K. Zioutas (2021). arXiv:2106.15408
22. V. Anastassopoulos et al., Nat. Phys. **13**, 584 (2017)
23. A. Arza, A. Kryemadhi, K. Zioutas, Phys. Rev. D **108**(8), 083001 (2023). arXiv:2212.10905
24. G. Cantatore, H. Fischer, W. Funk, M. Karuza, A. Kryemadhi, M. Maroudas, Y.K. Semertzidis, K. Zioutas (2022). arXiv:2012.03353
25. H. Fischer, X. Liang, A. Zhitnitsky, Y.K. Semertzidis, K. Zioutas, Phys. Rev. D **98**, 043013 (2018)
26. NASA, *Scientific Balloons*, Types (2022). https://www.nasa.gov/scientific-balloons/types-of-balloons
27. NASA, *Scientific Balloons*, Overview (2021). https://www.nasa.gov/scientificballoons/overview
28. P. Gorham, J. Anderson, P. Bernasconi, S. Chakrabarti, T.G. Guzik, W. Jones, C. Kierans, R. Millan, A. Vieregg, C. Walker, E. Young, A roadmap for scientific ballooning 2020–2030 (2022)
29. Columbia Scientific Balloon Facility Homepage. http://www.csbf.nasa.gov/
30. A. Zhitnitsky, JCAP **2003**(10), 10 (2003)
31. A. Zhitnitsky, Phys. Dark Univ. **22**, 1 (2018)
32. N. Raza, L. van Waerbeke, A. Zhitnitsky, Phys. Rev. D **98**, 103527 (2018)
33. D. Budker, V.V. Flammbaum, A. Zhitnitsky, Symmetry **14**, 459 (2022)
34. A.R. Zhitnitsky, Mod. Phys. Lett. A **36**, 2130017 (2021)
35. H.B. Nielsen, C.D. Froggatt (2022). arXiv:2205.08871
36. H.B. Nielsen, C.D. Froggatt (2020). arXiv:2012.00445
37. B.G. Borštnik, N.S. Mankoč, Does dark matter consist of baryons of new stable family quarks? Phys. Rev. D**80**, 083534.1 (2009). https://doi.org/10.1103/PhysRevD.80.083534
38. C. O'Hare, AxionLimits, Zenodo (Vol. 1, 2020). https://doi.org/10.5281/zenodo.3932430. https://cajohare.github.io/AxionLimits/

39. C.M. Adair et al., (CAST Collaboration), Search for DM Axions with CAST-CAPP. Nat. Commun. **13**, 6180 (2022). https://doi.org/10.1038/s41467-022-33913-6
40. A.A. Melcon et al., (CAST Collaboration), First results of the CAST-RADES haloscope search for axions at 34.67 μeV. J. High Energy Phys.**75** (2021). https://doi.org/10.1007/JHEP10(2021)075

Recent Physics Results from CMS

John Strologas

Abstract We present recently released physics results from the CMS experiment at the LHC. Most analyses use the entire 138 fb^{-1} integrated luminosity of Run 2 of LHC at $\sqrt{s} = 13$ TeV. We also report on the first measurement of top-pair production with the first Run 3 data of 1.21 fb^{-1} at 13.6 TeV.

Keywords CMS · LHC · Experimental HEP · Standard model · Physics beyond the standard model · Higgs boson · Top quark

1 Introduction

We present physics results from the CMS experiment [1] at LHC, concentrating on results released within a month from the conference and excluding supersymmetry (SUSY) results and jet studies in the standard model (SM) that are covered by separate presentations. We discuss SM first observations and searches for dark matter, heavy neutral leptons, Higgs boson pairs and SM-like extra Higgs bosons. We also present the first released result (top-pair production cross section) from Run 3 of the LHC at $\sqrt{s} = 13.6$ TeV. Unless otherwise stated, the presented analyses utilize the entire Run 2 dataset, collected at $\sqrt{s} = 13$ TeV, and corresponding to an integrated luminosity of 138 fb^{-1}.

2 First Observation of $t\bar{t}t\bar{t}$ Production

The production of $t\bar{t}t\bar{t}$ is one of the rarest SM processes. The measurement of its cross section provides sensitivity to the top quark Yukawa coupling, probes the existence of beyond-the-SM (BSM) Higgs bosons, and is an important input to effective field

John Strologas on behalf of the CMS collaboration.

J. Strologas (✉)
University of Ioannina, Ioannina, Greece
e-mail: strolog@uoi.gr

J. Strologas and L. Perivolaropoulos (eds.), *HEP2023*, Springer Proceedings in Physics 333 https://doi.org/10.1007/978-3-031-76856-9_24

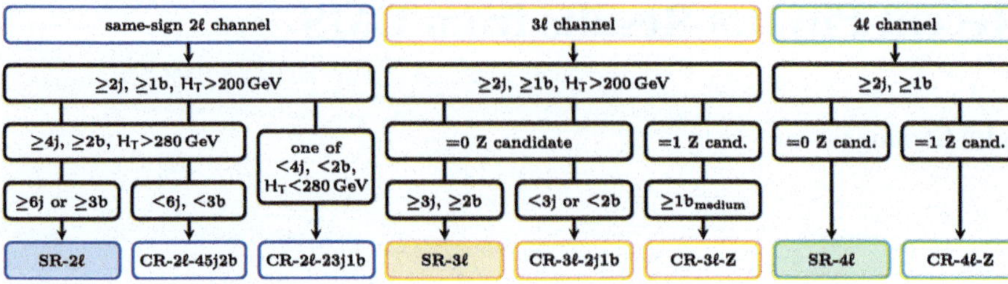

Fig. 1 The list of signal and control regions for the $t\bar{t}t\bar{t}$ analysis

theory interpretations. In the CMS discovery analysis [2], separate signal regions of two same-sign (SS), three, and four charged leptons (electrons and muons, transverse momentum $p_{\mathrm{T}} > \{25, 20, 10, 10\}$ GeV) with the presence of jets ($p_{\mathrm{T}} > 25$ GeV) some of them b tagged (with DeepJet [3]), are analyzed. Vetos are set on additional or low-p_{T} leptons. Figure 1 shows the object requirements for the signal and control regions studied. Multivariate discriminants are used to distinguish the signal process from the backgrounds, which are mainly $t\bar{t}W$ and nonprompt processes for the dilepton and trilepton channels, $t\bar{t}Z$ and nonprompt processes for the trilepton channel, $t\bar{t}Z$ for the 4-lepton channel, and $t\bar{t}H$ for all channels. Nonprompt background consists of genuine leptons from hadron decays or misidentification of jets as leptons, and it is estimated using data. The measurement is performed with a simultaneous fit of the boosted decision tree (BDT) scores of both signal and control regions. The main systematic uncertainties arise from luminosity, lepton-ID, trigger and b-tagging efficiencies, jet energy scale and nonprompt background estimation. The observed (expected) significance of the signal is 5.5 (4.9) standard deviations above the background-only hypothesis. The signal cross section is measured with a profile likelihood fit to be $\sigma_{t\bar{t}t\bar{t}} = 17.7^{+3.7}_{-3.5}$ (stat)$^{+2.3}_{-1.9}$ (syst) fb, in agreement with the best available theoretical prediction [4] (next-to-leading order (NLO) in QCD and EW theory, including next-to-leading-logarithmic (NLL) accuracy in gluon emission corrections), which is $13.4^{+1.0}_{-1.8}$ fb.

3 First Observation of $WW\gamma$ Production

Triboson production is sensitive to both triple and quartic gauge couplings. The first observation of $WW\gamma$ [5] utilized the $e^{\pm}\mu^{\mp}\gamma$ channel, with the requirement of missing transverse momentum $p_{\mathrm{T}}^{\mathrm{miss}}$. This process is characterized by low background, mostly QCD-initiated $V\gamma$ (where V is a gauge boson) and from nonprompt photons (measured using data). The transverse momenta of the electron, muon, and photon are required to be above 20, 25 and 20 GeV respectively, the $p_{\mathrm{T}}^{\mathrm{miss}} > 20$ GeV and $\Delta R_{\ell\gamma} > 0.5$. Additional-lepton and b-tag vetos reduce the $WZ\gamma$ and top backgrounds respectively. Low dilepton p_{T} and mass and WW mass

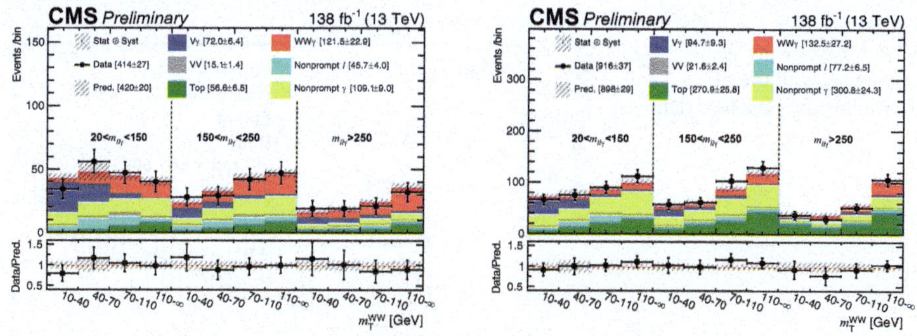

Fig. 2 The 2D m_T^{WW} versus $m_{\ell\ell\gamma}$ distributions for the $WW\gamma$ signal region with 0 jet (left) and ≥1 jet (right), after the fit to the data

are also excluded to reduce the QCD-induced background. The same-sign dileptons and b-tag control regions validate nonprompt and top backgrounds respectively. The signal region is divided into 0-jet and $\geq$ 1-jet, with the former demonstrating reduced top background. The final fit, 2D in m_T^{WW} and $m_{\ell\ell\gamma}$, includes both signal and control regions. Figure 2 shows the event yields for these signal regions. The observed (expected) signal significance from the fit is 5.6 (4.7) standard deviations. The measured cross section is $\sigma_{WW\gamma} = 6.0 \pm 1.0$ (stat) ± 1.0 (syst) ± 0.9 (theo) fb, where the SM prediction, based on the NLO MadGraph5_amc@nlo simulation is 4.61 ± 0.34 (scale) ± 0.05 (PDF) fb. For the $H\gamma$ search, the large lepton-lepton separation is excluded. Limits are set on the production of $H\gamma$ with subsequent decay to $H \to WW \to e\mu\nu_e\nu_\mu$ and on modified Higgs boson couplings to light quarks.

4 First Observation of $\eta \to \mu^+\mu^-\mu^+\mu^-$

The measurement of branching fractions of rare decays offers precision tests of the SM, a possibility of new discoveries and, in the case of multi-muon final states, a better understanding of the muon anomalous magnetic moment. The first observation of the decay $\eta \to \mu^+\mu^-\mu^+\mu^-$ by CMS [6] became possible with the use of dedicated low-p_T (down to 3 GeV), high-rate dimuon triggers. The analysis uses the decay $\eta \to \mu^+\mu^-$ (with known branching fraction $\mathcal{B}_{2\mu}$) as the normalization channel and measures the branching fraction $\mathcal{B}_{4\mu}$ as

$$\frac{\mathcal{B}_{4\mu}}{\mathcal{B}_{2\mu}} = \frac{N_{4\mu}}{\sum_{i,j} N_{2\mu}^{i,j} \frac{A_{4\mu}^{i,j}}{A_{2\mu}^{i,j}}}, \tag{1}$$

where i is the p_T bin of the η meson, j its pseudorapidity bin, $N_{4\mu}$ ($N_{2\mu}$) is the 4-muon (2-muon) signal yield and $A_{4\mu}^{i,j}$ ($A_{2\mu}^{i,j}$) the 4-muon (2-muon) acceptance

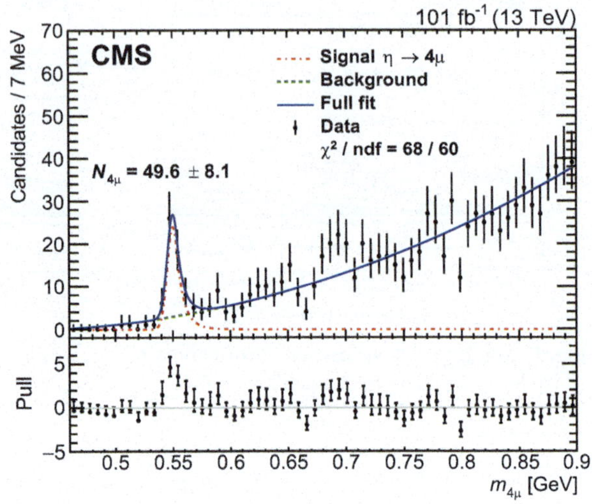

Fig. 3 The 4-muon mass spectrum, showing the η resonance, with the fit overlaying the observed data

times efficiency, measured in MC. The resulted value is $\mathcal{B}_{4\mu} = [5.0 \pm 0.8\ \text{(stat)} \pm 0.7\ \text{(syst)} \pm 0.7\ (\mathcal{B}_{2\mu})] \times 10^{-9}$, to be compared with the SM prediction [7] of $(3.98 \pm 0.15) \times 10^{-9}$. Figure 3 shows the observed 4-muon mass spectrum.

5 Search for Inelastic Dark Matter in Events with Displaced Muons

CMS searched [8] for inelastic dark matter, in models that postulate the existence of two coupled dark matter states (χ_1 and χ_2) as well as a dark photon (A') that provides the connection between the dark sector and the SM, needed to account for the thermal equilibrium in the early Universe. A process the models predict is the associated production of χ_1 and χ_2, followed by $\chi_2 \to \chi_1 A'$, and the subsequent decay of A' to two leptons. The inelastic coupling and the small mass gap between the two dark matter states increase the lifetime of the heavier χ_2, whose decay leads to displaced, low-p_T and collimated leptons. The produced two lighter χ_1 particles escape detection and lead to p_T^{miss}. Initial-state radiation produces at least one jet that balances the dark-matter system. Due to the higher CMS efficiency in reconstructing soft muons, the final signal signature investigated is $\mu\mu$+jet+p_T^{miss}. The range of theoretical parameters studied includes the mass difference of dark-matter states of $\Delta m = 10\%$ and $40\%\ m_{\chi_1}$, dark photon mass $3m_{\chi_1}$ and lifetime $c\tau = 1 - 1000$ mm. Triggers requiring p_T^{miss} and H_T^{miss} above 120 GeV are used and a b-tag veto is applied. A stand-alone algorithm that reconstructs displaced muons using only muon chambers and not the primary vertex achieves 90% efficiency for up to two meters of transverse displacement. Collimation of the muons is guaranteed by requiring $\Delta R_{\mu\mu} < 0.9$, while the dimuon system is azimuthally aligned with p_T^{miss}

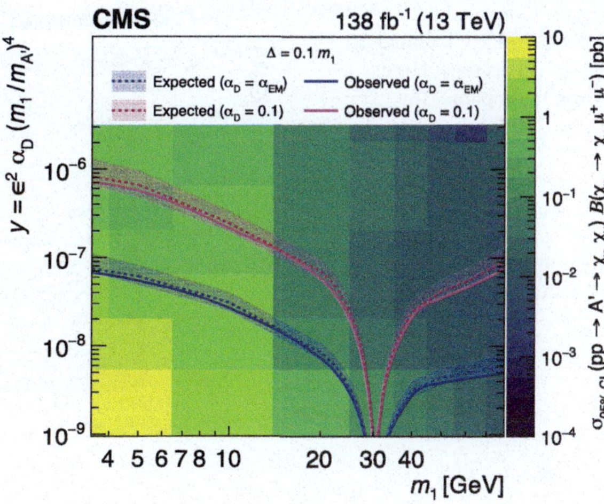

Fig. 4 Observed and expected 95% CL limits on the interaction strength of the dark photon to the dark matter in the displaced muons analysis, for $\Delta m = m_{\chi 1}/10$

($\Delta\phi_{\mu\mu^{\mathrm{MET}}} < 0.5$). The matching with the default CMS muons separates the events to the ones with 0 matched (highly displaced) and 1, or 2 matched (mildly displaced). Dominant background is QCD-induced real and misidentified displaced muons and W/Z+jets. They are determined using data in the "ABCD" manner in isolation versus impact parameter (for the 1- and 2-matched categories) and isolation versus $\Delta\phi_{\mu\mu^{\mathrm{MET}}}$ (for the 0 matched category). The data are consistent with the SM prediction in all categories and limits are set separately for $\Delta m = 10\%$ and 40% m_{χ_1}, for $\alpha_D = \alpha_{EM}$ and 0.1, where α_D of the coupling between the dark photon and the dark matter. This limit for $\Delta m = 10\%$ m_{χ_1} can be seen in Fig. 4.

6 Search for Dark Matter in WW+$p_{\mathrm{T}}^{\mathrm{miss}}$

The interaction of the dark sector with the SM could take place with a Z' mediator and an additional "dark" Higgs boson (s) [9] which also provides mass to the dark-matter particles. A CMS analysis [10] searched for the dark Higgs boson decay to WW, accompanied by $p_{\mathrm{T}}^{\mathrm{miss}}$ from the undetected lightest dark-matter state χ. Dilepton ($e\mu$, for reduction of Drell Yan (DY) background) and lepton+jets final states are investigated, collected with dilepton and single-lepton triggers. The requirement of isolated leptons, a b-tag veto and veto on more than the expected number of leptons are set. The dilepton channel is split into three boosted signal regions based on $\Delta R_{\ell\ell}$, and the observable is the transverse mass of the trailing lepton and the $p_{\mathrm{T}}^{\mathrm{miss}}$ in bins of the dilepton mass. In the lepton+jets channel, the dijet mass is required to be in the W mass window, and there are extra object-separation cuts and a cut on the lepton-$p_{\mathrm{T}}^{\mathrm{miss}}$ transverse mass. For this channel, the observable is the output of a BDT. For the dilepton (lepton+jets) channel, $p_{\mathrm{T}}^{\mathrm{miss}} > 20$ (60) GeV is required. The prompt lepton(s) background, estimated with MC, is WW and DY for dilepton, W+jets for

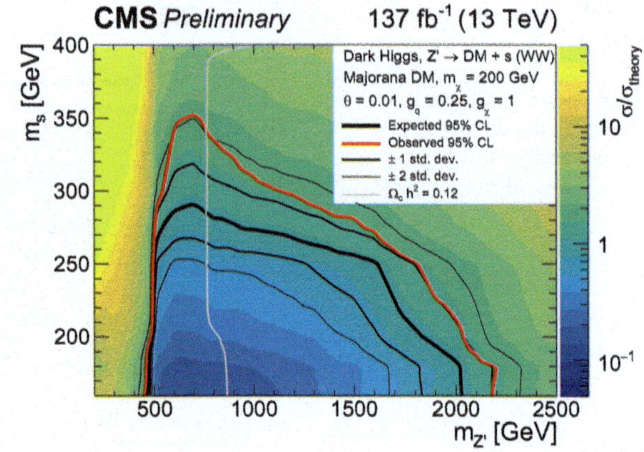

Fig. 5 Observed and expected 95% CL limits on the Z' versus dark Higgs boson mass plane, for $m_\chi = 200$ GeV

lepton+jets and tW, $t\bar{t}$ for both channels. The QCD-induced nonprompt background is estimated with application of lepton fake rates on events selected with looser cuts. Fits of the observables with systematics as nuisance variables lead to consistency with the SM, and limits on the plane of $m_{Z'}$ versus m_s are set for different masses m_χ of the dark-matter particle. Figure 5 shows the limit for $m_\chi = 200$ GeV.

7 Long-Lived Heavy Neutral Leptons

Long-lived heavy neutral leptons (HNL) can explain the small masses of neutrinos via the seesaw mechanism, account for the baryon asymmetry of the universe, and provide a dark-matter candidate. CMS searched [11] for HNL produced in association with a charged lepton and decaying further from the primary vertex to a charged lepton and a hadronically decaying W. Thus, the signature is a well-identified lepton, a displaced (as determined by the transverse impact parameter) lepton with looser identification requirements, and a displaced (as determined by a neural network) jet. Opposite-sign (OS) dileptons are used for studying Dirac HNL and SS for Majorana. The signal region requires $p_{\mathrm{T}}^{\mathrm{miss}} < 60$ GeV, a dilepton mass between 20 and 80 GeV and a total of 1–4 jets. The ΔR between the displaced jet and the displaced lepton should be less than 1.3. If it is less than 0.4 the event is called boosted, otherwise it is resolved. The control region of Z+jets with a dilepton mass above 80 GeV is used. The main background is Z/γ^*+jets for the OS dileptons and $W(V\gamma)$+jets for the SS, determined based on data. For greater sensitivity, the signal region is divided based on the lepton flavor combinations, OS/SS, resolved/boosted and impact parameter significance of the loose lepton. No excess of events in data over the expected SM background is observed. The best limit on the coupling strength is obtained for pure muon coupling scenarios excluding at 95% CL values of $|V_{\mu N}| > 5(4) \times 10^{-7}$ for Dirac (Majorana) HNLs with a mass of 10 GeV. Limits on the mass of the HNL are

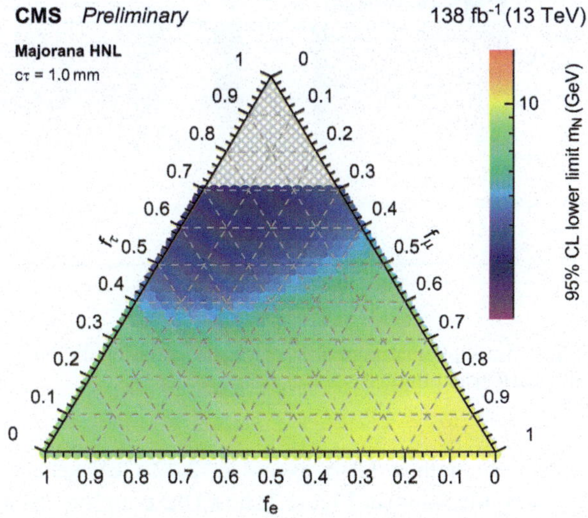

Fig. 6 Observed 95% CL limit on the mass of a Majorana heavy neutral lepton with lifetime 1 mm/c, as a function of the relative couplings to the three lepton generations

set for different values of coupling constants (and vice versa) and a great variety of the three coupling constant ratios, for Dirac and Majorana HNL cases separately. Figure 6 shows an example of such a limits plot, for Majorana HNL, with lifetime 1 mm/c.

8 Search for GeV-Scale Resonance Decaying to a Pair of Muons

CMS utilized high-rate, low p_T (> 3 GeV per muon) dimuon scouting triggers to search for new GeV-mass dimuon resonances in 96.6 fb^{-1} of data [12]. These are the same triggers used in the $\eta \to \mu^+\mu^-\mu^+\mu^-$ analysis described in Sect. 4. Possible BSM physics sources investigated are a dark photon produced by DY and decaying to a pair of muons with a coupling to SM processes suppressed by a factor, and an extra light pseudoscalar Higgs boson decaying to a pair of muons with coupling determined by the mixing angle θ and the ratio of vacuum expectation values of the Higgs bosons. The muons are reconstructed with only the high-level-trigger (HLT) information and use of a BDT. The search is conducted with the requirement of OS dimuons (in the mass range 1.1–2.6 GeV and 4.2–7.9 GeV, to avoid the Υ and ψ' resonances). The optimal mass resolution ($\approx$1.3%) is achieved for muons with $|\eta| < 1.9$. A small natural width of the signal is assumed, so this mass resolution determines the width of the resonance. The signal is extracted with simultaneous signal-plus-background fits to the dimuon mass distribution. The observation is consistent with the continuous dimuon spectrum and limits are set on parameters of the two models above. A local 3.2-σ (global 1.3-σ) excess at 2.41 GeV (high-p_T selection) is observed, interestingly

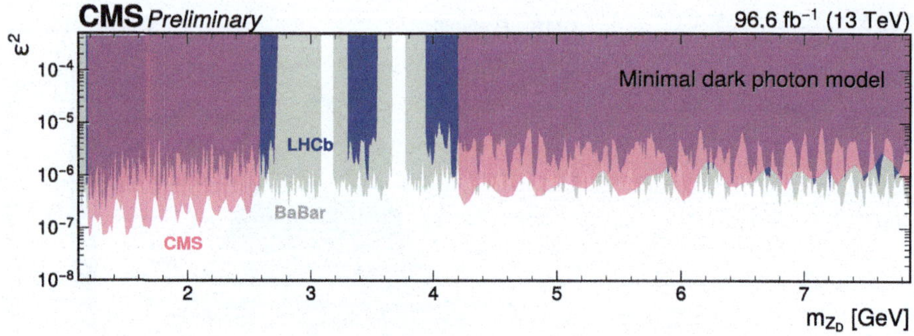

Fig. 7 Dark-photon coupling versus mass exclusion for the CMS analysis, compared to exclusion from LHCb and BaBar

consistent with a 3.1-σ excess at 2.42 GeV seen by the LHCb experiment in a comparable analysis. Figure 7 shows the coupling versus dark-photon mass exclusion compared to that of the LHCb and BaBar experiments.

9 Search for $HH \rightarrow W^+W^-b\bar{b}$

Double Higgs boson production at the LHC is achieved through gluon-gluon fusion or vector-boson fusion (VBF) and offers the opportunity to study the trilinear Higgs boson self-coupling strength modifier κ_λ and the $HHVV$ coupling strength modifiers κ_V, κ_{VV} (where $V = W$ or Z) respectively. A recent CMS analysis [13] looks for signals where one Higgs boson decays to $b\bar{b}$ and the other to W^+W^-. The two b quarks can be reconstructed as two small-radius ($R = 0.4$) separate jets (at least one b-tagged) or one large-radius ($R = 0.8$) jet. In the latter case two subjets should be reconstructed with at least one being b-tagged. The W bosons lead to either dilepton (DL) or single-lepton (SL) final states (electrons and muons are used). For the SL final state, either ≥ 3 small-radius jets (at least one b-tagged) or ≥ 1 small-radius jet accompanied with ≥ 1 large-radius jet (at least one large-radius b-tagged) are required. For the DL final state, exactly two leptons are required, combined with at least one large-radius or at least one small-radius jet, that is b-tagged. The dominant background is $t\bar{t}$ production for both channels, and, additionally, W+jets for the SL and DY for DL. Multijet events with at least one jet misidentified as a lepton is also a background for the two channels, mostly in the SL mode. A profile-likelihood fit is applied to distributions of deep neural network discriminators for all categories simultaneously. The observed (expected) upper limit on the inclusive cross section of HH production times the branching fraction for the $WWbb$ decay is 14 (18) times the expected SM value, as shown in Fig. 8. For VBF production, the respective results are 277 and 301. Limits are set on coupling-constant modifiers (κ_λ, κ_{VV} and

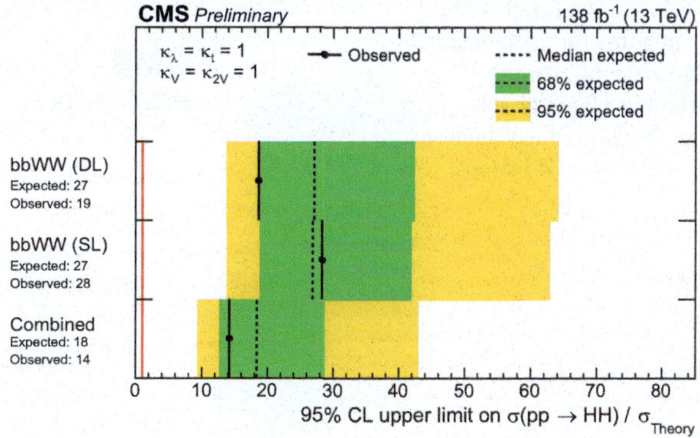

Fig. 8 The Observed and expected 95% CL upper limits on the SM HH production cross section, obtained for the DL and SL channels and from their simultaneous fit

others). If all other Higgs boson couplings are assumed to have the values predicted by the SM, the κ_λ trilinear self-coupling strength modifier is constrained in the range $[-7.2, 13.8]$ (expected $[-8.7, 15.2]$).

10 Search for $HH(\rightarrow b\bar{b}b\bar{b}) + W/Z$

A search for the rare process of Higgs boson pair production in association with a vector boson (W or Z) has been performed [14]. Only decays of both H to $b\bar{b}$ are considered, whereas the vector boson can decay either leptonically (including decay of Z to neutrinos) or hadronically. As a result, the channels used are labeled based on the vector boson decays as a) large p_T^{miss} (MET), b) single charged lepton (1L), c) double charged lepton (2L), d) fully hadronic (FH)—the latter used for the first time in such a search. All channels look for four small-radius jets (from the b quarks), whereas channels MET and 1L also look for two large-radius jets (because two boosted b quarks can be reconstructed as one wider jet). All leptonic channels require at least two b-tagged jets, whereas the fully-hadronic one requires at least three b tags. Single-lepton, dilepton, p_T^{miss} as well as multi-jet triggers are used. Signal regions are defined based on kinematics that enhance signal efficiency. Control regions are similarly defined and used to validate backgrounds. Machine-learning observables are used for all signal and control regions. The simultaneous fit to all the decay modes (inclusive) with a single signal strength modifier, and the individual fits per channel, with separate single strength modifiers, can be seen in Fig. 9. For a SM-like signal the observed (expected) significance is 2.6 (0.0) standard deviations with a fitted signal strength of 145 times the SM signal where all coupling modifiers are equal to 1. Limits are set on κ_λ, κ_{VV}.

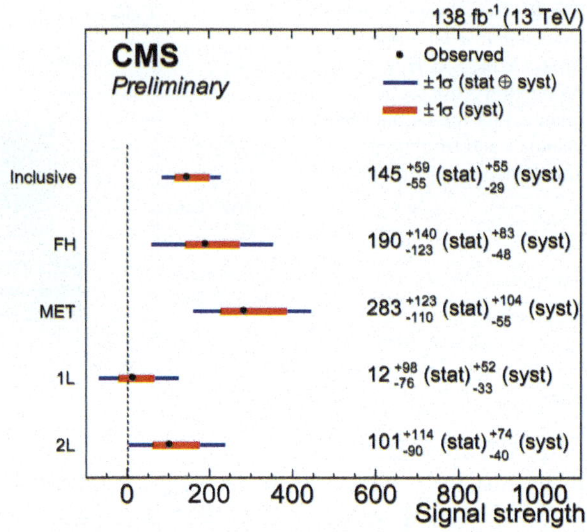

Fig. 9 The fitted signal strengths for each channel and overall, for the double-Higgs boson production in association with a W or Z boson

11 Search for a New Lower-Mass Standard-Model-Like Higgs Boson Decaying to Diphotons

Many BSM models predict two Higgs boson doublets and the existence of a lower mass Higgs boson. A recent CMS analysis [15] looked for diphoton resonances in the mass range from 70 GeV (to avoid HLT trigger turn-on effects) to 110 GeV. Asymmetric photon requirements are used, based on the diphoton trigger: for first photon $p_T^{\gamma_1} > 30$ GeV and $p_T^{\gamma_1}/m_{\gamma\gamma} > 0.47$ and for the second $p_T^{\gamma_2} > 18$ GeV and $p_T^{\gamma_2}/m_{\gamma\gamma} > 0.28$. The selection that is scaled by the diphoton mass prevents a distortion of the low end of the mass spectrum. The main backgrounds are the QCD-induced diphotons, jet+γ (where the jet fakes a photon), dijets (where both jets fake photons), and DY to dielectrons (where both electrons fake photons). Multivariate techniques are used for the photon identification and event classification. The number of event classes depends on the BDT score and is selected to cover the signal and background regions. Classes with extra jet activity are used to cover the VBF production channel. Parametric functions are used to model the diphoton invariant mass for signals corresponding to different Higgs boson masses. A fit to the mass spectrum leads to a cross section consistent with SM non-resonant backgrounds, with a local (global) excess of 2.9 (1.3) σ at the diphoton mass of 95.4 GeV, shown in Fig. 10.

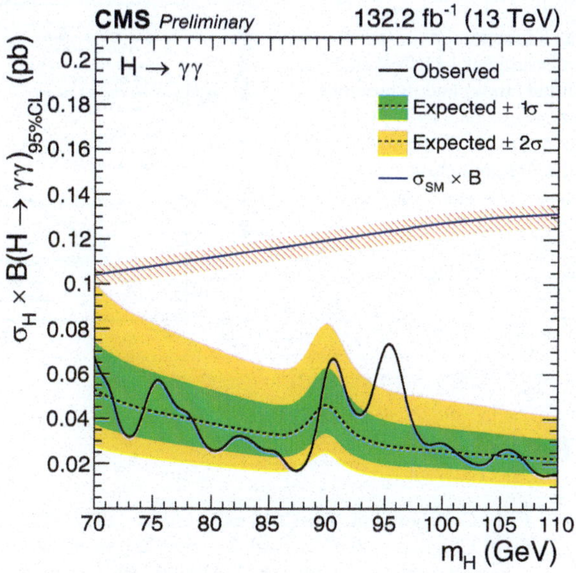

Fig. 10 Observed and expected 95% CL limits on the production cross section of an extra SM-like Higgs boson and its decay to diphotons as a function of its mass

12 First Measurement of the $t\bar{t}$ Production Cross Section in Proton-Proton Collisions at 13.6 TeV

CMS has measured the $t\bar{t}$ production cross section [16] at 13.6 TeV, using 1.21 fb^{-1} of data from the Run 3 of the LHC. Both the lepton+jets and dilepton channels are used. Data were collected with single-electron, single-muon, dielectron and dimuon triggers. In the lepton+jets channel, exactly one lepton with $p_T^\ell > 35$ GeV and at least three jets with $p_T^{jet} > 30$ GeV are required. Both leptons and jets have pseudorapidity $|\eta| < 2.4$. In the dilepton channel, exactly two OS leptons with $p_T^\ell > 35$ GeV are required. The b tagging is achieved with the DeepJet [3] neural network. In the ee, $\mu\mu$, and lepton+jets channels, 1 and 2 b-tag categories are used. In the $e\mu$ category, 0, 1, and 2 b-tag categories are used. The main background for the ee and $\mu\mu$ channels is the Z+jets (determined from MC, normalized with the use of data control regions, separately in b-jet categories), for $e\mu$ the single top, for lepton+jets the W+jets and the QCD multijets (the latter estimated with data). Major sources of systematic uncertainties are the luminosity, lepton and b-tagging efficiencies and the single-top background. The cross section is measured with a profile likelihood fit of event yields in all subcategories and it is found to be $\sigma_{t\bar{t}} = 882 \pm 23$ (stat + syst) ± 20 (lum) pb in agreement with the SM NNLO prediction for top mass equal to 172.5 GeV calculated with next-to-NLL resummation of soft gluons, which is 921^{+29}_{-37} pb. Figure 11 shows the CMS measurements of the $t\bar{t}$ cross section as a function of the LHC center-of-mass energy, including the 13.6 TeV measurement.

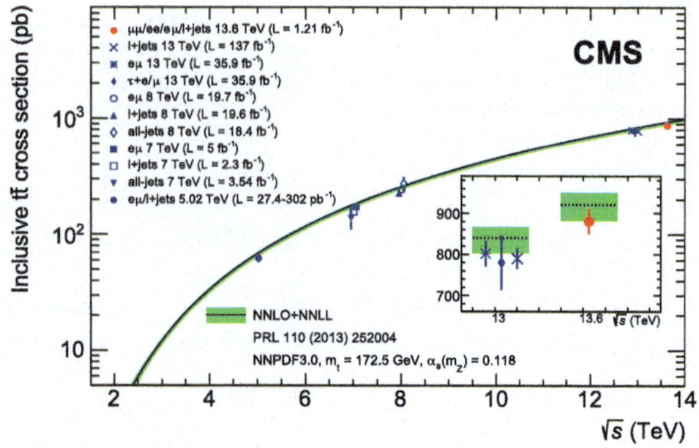

Fig. 11 The $t\bar{t}$ cross section as a function of $\sqrt{s}$, as measured at 13.6 TeV (red filled circle) and in previous measurements by the CMS experiment. Also shown is the SM prediction

References

1. CMS Collaboration, The CMS experiment at the CERN LHC. JINST **3**, S08004 (2008)
2. CMS Collaboration, Observation of four top quark production in proton-proton collisions at $\sqrt{s} = 13$ TeV. Phys. Lett. B **847**, 138290 (2023)
3. E. Bols et al., Jet flavour classification using DeepJet. JINST **15**, P12012 (2020)
4. M. van Beekveld, A. Kulesza, L. Moreno Valero, Threshold resummation for the production of four top quarks at the LHC (2022). arXiv:2212.03259
5. CMS Collaboration, Observation of $WW\gamma$ production and search for $H\gamma$ production in proton-proton collisions at $\sqrt{s} = 13$ TeV. Phys. Rev. Lett. **132**, 121901 (2024)
6. CMS Collaboration, Observation of the rare decay of the η meson to four muons. Phys. Rev. Lett. **131**, 091903 (2023)
7. R. Escribano, S. Gonzàlez-Solís, A data-driven approach to π^0, η and η^0 single and double Dalitz decays. Chin. Phys. C **42**, 023109 (2018)
8. CMS Collaboration, Search for inelastic dark matter in events with two displaced muons and missing transverse momentum in proton-proton collisions at $\sqrt{s} = 13$ TeV. Phys. Rev. Lett. **132**, 041802 (2024)
9. M. Duerr et al., Hunting the dark Higgs. JHEP **4**, 143 (2017)
10. CMS Collaboration, Search for dark matter particles produced in W^+W^- events with transverse momentum imbalance in proton-proton collisions at $\sqrt{s} = 13$ TeV. JHEP **03**, 134 (2024)
11. CMS Collaboration, Search for long-lived heavy neutral leptons with lepton flavour conserving or violating decays to a jet and a charged lepton. JHEP **03**, 105 (2024)
12. CMS Collaboration, Search for direct production of GeV-scale resonances decaying to a pair of muons in proton-proton collisions at $\sqrt{s} = 13$. JHEP **12**, 070 (2023)
13. CMS Collaboration, Search for Higgs boson pair production in the $b\bar{b}W^+W^-$ decay mode in proton-proton collisions at $\sqrt{s} = 13$ TeV. JHEP **07**, 293 (2024)
14. CMS Collaboration, Search for Higgs boson pair production with one associated vector boson in proton-proton collisions at $\sqrt{s} = 13$ TeV. JHEP **10**, 061 (2024)
15. CMS Collaboration, Search for a standard model-like Higgs boson in the mass range between 70 and 110 GeV in the diphoton final state in proton-proton collision at $\sqrt{s} = 13$ TeV. Phys. Lett. B **860**, 139067 (2024)
16. CMS Collaboration, First measurement of the top quark pair production cross section in proton-proton collisions at $\sqrt{s} = 13$. JHEP **08**, 204 (2023)

Correction to: HEP2023

John Strologas and Leandros Perivolaropoulos

Correction to:
J. Strologas and L. Perivolaropoulos (eds.), *HEP2023*, Springer Proceedings in Physics 333
https://doi.org/10.1007/978-3-031-76856-9

The original version of this book was inadvertently published without the Series Title as "Springer Proceedings in Physics" and the Volume Number as 333 which have now been included both in the book and also in the cover page. The book has been updated with the changes.

The updated version of this book can be found at
https://doi.org/10.1007/978-3-031-76856-9

J. Strologas and L. Perivolaropoulos (eds.), *HEP2023*, Springer Proceedings in Physics 333 https://doi.org/10.1007/978-3-031-76856-9_25

Author Index

J. Strologas and L. Perivolaropoulos (eds.), *HEP2023*, Springer Proceedings in Physics 333 https://doi.org/10.1007/978-3-031-76856-9

The manufacturer's authorised representative in the EU is Springer Nature Customer Service Centre GmbH, Europaplatz 3, 69115 Heidelberg, Germany. If you have any concerns regarding our products, please contact ProductSafety@springernature.com

Printed and bound by CPI Group (UK) Ltd, Croydon, CR0 4YY
30/09/2026
02225587-0004